T0275858

Offshore Electrical Engineering Manual

Notice From the Author

No book is a substitute for common sense and a thorough understanding of the particular installation design.

However, not all safety, reliability or maintenance problems can be detected or resolved easily and it is hoped that this book will be a useful tool in their detection and resolution.

BP's New FPSO Glen Lyon *(Courtesy BP)*

Offshore Electrical Engineering Manual

Second Edition

Geoff Macangus-Gerrard

Gulf Professional Publishing is an imprint of Elsevier
50 Hampshire Street, 5th Floor, Cambridge, MA 02139, United States
The Boulevard, Langford Lane, Kidlington, Oxford, OX5 1GB, United Kingdom

Library of Congress Cataloging-in-Publication Data
A catalog record for this book is available from the Library of Congress

British Library Cataloguing-in-Publication Data
A catalogue record for this book is available from the British Library

ISBN: 978-0-12-385498-8

For information on all Elsevier publications visit our website at
https://www.elsevier.com/books-and-journals

Publishing Director: Joe Hayton
Senior Acquisition Editor: Katie Hammon
Senior Editorial Project Manager: Kattie Washington
Production Project Manager: Surya Narayanan Jayachandran
Cover Designer: Miles Hitchen

Typeset by TNQ Books and Journals

Contents

PART 3 DESIGN PROJECT ORGANIZATION

PART 4 ELECTRICAL FAULTS AND PROTECTION DEVICES

PART 5 INSTALLATION

PART 6 SPECIFIC SYSTEMS AND VESSEL TYPES

PART 7 COMMISSIONING

PART 8 RELIABILITY, MAINTENANCE AND LOGISTICS

PART 9 STATUTORY REQUIREMENTS AND SAFETY PRACTICE

Preface to Second Edition

Many things have changed in the oil and gas industry since the first edition was published in 1992, four years after the Piper Alpha disaster.

Chief among these were the changes in the UK statutory legislation following on from Piper Alpha. Regrettably, I was too busy at the time to carry out the necessary revisions to this book, but now have the opportunity as I approach retirement to rectify that omission.

The change of name to 'Manual' reflects the intention to avoid covering theoretical material covered well in university courses, but to concentrate on the practicalities and precautions necessary to create an electrical design which has all the attributes expected of it in today's world, i.e., safe to use and maintain, robust in harsh conditions, reliable, efficient and environmentally sound.

I hope readers will appreciate the following significant content changes, which reflect the new offshore regulations, or have become necessary because of the modern information technology. These are as follows:

1. The addition of material related to the UK Offshore Safety Case and PFEER Regulations,
2. The addition of a section on offshore wind farms and their (offshore) substations,
3. The use of weblink references where possible,
4. Because of the constant revisions to standards, dates of standards are no longer quoted unless unavoidable,
5. At the time of writing, the IEE Recommendations for the Electrical and Electronic Equipment of Mobile and Fixed Offshore Installations, which has served us well over the years, is about to be superseded by a UK harmonised version of BS EN 61892, and this is reflected in the new text.

I look forward with interest as to what 'BREXIT' and the other global political upheavals create, but I am hopeful that the drive towards global harmonisation of standards will continue.

Acknowledgements

FIRST EDITION, OFFSHORE ELECTRICAL ENGINEERING (SEE REVISION NOTES)

The author wishes to repeat his thanks and gratefully acknowledge all those who provided material and advice for the production of the first edition, particularly the following:

Stephen Rodgers, John Brown Engineering Ltd, Clydebank (1)
Ian Stewart and Arlene Sutherland, BP Exploration Ltd, Aberdeen
Andrew White, Andrew Chalmers and Mitchell Ltd, Glasgow (2)
David Bolt, Ewbank Preece Ltd, Aberdeen (3)
Lynn Hutchinson, Ferranti Subsea Systems Ltd, Victoria Road, London W6 (10)
Hamish Ritchie, Geoff Stephens and John McLean, Foster Wheeler Wood Group Engineering Ltd, Aberdeen (4)
Gordon Jones, GEC Alsthom Large Machines Ltd, Rugby, Warwickshire (5)
Mr P.G. Brade, GEC Alsthom Measurements Ltd, Stafford (5)
Mrs M. Hicks, Publicity Department, GEC Alsthom Installation Equipment Ltd, Liverpool (5)
Pat Dawson, Hawke Cable Glands Ltd, Ashton-under-Lyne, Lancashire
Gordon Shear, Hill Graham Controls Ltd, High Wycombe, Buckinghamshire (6)
Richard Crawcour and Mr K.M. Hamilton, P&B Engineering Ltd, Crawley, Sussex
Sue Elfring, Crest Communications Ltd, for Rolls-Royce Industrial and Marine Ltd, Ansty, Coventry (7)
John Day, formerly with Shell UK Exploration and Production, Aberdeen
Jim Bridge and Keith Stiles, SPP Offshore Ltd, Reading
Ian Craig and Graham Sim, Sun Oil Britain Ltd, Aberdeen (8)
Prof. John R. Smith, University of Strathclyde, Glasgow
John Baker, GEC Alsthom Vacuum Equipment Ltd (5)
Mr John Hugill, Thorn Lighting Ltd, Borehamwood, Herts (9)

REVISION NOTES

1. Now part of Kvaerner
2. Now part of the Hubbell Group
3. Now part of Mott MacDonald
4. Now Wood Group PSN Ltd
5. GEC and Alsthom are now part of the US company General Electric (GE)

6. Hill Graham (Ansaldo) became part of the Robicon Corporation and acquired Siemens Energy and Automation Inc.
7. The Marine side of Rolls-Royce no longer produces Aero-Derivative Gas Turbine offshore packages
8. Sun Oil has pulled out of the United Kingdom and the Balmoral semisubmersible is operated by Premier Oil UK
9. Now part of the Zumtobel Group
10. Now part of Aker Solutions

SECOND EDITION, OFFSHORE ELECTRICAL ENGINEERING MANUAL

The author wishes to thank and gratefully acknowledge all those who provided material and advice for the production of the second edition, particularly the following:

Dr Luiz Fernando Oliveira, DNV-GL Brazil for provision of typical Orbit SIL printouts.
Scott Dean, BP plc, for provision of Glen Lyon photos
Gillian Ewan, DNV-GL Aberdeen Office
Paul Wilson, Nexen Petroleum UK (CNOOC Ltd), Aberdeen
James Sweeney, GenEx Design Ltd, Combermere, Whitchurch, Shropshire, UK
Keith Oliver MSc, BSc for his preparation of the Process Controls and Monitoring Systems chapter
Henryk Peplinski, for assistance with Kongsberg and marine integrated systems
Katie Ayres, SPP Pumps
Aimee Ironside, Aker Solutions
Siemens Gas Turbines, Lincoln, UK
GE Gerapid, USA
OTDS Transformers Ltd, Croydon, UK
P&B Relays, Manchester, UK
GE Grid Solutions, Stafford, UK
ETAP Inc., California, USA
Rob Dickson, CEE Relays (for SKM), Slough, UK
Bunmi Adefajo, TNEI (IPSA), Manchester, UK
Schneider Electric/EIG_Wiki
DEKRA Certification, B.V., Arnhem, Netherlands
Kongsberg Maritime, Kongsberg, Norway
Mike Lancashire, Abtech, Sheffield, UK, for provision of Appendix C1
Andy Smith, CML Certification, Ellesmere Port, Cheshire, UK for provision of Appendix C2

PART

Introduction and Scope

1

CHAPTER

The Offshore Electrical Dimension

1

INTRODUCTION

Designing for provision of electrical power offshore involves practices similar to those likely to be adopted in onshore chemical plants and oil refineries. However, other aspects peculiar to offshore oil production platforms need to be recognised. It is suggested that those unfamiliar with offshore installations read the brief guide in PART 6 of this book before continuing further.

The aspects which affect electrical design include the following:

1. The space limitations imposed by the structure, which adds a three-dimensional quality to design problems, especially with such concerns as
   a. hazardous areas
   b. air intakes and exhausts of prime movers
   c. segregation of areas for fire protection
   d. avoidance of damage to equipment due to crane operations
2. Weight limitations imposed by the structure which require
   a. The careful choice of equipment and materials to save weight.
   b. The avoidance of structurally damaging torques and vibrations from rotating equipment.
   c. The inherent safety hazards presented by a high steel structure surrounded by sea. Such hazards require
      i. Particular attention to electrical shock protection in watery environments
      ii. Good lighting of open decks, stairways and the sea surrounding platform legs
      iii. Protection of materials and components from the corrosive marine environment and avoidance of stray corrosion cells due to contact between dissimilar metals

MARINE ENVIRONMENT

Wave heights in the North Sea can exceed 20 m, with wind speeds exceeding 100 knots.

Offshore Electrical Engineering Manual. https://doi.org/10.1016/B978-0-12-385499-5.00001-7

HAZARDS OFFSHORE

GAS

Accumulation of combustible gas can occur on an offshore installation from various sources, including the following:

1. Equipment and operational failures such as rupture of a gas line, flame out of an installation flare, a gland leak, etc.
2. Gas compressor surge/vibration causing failure of pipe flanges, loss of compressor seal oil, etc.
3. Drilling and workover activities.
4. In concrete substructures, the buildup of toxic or flammable gases due to oil stored in caisson cells.

CRUDE OIL AND CONDENSATES

1. Equipment and operational failures such as the rupture of an oil line, a gland leak, etc.
2. The high pressures involved in some cases could cause spontaneous ignition because of the electrostatic effects.

OPERATIONAL HAZARDS

1. Apart from the fire and explosion hazard of process leaks, there is a hazard to personnel purely from the mechanical effects of the leak jet and the sudden pressure changes caused by serious leaks in enclosed compartments.
2. Care must be taken in the siting of switchrooms, generator sets and motor drives to minimise the risk of damage due to crane operations, especially if sited near drilling equipment areas where heavy pipes and casings are being frequently moved.

ELECTRICAL SYSTEM DESIGN CRITERIA

The purpose of any offshore electrical supply system is to generate and distribute electricity to the user such that

1. power is available continuously at all times the user's equipment is required to operate,
2. the supply parameters are always within the range that the user's equipment can tolerate without damage, increased maintenance or loss of performance,
3. the cost per kilowatt hour is not excessive taking into consideration the logistical and environmental conditions in which generation and distribution are required,

4. impracticable demands are not made on the particular offshore infrastructure, such as those for fuel or cooling medium,
5. the safety requirements pertinent to an offshore oil installation are complied with, in particular those associated with fire and explosion hazards,
6. the weight of the system is not excessive for the structure on which it is installed. In the case of rotating machinery, the effects of vibration and shock loads must be taken into account.

DESCRIPTION OF A TYPICAL SYSTEM

A single-line diagram of a typical offshore electrical system is shown in Fig. 1.1.1.

MAIN PRIME MOVERS

With the obvious availability of hydrocarbon gas as a fuel, and the requirement for a high power-to-weight ratio to keep structural scantlings to a minimum, gas turbines are the ideal prime movers for power requirements in excess of 1 MW. Below this value, reliability and other considerations, dealt with in PART 8 Chapter 1, tend to make gas turbines less attractive to the system designer.

Because of the complexity and relative bulk of gas turbine intake and exhaust systems, the designer is urged towards a small number of large machines. However, the designer is constrained by the need for continuity of supply, maintenance and the reliability of the selected generator set to an optimum number of around three machines. A variety of voltages and frequencies may be generated, from the American derived 13.8 kV and 4.16 kV 60 Hz to the British 11 kV, 6.6 kV and 3.3 kV 50 Hz. Many ships operate at 60 Hz, including all NATO warships and there is a definite benefit to be gained from the better efficiencies of pumps and fans running at 20% higher speeds.

KEY SERVICES OR SUBMAIN GENERATORS

On some platforms, smaller generators are provided to maintain platform power for services other than production. These are also normally gas turbine driven and can provide a useful black-start capability, especially if this is not available for the main machines.

MEDIUM-VOLTAGE DISTRIBUTION

The design of the distribution configuration at the platform topside conceptual stage is highly dependent on the type of oil field being operated and the economic and environmental constraints placed on the oil company at the time. The older platforms originally had little or no facilities for gas export or reinjection and therefore the additional process modules installed when these facilities were required have their own dedicated high-voltage switchboards. This is also the case if such a heavy-power

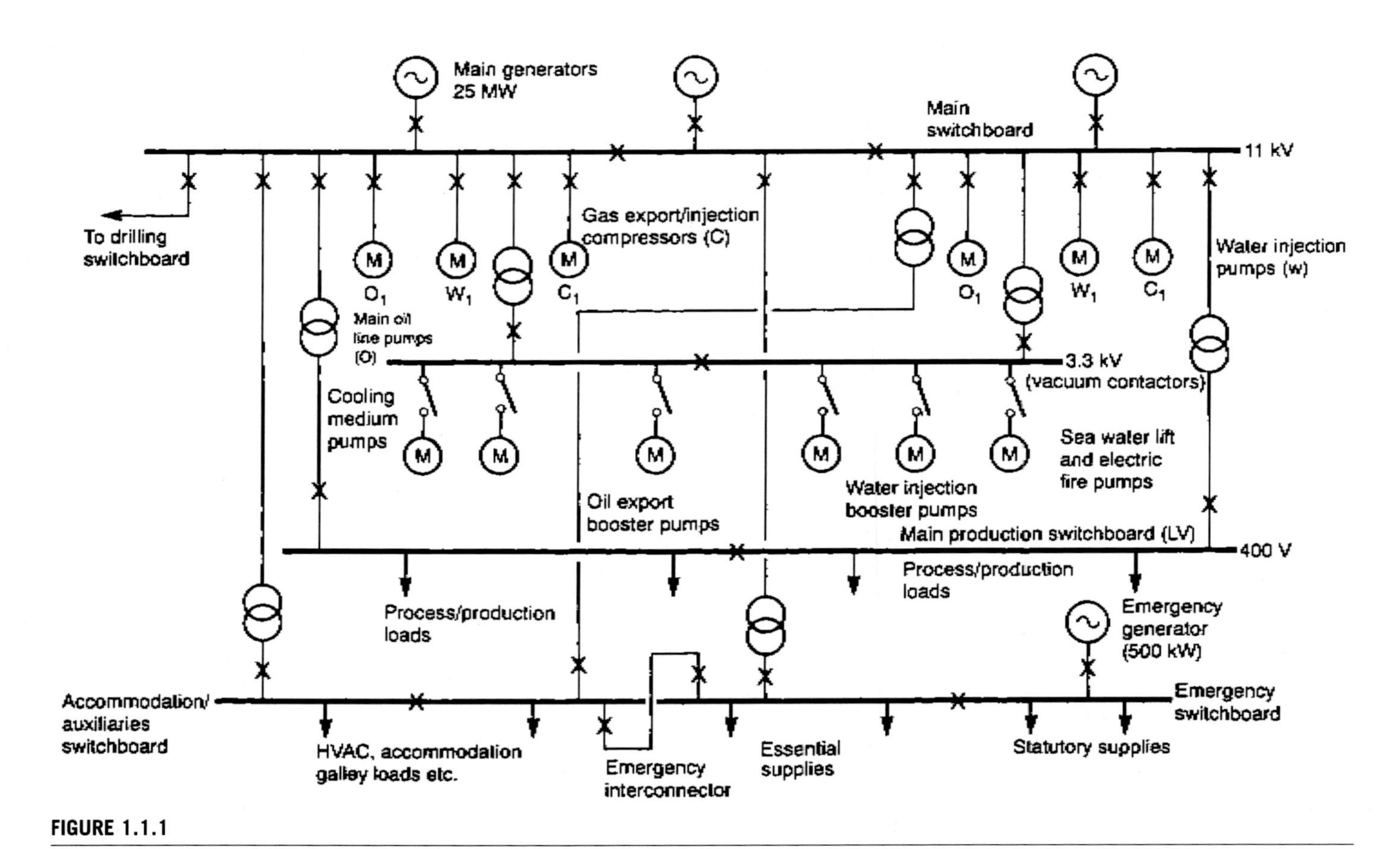

FIGURE 1.1.1

Single-line diagram of a typical offshore electrical system. *HVAC*, heating, ventilation and air conditioning.

consumer as seawater injection is underestimated for at the time of construction. In general, however, it is better to concentrate switchrooms in one area of the platform to avoid complications with hazardous areas and ventilation etc., as discussed in PART 1 Chapter 2.

With such relatively high generation capacities and heavy-power users within the limited confines of an offshore platform, calculated prospective fault currents are often close to or beyond the short circuit capabilities of the millivolt switchgear designs available at the platform topside design phase. Currently fault ratings of 1000 MVA are available, and with careful study of generator decrement curves etc., it is usually possible to overcome the problem without resorting to costly and heavy reactors. All the available types of millivolt switchgear are in use offshore. The use of bulk oil types, however, is questionable because of the greater inherent fire risk. Unlike land-based switchboards, there has been found to be a significant risk of earth faults occurring on the busbars of offshore switchboards and so some form of earth fault protection should be included for this. The platform distribution at medium voltage normally consists of transformer feeders plus motor circuit breaker or contactor feeders for main oil line pumps, seawater lift and water injection pumps and gas export and reinjection compressors. Depending on the process cooling requirements, the cooling medium pumps may also be driven by medium-voltage motors.

Operating such large motors on an offshore structure (i.e., three quarters of the way up a high steel or concrete tower) can lead to peculiar forms of failure due to the associated vibration and mechanical shock, almost unheard of with machines securely concreted to the ground. This has led to offshore platform machines being fitted with more sophisticated condition monitoring than that usually found on similar machines onshore.

Another problem, which will be discussed in more detail later in PART 4, is the transient effect on the output voltage and frequency of the platform generators with such large motors in the event of a motor fault or, for that matter, during normal large motor switching operations. Computer simulation of the system must be carried out to ensure stability at such times, both at initial design stage and also when any additional large motor is installed. Facilities such as fast load shedding and automatic load sharing may be installed to both improve stability and make the operator's task easier. This subject is discussed in PART 2 Chapters 12 and 15 and 15.

LOW-VOLTAGE DISTRIBUTION

Using conventional oil- or resin-filled transformers, power is fed to the low-voltage switchboards via flame-retardant plastic-insulated cables. Cabling topics are covered in PART 2 Chapter 8. Bus trunking is often used for incoming low-voltage supplies from transformers. Because of the competition for space, this is just as likely to be due to bending radius as to current rating limitations of cables, as bus ducting may have right angle bends. The type of motor control centre switchboard used offshore would be very familiar to the onshore engineer. However, the configuration of the low-voltage distribution system to ensure that alternative paths of supply are always available is usually much more important offshore. This is because although every effort is put into keeping it to a minimum, there is much more interdependence between systems offshore.

A few examples of small low-voltage supplies which are vital to the safe and continuous operation of the installation are

1. safe area pressurisation fans,
2. hazardous area pressurisation fans,
3. generator auxiliaries,
4. large pump auxiliaries,
5. large compressor auxiliaries,
6. galley and sanitation utilities for personnel accommodation,
7. uninterruptible power supply systems for process control and fire and gas monitoring,
8. seawater ballast systems (on tension leg platforms and semisubmersibles).

The topics of maintenance and availability are covered in PART 8 Chapter 1.

EMERGENCY OR BASIC SERVICES SWITCHBOARD

As a statutory requirement, every installation must have a small generator to provide enough power to maintain vital services such as communications, helideck and escape lighting, independent of any other installation utility or service. In the event of a cloud of gas enveloping the platform due to a serious leak, even this will need to be shut down as a possible ignition source.

FIRE PUMPS

Again as a statutory requirement, every installation must be provided with at least sufficient fire pumps with enough capacity to provide adequate water flow rates for fighting the most serious wellhead, pipeline riser or process fires. The numbers and capacities of these pumps have to take into account the unavailability due to routine maintenance and failure. These pumps may be submersible electrical, hydraulic powered or directly shaft driven from a diesel engine.

Typically, one pump arrangement could have an electrically driven 100% capacity pump supplied from a dedicated diesel generator set which is directly cabled to the pump, i.e., with no intervening switching or isolating devices, and has the advantages of increased reliability because of fewer components and 'soft' starting of the motor.

This kind of pump runs up to operating speed with the generator, in the same manner as a diesel-electrical railway locomotive would accelerate from start. The second 100% capacity pump could again be electrical but supplied from the platform distribution system in the conventional way. The purpose of this arrangement is to avoid failure of both pumps due to a common operational element, i.e., common mode failure. A third 100% capacity pump would be required to allow for maintenance down time. Details on the electrical design of diesel–electrical fire pump packages are given in PART 2 Chapter 12.

SECURE ALTERNATING CURRENT AND DIRECT CURRENT POWER SUPPLIES

On any platform, there are a large number of systems which require supplies derived from batteries to minimise the risk of system outage due to supply failure.

The following is a typical platform inventory:

1. fire and gas monitoring and protection,
2. process instrumentation and control,
3. emergency shutdown system,
4. emergency auxiliaries for main generator prime movers,
5. emergency auxiliaries for large compressors and pumps,
6. navigational lanterns and fog warning system,
7. emergency and escape lighting,
8. tropospheric scatter link,
9. line of sight links,
10. carrier multiplexing and voice frequency telegraphy equipment,
11. telecoms control and supervisory system,
12. public address,
13. general alarm system,
14. platform private automated branch exchange (PABX),
15. marine radio telephones,
16. aeronautical very-high-frequency (VHF) (AM) radio,
17. VHF (FM) marine radio,
18. aeronautical nondirectional beacon,
19. company high-frequency independent sideband and ultra-high-frequency (FM) private channel radios,
20. telemetry system,
21. satellite subsea well control systems.

The majority of these systems operate at a nominal voltage of 24V direct current (DC), and although it is not necessary for each of the above-mentioned systems to have separate battery and battery charger systems, the grouping criteria require more detailed discussion. These are covered in PART 2 Chapter 15, as is the need to provide dual chargers and batteries for certain vital systems.

In addition to the above-mentioned systems, there are, of course, switchgear tripping/closing supplies and engine start batteries which are dedicated to the equipment they supply. In the case of engines which drive fire pumps, duplicate chargers and batteries are also required. This subject is discussed in PART 2 Chapter 12.

DRILLING SUPPLIES

A typical self-contained drilling rig supply system's single-line diagram is shown in Fig. 1.1.2.

The usual arrangement is for two or more diesel generators rated at around 1MVA to feed a main switchboard which also has provision for a supply from the

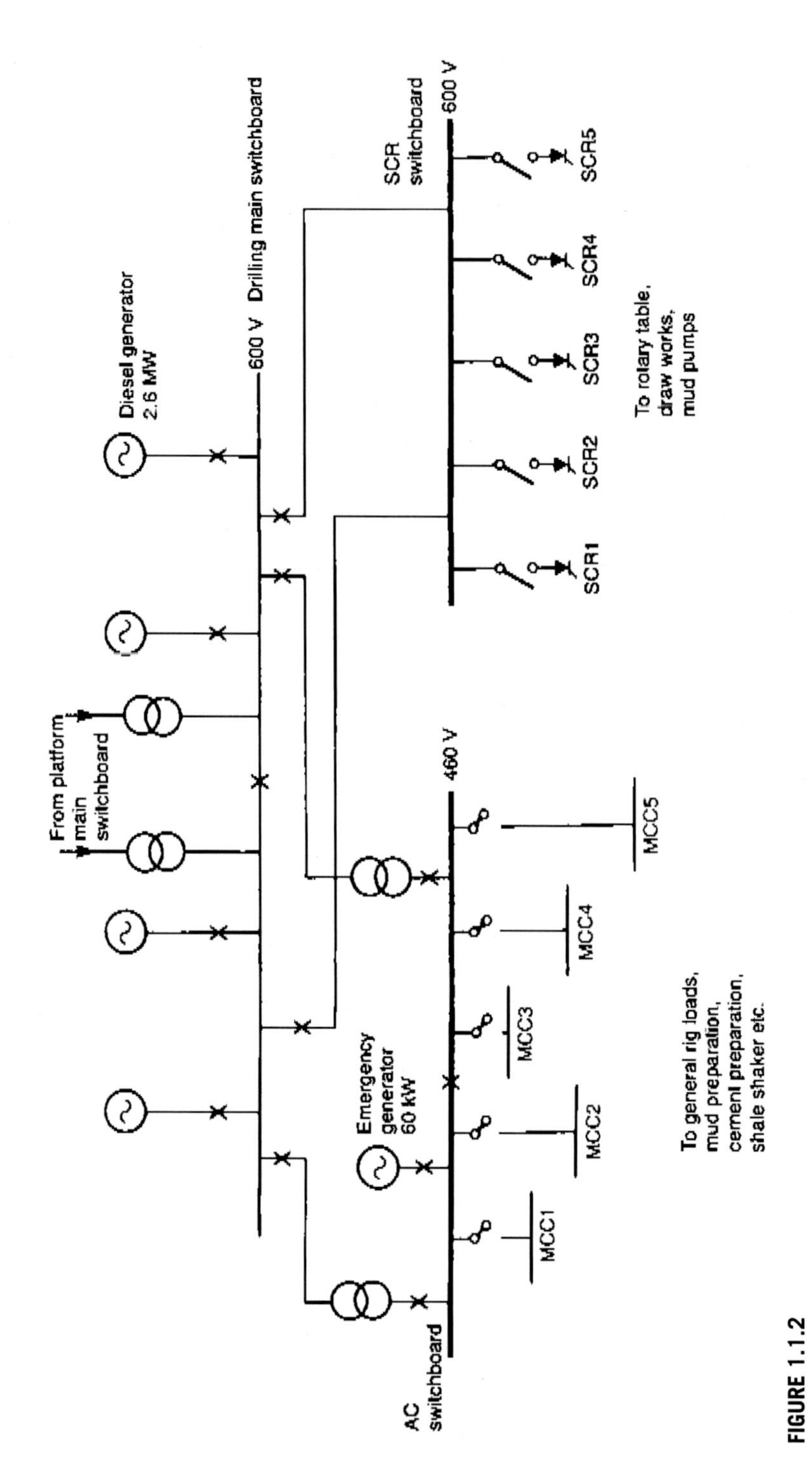

FIGURE 1.1.2

Single-line diagram of a typical drilling electrical system. *AC*, alternating current; *SCR*, silicon controlled rectifier.

platform generation and distribution system via transformers. This main switchboard then supplies a series of motor control centres, one or more of which contains a series of silicon controlled rectifier (SCR) DC variable speed drive controllers. By using an arrangement of DC contactors, these controllers may be assigned to various DC drive motors from the driller's console. As the SCR systems are phase angle controlled, a certain variable amount of harmonics is generated, depending on the kind of drilling operation being carried out. When the drilling rig is being fed from the platform supply, the harmonics may affect certain sensitive equipment, such as secure supply invertors. This subject is discussed in detail in PART 4 Chapter 9.

CHAPTER

The Offshore Electrical Environment

2

INTRODUCTION

Some background information on oil and gas production is given in Appendix A. This chapter discusses the more general criteria governing offshore electrical systems and equipment design.

SAFETY

The environment on an offshore installation is not inherently safe because of the heavily salt-laden atmosphere and the highly conducting nature of the structure and virtually all the equipment it contains. It must not be possible for personnel to come into contact with live or moving parts either by accident or while performing their normal duties.

Protection against electrical shock relies on the safe design and installation of equipment, training (making personnel aware of the dangers and ensuring that the necessary precautions are taken) and the use of special safe supplies for most portable equipment.

An electrical current of only a few milliamperes flowing through the human body can cause muscular contractions and, in some circumstances, will be fatal. The current may result in local burning or some involuntary reaction which in itself may lead to injury. Additionally, of course, varying degrees of pain will be experienced.

ISOLATED SITUATION

Except in the case of one or two installations, the electrical system is totally isolated from any other means of electrical supply. The system must be designed and configured in such a way that it is never dependent on one small component or electrical connection to continue in operation. This point may sound rather obvious, but it is the author's experience that hidden vulnerabilities may be designed into systems which are both costly and disruptive in their first effects and in their eradication. The following examples of actual occurrences illustrate the point.

Example 1 A platform reinjection compressor is driven by a 500 kW 3.3 kV motor having a lube oil system pressurised by auxiliary lube oil pumps driven by low-voltage motors. Both the lube oil pump motors and the control supply for the

Offshore Electrical Engineering Manual. https://doi.org/10.1016/B978-0-12-385499-5.00002-9

3.3 kV latched contactors are supplied from the same low-voltage switchboard. A spurious gas alarm occurs in the vicinity of the low-voltage switchroom and the low-voltage switchboard incoming circuit breakers are opened by the emergency shutdown system. The lube oil pumps then stop but the compressor control system is unable to open the main drive motor contactors and the motor runs to destruction.

Example 2 A platform has two low-voltage switchboards dedicated to provide the safe and hazardous ventilation necessary for continued safe operation of the platform. Depressurisation of any module would lead to process shutdown. The particular platform is a pumping station for oil from other platforms including those of other companies and therefore a considerable amount of oil revenue is at stake if the platform is shut down. Unfortunately, each switchboard is fed by a single incomer and the ventilation fan motor starters are distributed so that the majority of supply fans are on one switchboard and the majority of extract fans are on the other. This arrangement resulted in the export of oil from a number of large North Sea installations being dependent on the continuous operation of two small low-voltage switchboards.

The subject of reliability is dealt with in greater detail in PART 8 Chapter 1.

ENVIRONMENT

This topic is covered in greater detail in PART 5 Chapter 4 and is covered exhaustively in all the relevant standards, recommendations and codes of practice (see Bibliography). However, it is important to be very clear as to the fundamental reasoning behind all the regulations governing electrical installation offshore. Because both the safety and cost of an installation are highly sensitive to equipment selection, it is also important to have a clear understanding of the reasons behind the classification of hazardous areas and of the different methods employed by equipment manufacturers to make their equipment suitable for particular environments. Where this is practicable, electrical equipment is best installed in an environmentally controlled room located in an area that is not classified as hazardous with respect to hydrocarbon gas ignition risk, effectively sealed from the outside atmosphere and provided with a recirculating air conditioning system. Of course, this optimum scheme cannot be considered for equipment which

1. has to be located outside (such as navigational aids),
2. has to be located under or near water (such as seawater lift pump motors), or
3. is associated with some other equipment which may occasionally or does normally leak hydrocarbon gases (such as gas compressor drive motors).

Often the electrical equipment installed has to cater safely for a combination of all three of these situations and may also be required to operate at elevated pressures and temperatures.

WATER HAZARDS

Hazards due to water coming into contact with electrical equipment are similar to those experienced on ships but can be more catastrophic because on offshore installation, more power is often generated, at higher voltages, and with greater prospective fault ratings.

Water may leak from large bore water-carrying pipes routed over switchboards or generators. The following situations are two examples seen by the author where such pipes were routed over switchboards.

Example 1 A fire–water main routed over a 4 MW gas turbine generator. The fire main was flanged and valved directly over the alternator. When the valve was serviced while the generator was running, the pipe fitter inadvertently drained an isolated section of fire main over the alternator. The generator promptly shut down, because of the operation of the differential protection, with a stator winding fault.
Example 2 In an accommodation module a sewage pipe from the floor above was routed directly above a low-voltage switchboard and along almost its entire length. Although some minor leaking had taken place, it was fortunate that the only problem for the electrical maintenance staff was one of hygiene.

If routing of water-carrying pipes over switchgear is unavoidable, there should be no flanges in the section of the pipe over the switchboard.

HYDROCARBON HAZARDS

In the planning of platform superstructures, designers try to arrange to segregate the wellhead and process areas from the accommodation and other normally manned areas to the greatest possible extent. This involves not only horizontal and vertical segregation but also segregation of all piped or ducted services such as ventilation ducting and drains. Following the Piper Alpha disaster, the whole philosophy regarding the segregation of accommodation areas on offshore platforms has been rethought, and safety philosophies have become more 'goal setting' rather than prescriptive. As is common knowledge, 165 people lost their lives either as a result of the initial explosions, dense smoke and fire or following the ensuing riser fires which led to the loss of structural integrity and the falling of the Piper Alpha accommodation module into the sea. The Cullen Report gives over 100 recommendations, covering all aspects of offshore installation design, construction, operation and safety. In one of, in my view, the most important recommendations, Lord Cullen states that the operator should be required by regulation to submit to the regulatory body a Safety Case in respect of each of its installations. It is important that the safety aspects of each installation are considered uniquely so as to meet objectives, rather than impose fixed solutions which may or may not work on a particular installation. Safety Case Regulations have now been in force for almost as long as the first edition of this book has been published and these statutory regulations are considered in PART 9 Chapter 1.

Whatever further means of ensuring the survival of the particular installation and its personnel are considered in each installation Safety Case, it is certain to influence the design of the electrical system and equipment, particularly in minimising the risk of electrical ignition sources and in the provision of emergency secure electrical supplies, completely independent of normal platform supplies.

As an important means of minimising risk of ignition sources, the hazardous area boundary drawings produced during the platform process design stage are discussed in PART 5 Chapter 4 and represent the situation during normal operating conditions. However, it is also necessary to consider the situation during a major outbreak of fire or after serious gas leak; the so-called 'post red' situation.

There are three electrical systems which normally monitor, control and mitigate the extent and spread of oil and gas leaks and hence the safety of the platform:

1. the fire and gas monitoring system
2. the emergency shutdown system
3. safe and hazardous area ventilation systems.

These systems and devices are defined as the 'Safety Critical Elements' (SCEs) in the statutory regulations, the definition being

such parts of an installation and such of its plant (including computer programmes), or any part thereof–

(a) the failure of which could cause or contribute substantially to; or
(b) a purpose of which is to prevent, or limit the effect of, a major accident.

Emergency power, and the various communication systems, and on floating installations the ballast control system may also be included as electrical SCEs. All these systems will have some bearing on the design of the platform electrical system, either because they may include the facility to shut all or part of the electrical system down, or because a secure, or at least a more reliable electrical supply is needed to operate them. Every installation is different, and non-oil and gas installations such as offshore wind farm substation platforms will need their own safety cases at every stage including decommissioning and abandonment.

FIRE AND GAS MONITORING

Every installation must have, as a statutory requirement, a designated control point located in a nonhazardous area, capable of overall management of the installation and manned continuously.

All pertinent information from the production processes, drilling, utilities and firefighting systems need to be monitored at this control point, and emergency controls associated with these systems have to be available there to enable sufficiently effective control to be exercised in all operational or emergency conditions.

On normally manned installations the control point needs to be located in or adjacent to the accommodation area and may be in or adjacent to the Offshore Installation Manager's office and radio room. This control point also requires public address facilities to be close at hand, and as a Temporary Safe Refuge, it would normally be the last area to be vacated in an emergency; the room in which it is contained would be H120 fire rated, with a dedicated ventilation system (see PART 5 Chapter 5).

In the larger platforms, this is a limited repeat of a far more sophisticated monitoring system located in or near the process control room.

On normally unmanned platforms the basic control point may be located on an adjacent normally manned installation or even at a control centre onshore.

EMERGENCY SHUTDOWN SYSTEM

As in the case of a nuclear reactor or similar complex system, the continuing safe condition of the platform cannot be left solely to the human operators, as they would not always have sufficient time to investigate each abnormality and respond with the appropriate sequence of corrective actions in every case. Because of this, it is necessary to provide a system which either initiates the correct sequence of actions itself or provides a series of simple options (levels of shutdown) that the operator may take when a particular event occurs.

Every installation has start-up and shutdown systems of varying sophistication, which attempt to provide the greatest possible safety for personnel and equipment. These systems are interrelated, with the process control system being subordinate to the emergency shutdown system. The various levels of shutdown and their effects on the electrical system are discussed in PART 7 Chapter 6.

SAFE AND HAZARDOUS AREA VENTILATION

Most of the heating, ventilation and air-conditioning (HVAC) systems must run continuously during normal platform operation to ensure the following:

1. Acceptable working environments are maintained in process modules containing equipment or pipework which may leak hydrocarbon gas.
2. Comfortable environmental conditions are maintained within the accommodation modules and normally manned nonhazardous areas, and an acceptable working environment is provided in normally unmanned modules.
3. Positive pressurisation with respect to adjacent hazardous areas or the outside atmosphere is maintained in nonhazardous modules or rooms.
4. Potentially hazardous concentrations of explosive gas mixtures are diluted in, or removed from, hazardous area modules.
5. When fires occur or dangerous concentrations of gas are detected, individual areas are sealed from ventilation and the associated fans shut down in accordance with the logic of the emergency shutdown system.

6. Uncontaminated combustion, purging and normal ventilation air is available to prime movers.
7. Uncontaminated air supplies are available to personnel and emergency generator and fire pump prime movers and other essential service equipment are provided with combustion and ventilation air in times of emergency.

Ventilation systems, especially those associated with switchrooms and generator rooms, are discussed in more detail in PART 5 Chapter 5.

The three above-mentioned systems (i.e., fire and gas, emergency shutdown and HVAC) are interconnected and are often required to work in concert. An example of this would be if a fire occurred in a particular switchroom. The fire would be detected by smoke or heat detectors and the central fire and gas system monitoring the room would initiate the following actions:

1. Signals the ventilation system to seal the room by the closure of ventilation fire dampers and switch off associated fans.
2. Sounds an alarm in the switchroom to warn personnel that escape is necessary and that a fire extinguishant is to be released.
3. Depending on the system logic signals the emergency shutdown system to isolate the switchboards in the switchroom by opening the appropriate feeder circuit breakers, or even shuts down all main generators if the switchroom in question contains the main switchboard.
4. Releases the fire extinguishant (CO_2, water mist or halon gas) into the switchroom after a suitable time delay to allow for personnel to escape.

DISTRIBUTION CONFIGURATION

Figs 1.2.1–1.2.3 show various ways in which attempts have been made to obtain optimum availability from the platform electrical power system within the limitations of weight and conditions imposed by the situation.

Fig. 1.2.1 shows the system arrangement on an earlier installation, consisting of a main 6.6 kV switchboard with two 15 MW gas turbine generators and a smaller 6.6 kV switchboard with two 2.5 MW gas turbines. In this configuration the smaller 'utilities' switchboard provides supplies for all cooling seawater pumps and also feeds all low-voltage distribution transformers except those in a water injection module which was added to the platform at a later date. Should a fault occur affecting the whole of the utilities switchboard, for example, a fault in the bus sections or sectioning switch leading to busbar damage, then only the emergency generator is left available. This is because there is insufficient spare capacity in the emergency generator to run the main generator auxiliaries, and in this particular case the main generators require seawater for cooling, which is normally available from the seawater lift pumps powered from the utilities switchboard.

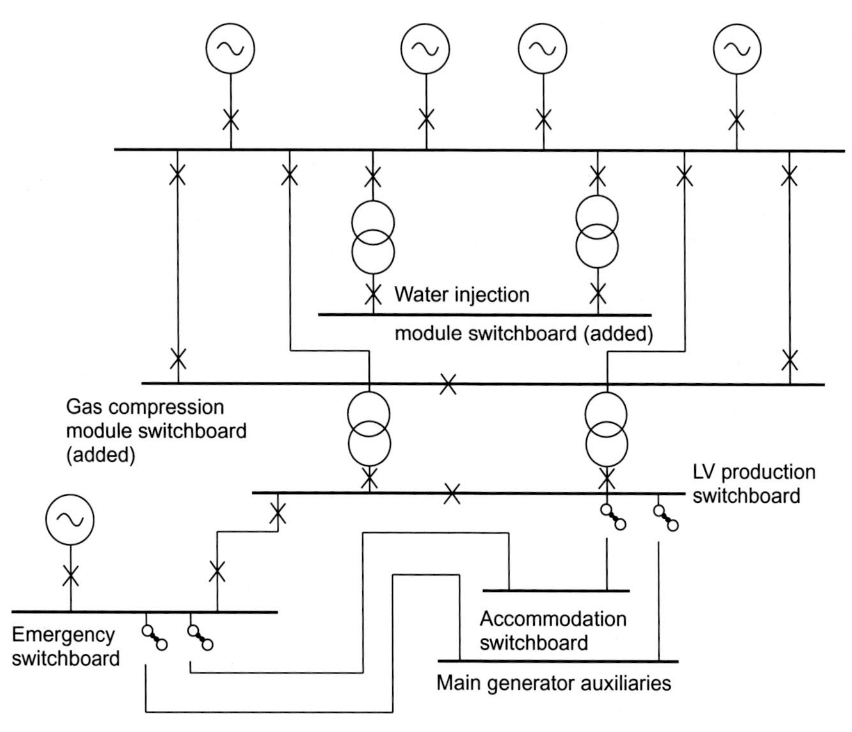

FIGURE 1.2.1

System arrangement on an earlier installation, consisting; main 6.6 kV switchboard; two 15 MW gas turbine generators; smaller 6.6 kV switchboard with two 2.5 MW gas turbines.

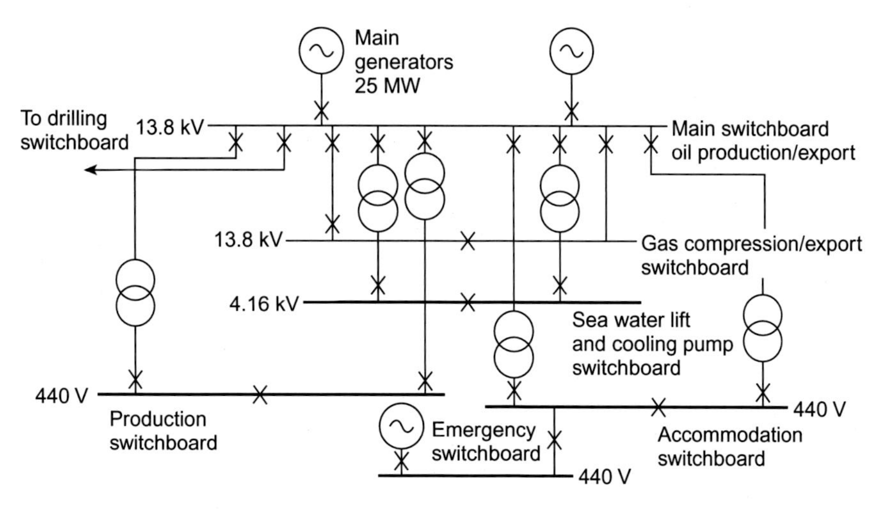

FIGURE 1.2.2

Scheme where the main switchboard consists: single section switchboard fed by two 24 MW gas turbine generators, each capable of taking the entire platform load including drilling.

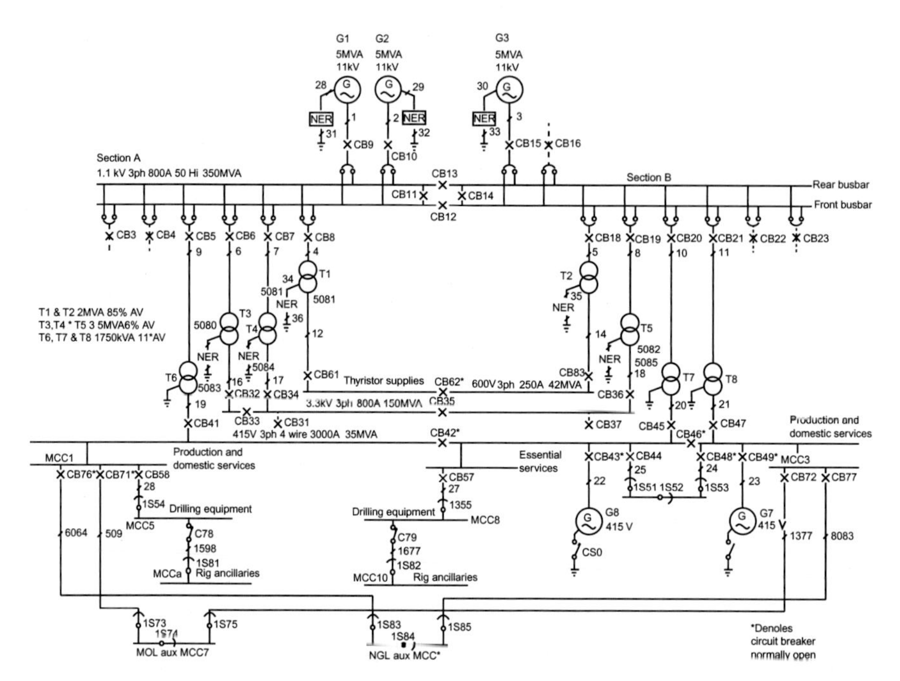

FIGURE 1.2.3

Most favoured main switchboard arrangement for offshore use since, assuming supplies are maintained to generator auxiliaries, it requires several failures to occur before production is affected.

The probability of water ingress, the most likely cause of such a failure, would be reduced if the switchroom housing the utilities switchboard was surrounded by other rooms or modules.

Fig. 1.2.2 shows a scheme where the main switchboard consists of a single-section switchboard fed by two 24 MW gas turbine generators, each capable of taking the entire platform load including drilling.

This medium-voltage switchboard not only directly supplies the platform with large motor drives such as main oil line pumps and gas compressors but also feeds the majority of lower-voltage switchboards via transformers.

There are two points at which connections are made to the drilling electrical system. One is a supply at medium voltage and the other is an alternative supply to the emergency switchboard at 440 V. The drilling system is also supplied by two dedicated 1.5 MW diesel generators; however, these are much too small to sustain any platform production but could maintain some utilities.

It can be seen that production is dependent on the continued functioning of the main switchboard, a single fault in which causes the majority of the installation to be blacked out. The insertion of a bus section switch in this particular case may do little to improve the system integrity, as the reliability of the switch is likely to be less than that of the busbars.

The configuration shown here for the main switchboard is now the most favoured for offshore use, as, assuming supplies are maintained to generator auxiliaries, it requires several failures to occur before production (and hence revenue) is affected.

MODULAR CONSTRUCTION

Provided lifting and transport facilities of sufficient capacity are available at an economically viable cost, it is invariably better to build a complete module containing the generators, switchgear and transformers that are completely fitted out, tested and commissioned at a suitable fabrication yard than to carry out any of the construction on the platform.

Apart from fuel, cooling air and combustion air, it is preferable to make the 'electrical power module' totally independent from the rest of the installation. This also means that it has integral engine-starting facilities as well as engine auxiliaries which do not require external low-voltage electrical supplies. This is not always possible because of the weight of the extra transformers required. The need for seawater cooling for alternator heat exchangers may be unavoidable because of the bulk of air-cooled units. The optimum independence of the module has the added advantages of

1. minimal hook-up requirements during offshore installation,
2. minimal service requirements during test and commissioning at the module fabrication yard.

SUBSEA CABLE VERSUS ON-BOARD GENERATION

The cost of laying subsea power cable is presently in the order of £2 to £5 million/cable kilometre. This cost includes that involved in mobilising a suitable cable-laying vessel. At first sight, this may appear prohibitively expensive but is worth investigation if most of the following conditions are met:

1. The source of supply is conveniently at hand, i.e., on the mainland or another installation. Subsea cable lengths greater than 50 miles are unlikely to be viable.
2. The cable route is not through a busy shipping lane or anchorage where there is high-risk anchor damage.
3. The cable route does not pass through a popular fishing ground where there is the consequent risk of cables being dredged up by trawl nets or gear.
4. The fuel gas supply on the installation to be supplied by the cable is unreliable, exhausted or of too small a capacity for the required prime mover. Of course, the converse must be true of the supplying installation.
5. The transmission route is through reasonably sheltered waters where it is probable that cable repairs could be carried out throughout the year. If this is not the case, then the expense of laying duplicate cables over separate routes becomes more attractive, as, in winter, it could be 3 months or more before a suitable weather window becomes available to repair a cable fault.

6. The sea floor is suitable for burial of the cable as a means of protection. Drifting sands, solid rock or strong tidal currents would militate against this.

In listing these conditions, it has been assumed that the import of diesel fuel for normal operation is not a viable option because of the high cost involved. However, this may depend on commercial considerations such as the revenue value of the gas to the installation operators.

For example, Centrica's Morecambe Bay complex has a mixture of subsea cable, gas turbine and diesel-generator-powered platforms. If it becomes necessary to supplement the available power on a particular platform, then the additional weight of supplementary generator may be too great for the platform to bear without very costly modifications. Even if weight is not a problem, it is not always possible to find a sufficiently spacious location on the installation either. Shore supplies may be of the wrong frequency for use on the particular installation, and it may be necessary to install a motor generator set. This has the additional advantage of improving the motor-starting capability of the supply, as the generator impedance will be much lower than a series of transformers and long subsea cables. The transmission voltage required will vary depending on the length of the subsea cable but is likely to be either 11 or 33 kV. The weight and space taken up by the transmission transformers and the associated extra switchgear needs to be taken into consideration whenever subsea cable options are proposed.

If there is a group of several small installations separated only by a few kilometres of water, it may be economic to supply all their main power requirements from one central platform. This is more likely to be the case if centralising the main generation allows gas turbines of 1 MW or more to be considered (see discussion on prime mover selection in PART 2 Chapter 4).

Finally, it is advisable to carry out some form of reliability analysis to numerically rank the reliabilities of various supply and generation schemes before making the final decision.

Reliability topics are discussed in PART 8 Chapter 1.

PART 2

Offshore Electrical Systems and Equipment

CHAPTER

1 Insulation and Temperature Ratings of Equipment

INSULATING MATERIALS

Insulating materials may be grouped into classes according to their properties and the working temperature for which they are suitable. Further information is available in BS EN 60085:2004. The main insulation classes in use today are listed below:

1. Class A insulation consists of materials such as cotton, silk and paper when suitably impregnated or coated or when immersed in a dielectric liquid such as oil. Other materials or combinations of materials may be utilised for this class if by experience or tests they can be shown to be capable of operation at the Class A temperature.
2. Class E insulation consists of materials or combinations of materials which by experience or testing can be shown to be capable of operation at the Class E temperature; i.e., materials possessing a degree of thermal stability allowing them to be operated at a temperature 15°C higher than that for Class A materials.
3. Class B insulation consists of materials or combinations of materials such as mica, glass fibre etc., with suitable bonding, impregnating or coating substances. Other materials or combinations of materials, not necessarily inorganic, may be included in this class if by experience or testing they can be shown to be capable of operation at the Class B temperature.
4. Class F insulation consists of materials or combinations of materials such as mica, glass fibre, etc., with suitable bonding, impregnating or coating substances. This class also includes other materials or combinations of materials, not necessarily inorganic, which by experience or testing can be shown to be capable of operation at the Class F temperature; i.e., materials possessing a degree of thermal stability allowing them to be operated at a temperature 25°C higher than that for Class B materials.
5. Class H insulation consists of materials such as silicone elastomer and combinations of materials such as mica, glass fibre etc., with suitable bonding, impregnating or coating substances, such as appropriate silicone resins. Other materials or combinations of materials may be included in this class if by experience or testing they can be shown to be capable of operating at the Class H temperature; i.e., materials possessing a degree of thermal stability allowing them to be operated at a temperature 25°C higher than that for Class F materials.

Offshore Electrical Engineering Manual. https://doi.org/10.1016/B978-0-12-385499-5.00003-0

6. Class C insulation consists of materials or combinations of materials such as mica, porcelain, glass, quartz with or without an inorganic binder. Other materials or combinations of materials may be included in this class if by experience or testing they can be shown to be capable of operating at above the Class H temperature limit. The temperature limits for specific materials in this class will be dependent on their physical, chemical and electrical properties.

In each class, a proportion of materials of a lower temperature class may be included for structural purposes only, provided that adequate electrical and mechanical properties are maintained during the application of the maximum permitted temperature.

An insulating material is considered to be 'suitably impregnated' when a suitable substance such as varnish penetrates the interstices between fibres, films, etc., to a sufficient degree to bond components of the insulation structure adequately and to provide a surface film which adequately excludes moisture, dirt and other contaminants.

For some applications, compounds and resins without solvents may be used which may substantially replace all the air in the interstices. In other applications, varnishes or other materials containing solvents may be used which provide reasonably continuous surface films and partial filling of the interstices with some degree of bonding between components of the insulation structure.

An insulating material is considered to be to be 'suitably coated' when it is covered with a suitable substance, such as varnish, which excludes moisture, dirt and other contaminants to a degree sufficient to provide adequate performance in service.

It will be seen that Class E is intermediate between Classes A and B, and Class F covers Class B materials with bonding substances which make them suitable for an additional 25°C. Generally speaking, Class C is not appropriate for rotating machines but is used for static plants such as transformers.

The endurance of insulation is affected by many factors such as temperature, electrical and mechanical stresses, vibration, exposure to deleterious atmospheres and chemicals, moisture and dirt. For example, some varnishes tend to harden with age to such an extent that cracks are formed and moisture is then admitted.

The majority of marine apparatus is now insulated with class B or F materials. Class Y, being without impregnation, is hygroscopic and therefore unsuited to marine conditions.

HOT-SPOT TEMPERATURES

When considering suitable operating temperatures, it is the temperature at the hottest point that is important, and this is referred to as the 'hot-spot' temperature. In a field coil, for instance, the hot spot is somewhere near the centre of the winding and there is a temperature gradient from there to the surface, so that the temperature is not uniform throughout the coil. The temperature of this spot can be measured by embodying a thermocouple in the winding, but this is not practicable in small production machines but is very common in larger machine stators. In small machines the only means available in practice may be to determine the temperature either by the change in resistance of the winding, or by measuring the surface temperature by thermometer. The surface temperature will obviously be less than that of the hot spot, and to a certain extent the difference will depend on the depth of winding. From previous research and experience, values of surface temperature corresponding to specified hot-spot temperatures have been determined and are now universally accepted. If the resistance method is used, it is evident that as there is a temperature gradient, the resistance must lie somewhere between what it would be if the whole coil was at hot-spot temperature, and what it would be at surface temperature. The temperature determined by resistance is therefore higher than the surface temperature, and the accepted difference will be noted in the standard tables.

For Classes A and B, hot-spot temperatures of 105 and 130°C respectively have been universally accepted for very many years. These figures correspond roughly to about 20 years' working life under average industrial conditions. It must be remembered, however, that in industry there are usually peak periods of loading interspersed with off-load or reduced load periods for meal breaks, etc., and these periods of rest have considerable influence on the life of machines. Under marine conditions some machines may run for days at a constant load, and experience has shown that under these conditions the life may be reduced to about 15 years. No hard and fast rule can be made because conditions vary, but taking into account the necessity for the utmost reliability in marine installations the need for a more conservative approach to this problem is indicated. It is generally accepted that insulation life is approximately halved for each 10°C increase above the accepted hot-spot temperature limits for Class A and Class B materials.

TEMPERATURE RISE

A continuously rated machine will eventually reach a steady temperature at which the heat in the windings and magnetised cores and the heat arising from frictional losses will be dissipated at the same rate as they are generated. The difference between this steady temperature and that of the incoming cooling air is the temperature rise. For all practical purposes, other parameters being equal, this rise is always the same regardless of the temperature of the cooling air. For example, if a machine is tested in an ambient or cooling temperature of 20°C and a machine temperature of 55°C is recorded, the rise is 35°C; when the same machine is in the tropics and the cooling air is at 45°C the rise will still be 35°C, giving a total machine temperature of 80°C.

When the commutators' sliprings or bearings of machines provided with water coolers are not in the enclosed air circuit cooled by the water cooler but are cooled by the ambient cooling air the permissible temperature-rise above the ambient cooling air should be the same as for ventilated machines.

The appropriate hot-spot temperature for a given class of insulation is determined, and from that the surface temperature. Then the permissible temperature rise is arrived at by deducting the maximum ambient air temperature under which the machine will be called on to operate. When carrying out temperature tests on machines it is important to remember that the surface temperature of windings is affected by windage, and the temperatures recorded while the machine is rotating must not be taken as the maximum for determining temperature rise. After the machine has come to rest, a further rise will occur and the thermometers must be observed for several seconds after stopping, until the temperature reaches its maximum. (Note that this is not to be confused with the effects of reduced windage in variable speed machines, which is considered at the end of this chapter).

Temperature rise limits vary according to the application, e.g., rotating machinery, transformers, contactor coils. Reference should be made to the appropriate standard and/or regulation for a particular application. Table 2.1.1 is an example of limits for rotating machines on vessels for unrestricted worldwide service. It is based on the ambient air temperature of 45°C and cooling water at 30°C. The limits for Classes F and H insulation are 20 and 40°C, respectively, higher than Class B.

AMBIENT AIR TEMPERATURES

For ocean-going vessels and mobile oil installations it is required that temperature rises should be based on tropical conditions. Whatever the intentions of the first owner, such vessels sometimes change hands and must therefore be suitable for service in any part of the world.

Temperatures recorded in various positions in many types of ship have been tabulated and studied. It has been shown for instance that certain locations are almost invariably warmer than other parts of the same machinery space, and it is logical therefore to accept lower temperatures for switchboards, for instance, than for generators and motors.

BASIS OF MACHINE RATINGS

[Reference should be made to BS EN 60034-1:2004: Rotating electrical machines – PART 1: Rating and performance]

Insulation and temperature ratings of equipment: Momentary overloads (for which, for test purposes, a duration of 15 s is recognised) of 50% in current for generators and of varying amounts in torque for motors according to type, size and duty are recognised. Continuous Maximum Rating (CMR) does not mean that machines are incapable of carrying moderate overloads of reasonable duration, but that the makers are not required to state either magnitude or duration or to submit machines to an overload test. In actual fact, electric motors, generators and particularly transformers have an inherent capacity sufficient for average requirements.

For exceptional applications where overloads are anticipated in normal service, the purchaser should seek the advice of the manufacturer or select a standard motor of higher rating. Such cases might arise with motors coupled to oil pumps, where the load may be increased for short periods while pumping cold oil.

Table 2.1.1 Limits of Temperature Rise in Degree Celsius for Rotating Machines

Item	Part of Machine	Temperature Measurement Method	Air-Cooled Machine Class			Water-Cooled Machine Class		
			A	E	B	A	E	B
1(a)	AC windings of turbine-type machines having output of 5000 kV a or more	ETD or R	50	60	70	70	80	90
1(b)	AC windings of salient-pole and of induction machines having output of 5000 kV a or more							
2(a)	AC windings of machines smaller than in item 1							
2(b)	Field windings of a.c. and d.c. machines having d.c. excitation other than those in items 3 and 4	R T	50 40	65 55	70 60	70 60	85 75	90 80
2(c)	Windings of armatures having commutators							
3	Field windings of turbine-type machines having d.c. excitation	R	–	–	80	–	–	100
4(a)	Low-resistance field windings of more than one layer, and compensating windings	T, R	50	65	70	70	85	90
4(b)	Single-layer windings with exposed bare surfaces	T, R	55	70	80	75	90	100
5	Permanently short- circuited insulated windings	T	50	65	70	70	85	90
6	Permanently short-circuited windings, uninsulated	T	The temperature rise of these parts shall in no case reach such a value that there is a risk of injury to any insulating or other material adjacent parts					
7	Iron core and other parts not in contact with windings	–						
8	Iron core and other 90 parts in contact with windings	T	50	65	70	70	85	90
9	Commutators and sliprings, open or enclosed	T	50	60	70	70	80	90

ETD, *embedded temperature detector;* R, *resistance method;* T, *thermometer method.*

Class E insulation was introduced to enable full advantage to be taken of the then new synthetic resin enamels, and similar materials now in use, which are suitable for higher working temperatures than the oleo-resin enamels which were available when the present limits for Class A were established. This has now largely been displaced by Class B and F materials.

METHOD OF MEASUREMENT OF TEMPERATURE RISE BY RESISTANCE

If it is desired to measure temperature rise by resistance, it is important to obtain a reliable reading of the cold resistance and corresponding temperature. The machine should have been standing at atmospheric temperature for some time so that it is at a uniform temperature. The initial resistance and the temperature, at which this is read, as measured by a thermometer on the winding, should be recorded simultaneously.

The hot temperature is determined from the following formula:

$$t_2 = [R_2/R_1\,(t_1 + 235)] - 235,$$

where R_1 is the resistance of the winding cold, R_2 the resistance of the winding hot, t_1 the temperature of the winding cold (°C), and t_2 the temperature of the winding hot (°C).

Since the increase in resistance of copper per degree Celsius is about 0.4%, special care must be taken to use calibrated instruments, accurate recording and rapid measurement after shutting down.

A quick estimation of temperature rise can be made on the basis of 0.4% increase in resistance per degree Celsius. For instance, if a 20% increase in resistance is recorded the temperature rise is approximately 20/0.4 = 50°C.

When it is intended to use embedded temperature detectors (ETDs) they must be built into the machine during construction. They may take the form of either thermocouples or resistance thermometers. This method is generally used for large alternators to record the temperature of stator windings, in which case at least six detectors are suitably distributed around the stator and placed where it is expected that the highest temperatures are likely to occur. They are usually placed between upper and lower coil sides, within the slots and midway between radial ducts, if any.

MEASUREMENT OF AMBIENT AIR TEMPERATURE

The cooling air temperature should be measured at different points around and halfway up the machine, and at distances 1–2 m away from it. The thermometers should indicate the temperature of the air flowing towards the machine, and should be protected from heat radiation and stray draughts.

The value to be adopted to determine temperature rise is the average of these temperatures from readings taken at the beginning and end of the last half hour of the test. If the conditions are such that parts of a machine are in a position in which ventilation may be impeded, e.g., in a pit, the air temperature in such a restricted area is deemed to be the ambient temperature. If the air is admitted into a machine through a definite inlet opening or openings, the temperature of the cooling air should be measured in the current of air at or near its entry to the machine.

TYPICAL ALTERNATOR HEAT RUN PROCEDURE

The test is carried out at zero power factor lagging to avoid the need for a large prime mover. The normal safety precautions for testing electrical machinery should be taken.

1. Test Equipment
   a. An ac voltmeter is required to be connected across the alternator output terminals.
   b. Two ac ammeters with suitable current transformers (CTs) are required for monitoring the alternator output current.
   c. Two single phase wattmeters with suitable voltage and current inputs are required.
   d. A Tachometer suitable for the generator speed at operating frequency is needed.
   e. A dc ammeter placed in series with the exciter field is required
   f. A dc voltmeter with fused leads, placed across the exciter field, is needed.
   g. Portable (digital probe) thermometers for internal temperature measurement are needed.
   h. Portable thermometers for air temperatures
   i. A driving motor capable of maintaining the rated speed at full load (zero p.f.), 150 & load for 15 s and 25% overspeed is required.
   j. Adjustable 3 phase reactors are needed
   k. Load circuit breaker is needed.
2. Method
   a. Connect the above list of equipment.
   b. Couple the alternator to the drive motor and fit an adequate guard over the coupling and exposed shaft.
   c. Check that the alternator enclosure is as specified by the customer, is complete and fitted correctly.

If terminal box lids connections cannot be fitted after making test connections, seal box and cable outlets with paper and masking tape.

Calculate the expected ammeter and wattmeter readings from the CT ratios.

Run generator at rated speed (if possible, plus 4.0%) and record no load reading using the appropriate Test Sheet. Apply the load and continue the run, recording all readings every 60 min for at least 4 h or until thermal stability is reached.

DO NOT READJUST AVR DURING HEAT RUN

Portable thermometers are attached to the air intake and outlet; in the latter case it is advisable to run generator for a short while and determine the hottest spot by hand before attaching the thermometer. Air thermometers must be no more than 2m from generator.

When thermal stability is reached, take final hot load readings and running temperatures and then shut down as quickly as possible. No specific time for shutdown and taking hot temperatures is given in marine specifications.

With digital thermometer check spot temperatures and record as follows:

Stator core	(usually accessible through the terminal box)
Stator coils	(end windings)
Rotor coils	(At hottest spot, usually near coil supports)
Exciter armature	(Windings)
Exciter field	(Windings)
Bearings	(D. E. and N. D. E. temperatures to be recorded)

Take hot resistances of stator and exciter field windings.

On large generators, take rotor winding resistances as soon as possible after generator has stopped.

As mentioned earlier in this chapter, temperature rise by resistance is given by the following formula:

$$t_2 = [R_2/R_1\ (t_1 + 235)] - 235$$

where R_1 is the resistance of the winding cold, R_2 the resistance of the winding hot, t_1 the temperature of the winding cold (°C), and t_2 the temperature of the winding hot (°C) (Fig. 2.1.1).

RATINGS FOR VARIABLE SPEED MOTORS

Where motors are being supplied from converters, phase-angle controlled thyristor drives or similar systems where the machine is subjected to prolonged running at significant load but reduced speeds, a larger de-rated machine may be required because the cooling effects of shaft-driven fans and windage will be reduced. Alternatively, an independently driven cooling fan will be required. Because of this, it is preferable to purchase the drive and motor as a unit with the appropriate ATEX Declaration of Conformity from the manufacturer.

THERMAL OVERLOADS AND MOTOR THERMAL PROTECTION

During starting, both in the winding of the rotor and in the winding of the stator, currents may be present that are well above regular on-load currents. A fast trip must be initiated when a fault occurs during startup, for example, in the event

(A) Temperature Probes [Temperatures in Deg C]

Time	Alternator Frame	Exciter Frame	Air Out	Air Out	ND/E Bearing	D/E Bearing	Ambient Temp.
8:30	17.5	17.5	17	19	18	17	16.75
9:30	37	36.5	19.5	30	33	35.5	19.5
10:30	43.5	41.5	21.5	33	37.5	44	21.25
11:30	46	43.5	23	35	39	47	22
12:30	47.5	44.5	21	35	40	50	21.25
13:30	46.5	43.5	21	34	38.5	49	21.5
14:30	46.5	43.5	21	34	38.5	48.5	21.5

(B) Winding Thermocouples [Temperatures in Deg C]

°C	1	1	2	2	3	3	4	4	5	5	6	6
	Ohms	°C	Ohms	°C	Ohms	°C	Ohms	°C	Ohms	°C	Ohms	°C
08:30	105.42	11.5	105.37	11.5	105.61	12	105.32	11	105.45	11.5	105.38	11.5
09:30	129.2	75.5	133.54	87	130.6	79.5	130.9	80	132.6	79.5	128.2	73
10:30	132.5	84	136	93.5	134.45	89.5	134.6	89.5	135.8	93	132.5	84
11:30	124.6	89	137.7	98	135.7	92.5	136.2	94	137.7	98	132.6	84.5
12:30	134.6	89	138.3	99.5	135.8	93	136.3	94	137.9	98.5	132.8	85
13:30	134.15	88.5	137.9	98.5	135.4	92	135.9	93	137.6	98	132.75	85
14:30	134.15	88.5	137.9	98.5	135.5	92	135.9	93	137.6	97.5	132.3	83.5

FIGURE 2.1.1

Typical alternator temperature rise test result. (A) Temperature probes (temperatures in °C). (B) Winding thermocouples (temperatures in °C).

of a rotor block or motor start under heavy driven-machinery load conditions. In order to allow for cooling after each load attempt, the number of starts needs to be limited.

For starting, motor thermal protection should include the following elements:

- Blocked rotor protection
- Motor start protection
- Number of starts protection

To protect the motor during running, protection should include:

- Thermal overload protection – this uses a device or electronic simulation which replicates the heating characteristic (thermal replica) and will trip the motor if the replicated temperature exceeds the insulation maximum approved value for the insulation class. Machines may be rotor critical, where the rotor heats up faster than the stator or stator critical where the stator heats up faster.
- Unbalanced load protection or positive sequence monitoring – this is caused by a broken phase connection, which heats up the laminated rotor core, and may be detected by a bimetal device (in older designs of relays, e.g., P&B Golds) or a negative sequence monitor in electronic relays. For small motors a simple positive sequence monitor can be utilised.

ATEX EX 'E' INCREASED SAFETY CERTIFIED MOTORS

Thermal overload devices must be certified with the motor and form part of the motor certification.

For further hazardous area topics see **PART 5** Chapter 4.

Other useful references include:

- *Marine Electrical Practice*, Sixth Edition, Revised by BP Shipping Ltd., G. O. Watson, CEng, FIEE, FIEEE, FIMarE
- *Marine Electrical Equipment and Practice*, Second edition, H. D. McGeorge, CEng, FIMarE, MRINA
- *Dangerous Substances and Explosive Atmospheres Regulations 2002* – Approved Code of Practice and guidance
- *GAMBICA ATEX and Power Drive Systems User Guide* No. 4 Second Edition – Application of the ATEX Directives to Power Drive Systems

For further electrical system protection topics please refer to **PART 4**.

ACKNOWLEDGMENTS

My thanks to George Watson's 'Marine Electrical Practice' (Butterworth Heinemainn) for some of the material in this chapter.

CHAPTER

Alternating Current Synchronous Generators

2

INTRODUCTION

The intention here is not to produce yet another alternator theory text, but to provide the reader with some practical guidance on the installation and operation of alternating current (AC) generators offshore.

AC generators on ships and offshore installations are of the normal commercially available type of construction, with the field system in the rotor, either of salient-pole or cylindrical rotor design, and the armature windings in the (stationary) frame. There are still some slip ring excitation systems in use offshore, but by far the majority are now of brushless design, avoiding the need for regular brush replacement and the buildup of carbon dust.

Note: wind turbine generators are not normally synchronous machine designs and are covered in PART 6 Chapter 6.

PRINCIPALS OF OPERATION

Referring to the following phasor diagrams, the terminal voltage V is taken as the phase reference. The resistive and reactive volt drops are then added, giving the open circuit voltage or induced electromotive force E.

Hence this gives the equation $E=V+I_AR_A+jI_AX_S$, where E, V and I are phasors rotating at the synchronous speed and the subscripts A and S indicate armature and synchronous values. The angle δ is the load angle of the machine (see also Fig. 2.2.3). I_AR_A and jI_AX_S are volt drops due to losses, where the synchronous reactance X_S is a combination of armature winding leakage reactance and a reactance used to represent armature reaction.

The three diagrams in Fig. 2.2.1 correspond to the lagging, unity and leading power factor conditions, respectively.

From the design viewpoint, the size of the generator is governed mainly by the kilovolt-ampere rating and rotating speed (i.e., number of poles required). However the actual size used offshore will also depend to some extent on the prime mover ratings commercially available, as well as the reactive power required for the starting of large motors used for driving major oil or gas process rotating machinery such as export gas compressors. Oversising of the generator in relation to the prime mover is common in oil installations because the increased copper and excitation losses of the

Offshore Electrical Engineering Manual. https://doi.org/10.1016/B978-0-12-385499-5.00004-2

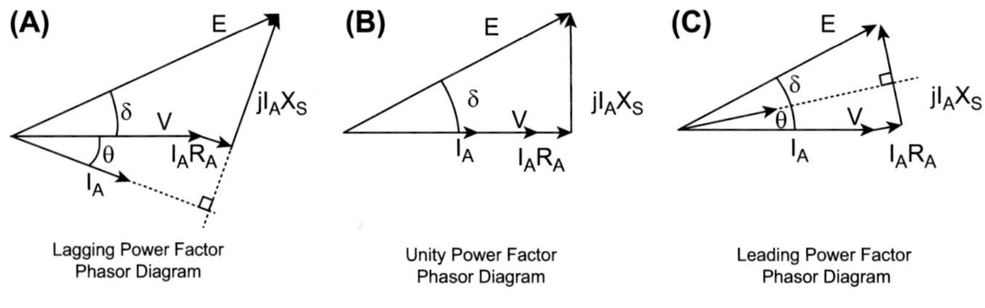

FIGURE 2.2.1

Phasor diagrams: (A) lagging power factor, (B) unity power factor and (C) leading power factor.

larger machine is a small price to pay for the increased flexibility in power management that the extra reactive power capacity gives.

SPEED

For large gas turbine machines on offshore platforms (30 MW+) AC generators are generally designed for two-pole operation, i.e., 3600 rpm for a 60 Hz output. The high speed gives greater efficiency of power transfer from the prime mover, and 60 Hz gives better efficiency for platform utilities such as pumps, as well as reduces the weight and size of transformers over 50 Hz units.

For the smaller generators driven by high-speed diesel engines, four-pole designs are the norm. For large ships with slow-speed diesel prime movers or propeller shaft driven generators, more poles may be necessary depending on the gearbox design available.

GENERATING VOLTAGE

For large offshore installations a medium voltage of 11 kV or higher may be used. The value actually used should reflect the economics of available designs from manufacturers, i.e., by avoiding significant restriction in the number of vendors that can be considered during tendering. Manufacturers should have no problems up to 20 kV.

SUBTRANSIENT REACTANCE

In prospective fault current and motor starting studies, generator subtransient reactance plays a vital role in optimising the power system design. The natural subtransient reactance for a generator of about 35 MVA would be around 15%, but this figure would produce very high fault levels.

A practical maximum subtransient reactance is around 22%, limiting the fault level to manageable values for the switchgear designs available commercially.

Actual fault levels in the system will depend on the number of generators in parallel, transformer reactances and general topology of the system and this will need to be optimised by computer simulation, using manufactures' data.

TRANSIENT REACTANCE

Unfortunately, increasing the subtransient reactance also increases the transient reactance and this will adversely affect the power system performance under motor-starting conditions. Therefore, there needs to be a trade-off between the two reactance values to ensure starting of the largest motor with the available spinning reserve, whilst avoiding intolerable fault levels.

POWER FACTOR

The power factor of the system is dependent on the type of load. There is little filament lighting these days, but normally trace heating, switched heating elements for heating the accommodation and for hot water will provide some loads at unity power factor.

Motors, transformers, fluorescent lighting and motors will provide loads at less than unity power factor. In addition, there will be nonsinusoidal, semiconductor controlled loads such as some forms of vessel propulsions (i.e., thrusters), process heating and variable-frequency and variable-speed process drives.

The nominal power factor for generator rating on both ships and offshore installations is taken as 0.8 lagging.

THE BRUSHLESS ALTERNATOR

In this typical offshore machine, slip rings and brushes are eliminated and excitation is provided not by conventional direct current exciter but by a small alternator. The AC exciter has the unusual arrangement of three-phase output windings on the rotor and magnetic poles fixed in the casing. The casing pole coils are supplied with direct current from an automatic voltage regulator (AVR) of the type described in the previous section. Three-phase current generated in the windings on the exciter rotor passes through a rectifier assembly on the shaft and then to the main alternator poles. No slip rings are needed.

The silicon rectifiers fitted in a housing at the end of the shaft are accessible for replacement and their rotation assists cooling. The six rectifiers give full-wave rectification of the three-phase supply (Fig. 2.2.2).

GENERATOR CAPABILITY DIAGRAM

The physical capability of the AC generator is limited by the following:

1. The real power available from the prime mover – if this is a gas turbine, as is often the case for the larger generators offshore, the power available will depend on the fuel in use (diesel or fuel-gas) and the ambient temperature.
2. The field heating limit – this will limit the maximum current in the rotor field.

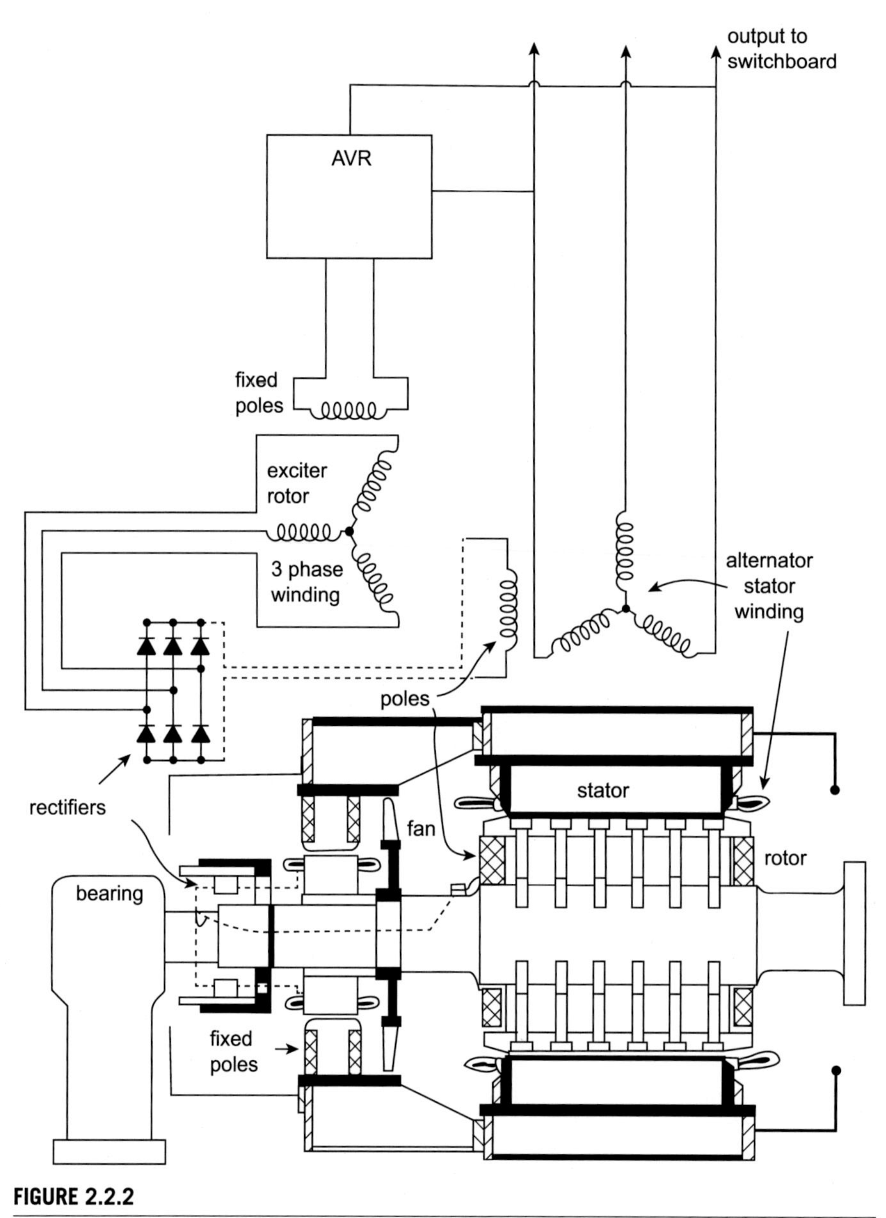

FIGURE 2.2.2

Diagram of a typical brushless alternator. *AVR*, automatic voltage regulator.

3. The stator and rotor heating limits – this is an alternator design limit associated with the type of insulation and the avoidance of damage.
   Considering the rotor, the major cause of losses in the rotor is the field winding resistive loss. If the field current is above a certain level, the I^2R losses will be too high, possibly causing insulation failure in the field winding. To manage thermal conditions, field current is limited; therefore, the induced excitation in the armature winding will be limited.
   Considering the stator, excessive armature currents will cause the stator temperature to rise beyond a safe level.
4. The underexcitation limit – this is an electromagnetic limit; underexcitation will weaken the field, causing instability and leading to loss of synchronism.

It is vital to operate the generator within the locus of the diagram in Fig. 2.2.3 to maintain a reliable supply and to avoid reduction in the operational life or damage to the generator.

PARALLEL OPERATION OF GENERATORS

When an AC generator is operating alone (often called 'Island' mode) the power factor is determined only by the load, and altering the excitation only changes the voltage (although changing the voltage will itself change the load and the fault level). Similarly, increasing the power input from the prime mover will increase the generator speed and hence the frequency.

When the offshore power system has more than one generator running on it in parallel, the power and reactive kilovolt-ampere contributions from each generator may be adjusted separately and it is possible to adjust the power factor of individual machines in this way.

A set of paralleling conditions must be met to avoid failure to synchronise and possible damage to both generators and switchgear:

1. The two generators must have the same root mean square line voltages.
2. The phase sequence must be the same in the two generators.
3. The two 'a' phases must have the same phase angles (assuming phases are a, b and c).
4. The frequency of the oncoming generator must be slightly higher than the frequency of the running system.

If the sequence in which the phase voltages peak in the two generators is different, then two pairs of voltages are 120 degrees out of phase and only one pair of voltages (the a phase) is in phase. If the generators are connected in this configuration, large currents will flow in phases b and c, causing damage to both machines and possibly the synchronising circuit breaker.

This would be a commissioning issue and corrected by swapping the connections on any two of the three phases on one of the generators.

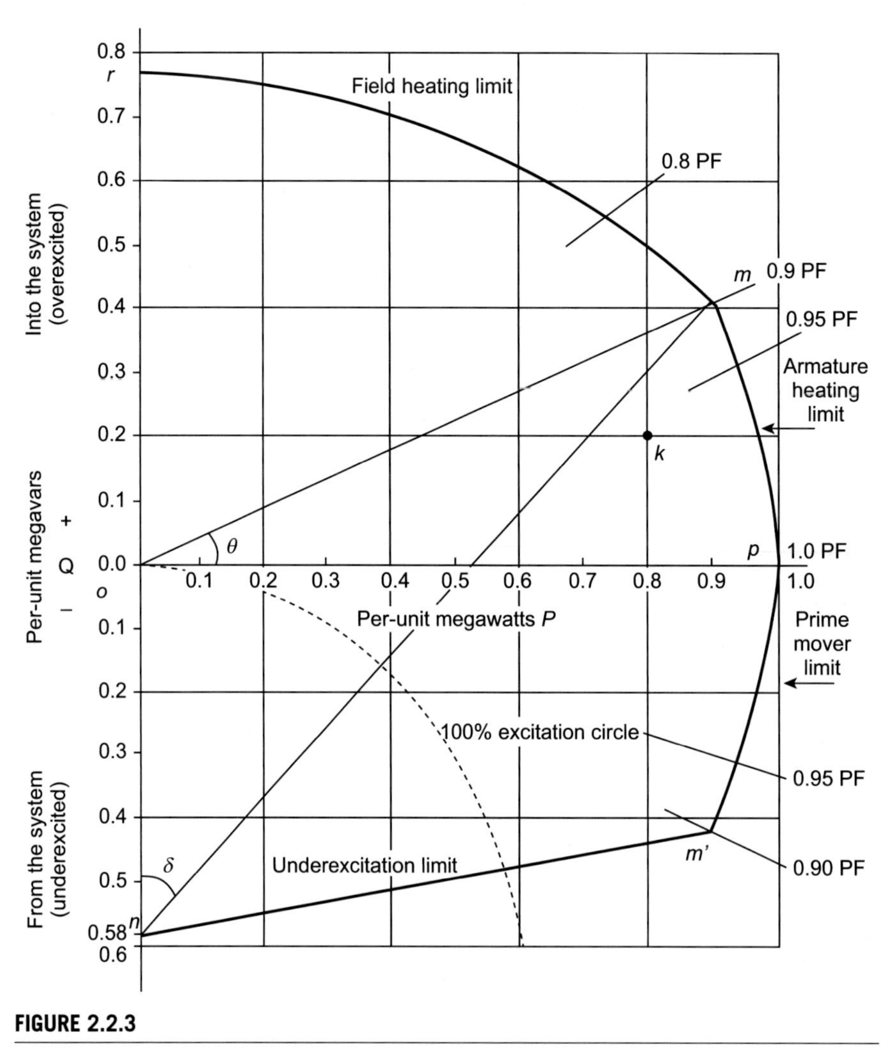

FIGURE 2.2.3

Alternating current generator capability diagram. *PF*, power factor.

If the frequencies of the power supplied by the two generators are not almost equal when they are connected together, large power transients will occur until the generators stabilise at a common frequency. The frequencies of the two generators must differ by a small amount so that the phase angles of the oncoming generator will change slowly relative to the phase angles of the running system. The angles between the voltages can be observed with suitable instrumentation, e.g., synchroscope, and the synchronising circuit breaker can be closed when the systems are exactly in phase.

LOAD SHARING OF GENERATORS

Unlike an 'infinite busbar' system as experienced in grid systems, where the frequency is maintained by a large number of generators and the output of each generator is controlled mainly by controlling the prime mover power output (governor), in the offshore or ship system the frequency is dependent on not only individual generator governors but also the type of load-sharing system operating. With a simple 'droop characteristic' control where each machine's speed decreases for the increasing kilowatt loading, if there is a tendency for the load on one machine to increase, it will tend to drop its speed thereby shedding some of its load. This arrangement works well with machines of equal size and characteristics (and with compounding when AVRs are utilised), but an electronic active load-sharing scheme will be required if machines of different size and characteristics are to share load successfully in most operating conditions. Such systems are available commercially as a modular part of a full multigenerator power management system (see PART 4 Chapter 8).

Once the generators have been paralleled, the power outputs of the two machines should be equalised by operating in governor droop mode, or by a load-sharing system. Reactive power is equalised by adjusting excitation. It is possible to parallel dissimilar machines, but there is a tendency for the differences in governor and AVR characteristics to cause one machine to take the load until the other is tripped on reverse power.

NOTE ON COMPOUNDING

To share kilovolt-ampere reactive between paralleled machines, a drooping voltage characteristic is required. This is provided inherently with hand regulation but machines under AVR control maintain virtually constant voltage (i.e., no droop), and instability will result unless the characteristic is modified. This can be achieved by using a compounding or quadrature droop current transformer to develop a voltage across a resistor. This voltage is added vectorially to the system voltage and the resultant applied to the AVR. The effect is to reduce the excitation of any machine supplying excessive kilovolt-ampere reactive.

Hand regulators should always be provided for testing/commissioning purposes. If part of the automation fails, it may also be necessary to operate one of the machines in hand regulator control, with other machines on AVR control providing for any load variations.

LOAD–VOLTAGE CHARACTERISTICS (REGULATION)

In machines of normal design, the drop between no load and full load voltage with a power factor of 0.8 will be of the order of 65%–75%, assuming constant excitation. At unity power factor the voltage drop is caused by stator resistance and by distortion

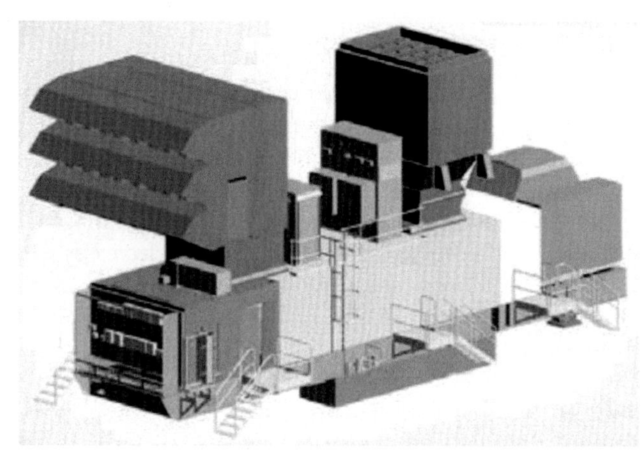

FIGURE 2.2.4

A typical gas turbine generator package (computer generated three-dimensional image).

in the flux path. As the power factor decreases, other voltage-reducing losses come into play caused by stator leakage reactance and by the component of the stator magnetomotive force which has a directly demagnetising action.

In practice the voltage is restored to normal by increasing the excitation, and it will be seen that the lower the power factor, the more excitation is required, thus increasing the amount of heat dissipated in the field windings. A typical gas turbine generator package is shown in Fig. 2.2.4.

CHAPTER

Emergency Generators

3

EMERGENCY GENERATION

Most offshore production installations have three or four main levels of operation which are reflected in control systems such as the emergency shutdown (ESD) system (PART 2 Chapter 15). If, however, there is a very large gas leak such that the installation is enveloped in a gas cloud, it would be necessary to isolate all forms of electrical power capable of igniting the gas, including in some instances, the direct current secure supply batteries. Assuming this dire situation (usually called Level Zero) has not occurred, the first level of operation is on battery power only and is considered in the ESD section in PART 2 Chapter 15.

The next level of operation is with the emergency generator only running. The provision of an emergency generator is a statutory safety requirement and as such should be designed to provide reliable power for statutory communications equipment, navigational aids, fire and gas monitoring, ballast systems for floating units and, although not a statutory requirement, accommodation cooking, drinking water, laundry and sanitation facilities. As this generator must not be dependent on the platform production processes for fuel, it is invariably diesel driven. Storage of petrol or propane on the platform would be considered a hazard, which would rule out the use of a spark ignition engine for this purpose. Emergency generators are usually designed to be automatically started on failure of other larger generators in the installation, by use of 'dead bus' relays. Again, there is a statutory requirement that the starting equipment for this generator is capable of at least six start attempts, frequently with a second bank of batteries on a manual changeover switch.

This generator should be located in a 'safe' area, close to the accommodation, radio room and process control room. A 'day tank' is required near the generator, big enough to run the machine for the time specified in the relevant statutory regulation. The time will vary depending on other installation conditions such as whether it is regarded as a 'manned installation' or 'normally unmanned installation' but may be 24, 48 or even 96 h.

The following points are often overlooked in specifications for emergency generator sets:

1. Despite the small size of the prime mover, air intakes must still be provided with spark-arresting devices and overspeed flap valves to prevent ingestion of gas, and exhausts with spark arrestors.

Offshore Electrical Engineering Manual. https://doi.org/10.1016/B978-0-12-385499-5.00005-4

2. Interlocking facilities must be provided to ensure that the generator circuit breaker cannot close on to an existing fault when the generator is automatically started.
3. In the event of a fault, means should be provided to maintain the generator output current for long enough to operate protection devices, where this is possible with the limited magnitude of fault current available from such a machine. Leaving the machine running with collapsed excitation is dangerous, as the fault may disappear followed by a sudden and possibly unexpected reappearance of full voltage on the system. Fig. 2.3.1 shows a photograph of the typical current design of an emergency diesel generator set.
4. The generator container should satisfy the following:
   a. Be pressurised, with automatic shutdown on loss of pressure, or alternatively all equipment within the container (including the prime mover) would need to be approved for use in a Zone 2 area (as per BS-EN 1834).
   b. Be fitted with overspeed protection on the diesel engine.
   c. Be fitted with appropriate noise suppression for its location.
   d. Be fitted with vibration reducing measures and sound isolating deck/structural connections/supports.
   e. The exhaust pipe on the engine to be fitted with an approved spark arrester.
   f. The electrical start battery to be fitted with a circuit breaker for 'Level Zero' installation ESD.
   g. A drip tray for collecting any oil or diesel leaks to be placed under the engine.

FIGURE 2.3.1

A typical packaged generator set.

Courtesy Gen Ex Design Ltd.

h. All oil and diesel lines to be made of hydrocarbon-resistant material (reinforced hose or piping).
i. Diesel tanks that are located such that the diesel is gravity-fed to the engine to be fitted with a manual shut-off valve.
j. Air intakes to be fitted with gas detection, which is part of the installation's central monitoring system.
k. Diesel engines to have an emergency stop on the outside of the container.
l. Machinery designed for unmanned operation to be equipped with monitoring facilities in the installation manned control room.
m. A permanent fire extinguishing system with automatic release to be installed in addition to a manual release located on the outside.
n. Be provided with protective earthing.
o. Be provided with equipotential bonding facilities.
p. Intrinsically safe (IS) instrumentation and telecommunications in the container will require an external separate reference earth.

5. The above-mentioned list is *not* exhaustive and will vary because of national standards, the installation's safety case, the location on the installation and the geographic location.
6. If for some reason (e.g., located in open air) the diesel engine requires to be approved for Zone 2 operation, then it will be necessary to comply with BS-EN 1834.

Because of the safety criticality of the emergency generator, a robust well-proven design should be utilised and rigorous factory inspection and test procedures must be applied during manufacture.

CHAPTER

Prime Mover Selection Criteria

4

INTRODUCTION

In this chapter the criteria for the selection of prime movers and generators for various applications are addressed. Before discussing the particular criteria for selecting an engine, the prime mover types are described.

GAS TURBINES

Although there are some smaller machines installed, it is not common practice to install gas turbines of less than 1 MW offshore. This is because

1. gas turbine reliability generally tends to improve at around this size for machines in continuous operation,
2. the bulk of intake and exhaust ducting involved to handle the large volume of air required, to reduce noise to acceptable levels and to protect the engine from the marine environment tends to make diesel or gas ignition engine prime movers more attractive up to this size.

There are two forms of gas turbine in use:

1. The aero engine derived type. This consists of a modified aeronautical jet engine known as the gas generator which exhausts into a separate power turbine. This combination often produces a unit with very good power-to-weight ratio, as the gas generator is lighter than the integral unit on the equivalent industrial machine. However, the unit may require better protection from the environment and in some cases shorter intervals between servicing. Examples of this type of machine are Rolls Royce Avon and RB211 based sets. Fig. 2.4.1A and B illustrates a typical example.
2. The industrial type. These are purpose-built engines which incorporate the gas generator and power turbine in a single design. The older machines were less fuel efficient than the equivalent aero-derived types but have a good reputation for reliability and for the toleration of fuel supply or load abnormalities. Examples of this type of machine are the General Electric/John Brown Frame 5 and the Ruston TB series. Fig. 2.4.2 illustrates a typical example.

Offshore Electrical Engineering Manual. https://doi.org/10.1016/B978-0-12-385499-5.00006-6

(A)

(B)

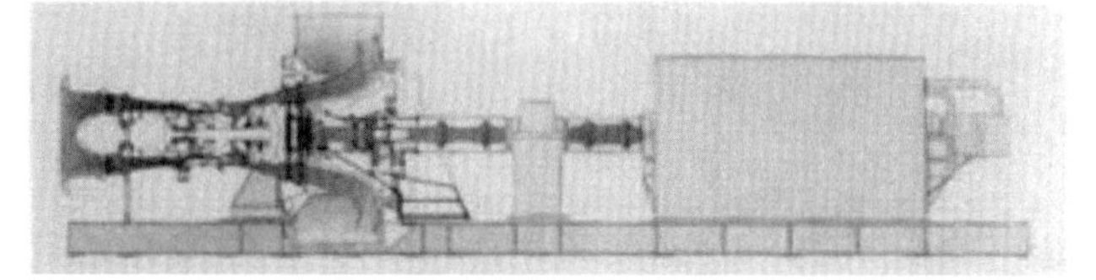

FIGURE 2.4.1

Aero engine derivative generator set. (A) Rolls Royce industrial Avon powered compression set being installed on the Brunei Shell Petroleum Company platform and (B) sectional drawing of set shown in (A).

Courtesy Crest Communications Ltd.

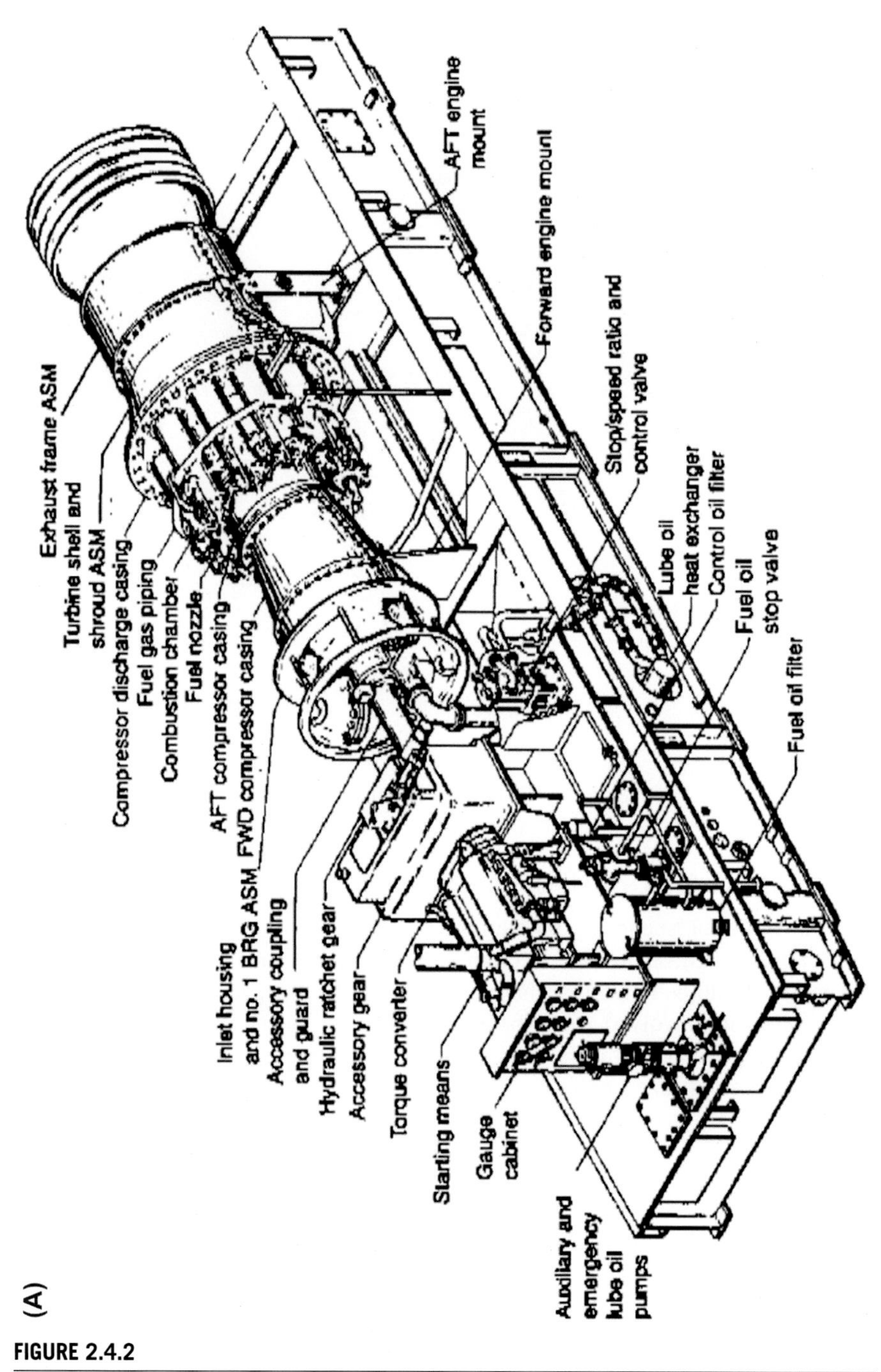

FIGURE 2.4.2

Industrial engine-based generator set: (A) engine and (B) generator.

(A) Courtesy John Brown Engineering (to GE (USA) design); (B) Courtesy Hawker Siddeley Electrical Machines Ltd.

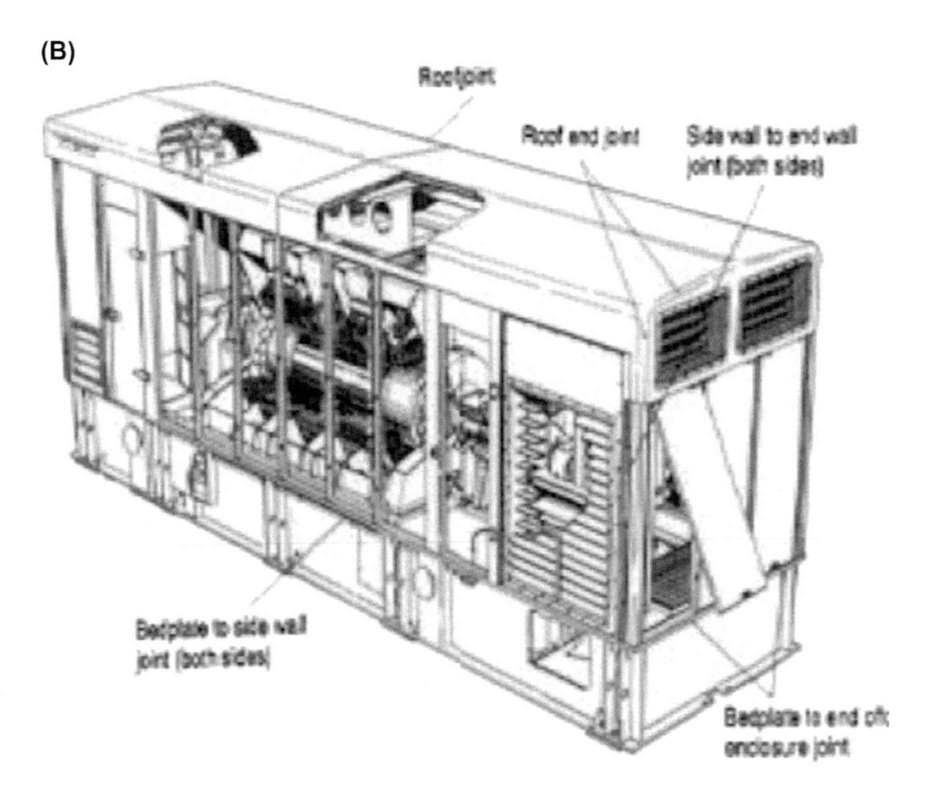

FIGURE 2.4.2, cont'd

GAS TURBINE APPLICATIONS

The following can be applied to both forms of gas turbine and is designed to assist the electrical engineer in the selection and application of gas turbines as prime movers. It should be noted that there are many other considerations, beside those in the following, involved when installing a turbine, but as they are of no direct concern to electrical engineers, they are beyond the scope of this book. However, some reference is made to firefighting facilities in generator rooms in PART 2 Chapter 15.

FUEL GAS SUPPLY DISTURBANCES

In most offshore situations (with the exception of storage and pumping stations) the gas is being produced via the process plant from a production well. Changes in well gas content, calorific value, pressure, etc. can have significant effects on engine power output, which may in turn affect the production process and the fuel supply. Slugs of condensate may also appear in the fuel gas supply, having a similar effect.

Although these unpredictable phenomena are avoided as much as possible by good process design, it is worth considering some means of catering for them such as one or more of the remedies described in the following.

1. A small separator or knockout pot may be located close to the engine intake to remove condensate. If the supply pipe is long, it will be necessary to provide this anyway in order to remove condensate which has condensed in the pipe.
2. The engine combustion system may be modified for dual fuel operation. When set up properly, this system can automatically transfer from gas to diesel combustion during fuel gas disturbances with negligible effect on engine power output.
3. If there is a problem with fluctuations in the calorific value of the gas to the extent that the electrical load cannot be met on a downward fluctuation, it may be necessary to install a fast-acting automatic load shedding system, as described in PART 2 Chapter 15.

TURBINE TEMPERATURE LIMITS

Gas turbines depend for their cooling on the vast quantity of air which passes through them, in other words high air mass flow. A basic design limitation on the operation of any gas turbine is the operating temperature of the power turbine blades. Disastrous changes in their mechanical properties will occur should they get too hot. By monitoring the exhaust gas temperature using a sophisticated control system, such as the GE 'Speedtronic' or the Ruston 'Rustronic' governor, it is possible to bring the engines up to power automatically and to continuously control the supply of fuel in such a way that the maximum exhaust temperature is never exceeded.

As the cooling effect of the incoming air is proportional to its density and temperature, the altitude and ambient air temperature have a very significant effect on the available output power. Altitude, of course, is fixed, but if the engine is running close to its rated power output then a small increase in ambient air temperature may cause the engine control system to limit the fuel supply in order to prevent the maximum exhaust temperature being exceeded. As it is unlikely that the generator load would have been reduced at the same time, the effect of this fuel supply reduction is for the generator set to begin slowing down until an underfrequency trip occurs. Even in the North Sea, warm weather conditions can sometimes reduce turbine power output capability below the required rating. The problem is often exacerbated by the poor location of combustion air intakes and exhausts.

The following must be considered:

1. Location of intakes away from any engine exhaust. Give these, including that of the engine under consideration, as wide a berth as possible.

2. The effects of any process flares must be taken into consideration, both for hot gas and radiation.
3. The effect of wind on all the various platform exhausts must also be considered. Although the prevailing wind is the most important consideration, the turbine must be able to develop sufficient power in any wind condition, and with any combination of other engines installed on the platform, working at their normal outputs. If the exhaust plume from another engine causes the engine in question to govern down, this may be overcome by the use of extra ducting or a water curtain installed around the exhaust of the other engine.

Airflow through the engine, and hence cooling, may be improved by cleaning the compressor section of the engine after a few months of operation. The improvement in output power after routine cleaning is usually significant and can be as much as 10% of its rated power. The selectors of the prime mover must take this into account in their rating calculations.

If it is necessary to install a load shedding system because of engine power limitations or increasing electrical demand, as mentioned earlier the system should take turbine exhaust temperature into account. If the load shedding system tripping level is based purely on monitoring electrical load for a fixed worst-case value, production operations may be unnecessarily curtailed. In colder weather conditions, as much as 15% of extra engine capacity would still be available. This may amount to several megawatts with a 25 MW generator set. The converse is also possible when exceptionally warm weather conditions may reduce engine capacity to below the load shedding system tripping point causing the generator set to trip on underfrequency.

GAS INGESTION FROM LEAKS

A large gas turbine generator set may take in the region of 30 s from the initiation of an emergency stop signal to slow down to a speed at which reacceleration is no longer viable without cranking. During this period of deceleration, it is vulnerable to ingestion of gas from serious process gas leaks on the platform. Such ingestion of gas in sufficient quantity, bearing in mind that all electrical loads would have probably been removed by the emergency shutdown system by this time, may cause the machine to reaccelerate and overspeed to destruction.

The risk of this occurring must be minimised by the following:

1. Carefully siting gas detectors to ensure that serious gas leaks are detected as soon as possible. A '2-out-of-N' voting system should be used to minimise spurious generator trips due to faulty detectors.
2. Governor response to an emergency engine stop signal should be as fast as possible within the metallurgical limitations imposed by turbine blade cooling rates etc.

RECIPROCATING ENGINES

DIESEL ENGINES

By far the most common engine for small- to medium-power requirements offshore, as well as prime movers for alternators, is the diesel engine and will be found directly driving anything from cranes to fire pumps. Diesel engines larger than 2 MW are rarely found on fixed platforms, however, for the following reasons:

1. Weight and vibration problems may be encountered in the platform structure.
2. If the engine is driving an alternator which provides normal production, i.e., nonessential supplies, and gas of sufficient quantity and quality is available during production, then importing large quantities of relatively expensive diesel oil is likely to be commercially unattractive. It would also require heavy storage tanks located in a site that would not constitute a fire hazard, which could well be a problem in the case of a steel structure.

It is possible to reduce the quantity of diesel oil consumed by burning a mixture of diesel oil and fuel gas. The ratio of gas to diesel is limited to approximately 90%, the limitation is because of the amount of diesel required to be injected as a pilot fuel to preserve the compression ignition action of the engine. This limitation has the benefit that a governed diesel engine is less likely to overspeed because of the ingestion of gas from a process gas leak. The gas supply has to be at very high pressure to enable injection at diesel compression cycle pressures, and therefore, fuel gas booster compressors are required for this duty.

GAS-IGNITION ENGINES

If it is required that the engine should run on fuel gas only, then an ignition system will be required similar to that found on petrol engine. It is normal, however, in order to improve reliability, to use a low-tension distribution system with individual coils mounted over the spark plugs on each cylinder, known as a 'shielded system'.

Reliability is improved by the following:

1. Reduced line loss in the coil secondary circuit, as it is very short. Plugs and coils screw together to form an integrated Ex certified unit.
2. The probability of earth faults occurring in the secondary circuit is substantially reduced, as there are no high-tension cables routed close together.
3. Common routing of high-tension cables can also lead to transformer effects which lead to ignition voltages appearing in the wrong cable at the wrong time, engine misfiring and loss of power.

FUEL GAS PRESSURE

As with a conventional petrol engine, the fuel–air ratio is controlled by means of a fuel injection engine management system, with the fuel gas pressure of a few pounds per square inch being accurately controlled.

FUEL GAS TEMPERATURE AND COMBUSTION KNOCK

Natural gas is predominantly methane which has a high octane rating and is therefore quite resistant to combustion knock ('pinking'). However, factors which increase charge temperature will also increase the likelihood of knock occurring. Such factors are

1. compression ratio
2. fuel gas temperature
3. high ambient air temperature
4. charge air temperature output from the turbocharger intercooler (where this is fitted).

There are other factors associated with the design of the engine which would affect how prone the engine is to combustion knock, and it is important to ensure that a manufacturer's warranty is provided, which states that the engine will run at the rated power output without any shortening of operational life, for the fuel and conditions expected on the platform. It would be prudent to provide the manufacturer with a recent platform fuel gas analysis if this is available.

SPARK DURATION AND VOLTAGE

As with most spark-ignition engines, it is important that the sparks are of sufficient voltage and duration to ensure good combustion. If the ignition engine generator sets poor sparking, the consequent poor combustion will give erratic speed control, with the effect being noticeably different from the regular hunting associated with governor control problems in that it appears as a random speed wandering over a few tens of revolutions per minute. The problem will tend to be worse at lower loadings and can make it difficult to obtain satisfactory operation from autosynchronising equipment.

FUEL GAS DISTURBANCES

Fuel gas is tapped from the platform process separators at often very high temperatures and pressures, and these must be reduced to values within the operating envelope of the prime movers being supplied. Instrumentation will be required to ensure that gas at the wrong temperatures and pressures cannot reach the engine. This would mean the automatic operation of isolation valves and shut down of generator set.

Occasionally, there may be fluctuations in the quantity of condensate that pass through the process system and reach the engines. 'Slugs' of condensate, as discussed earlier, are more likely to cause speed fluctuations in a gas turbine, but with a gas-ignition engine, significant quantities of 'incompressible' liquid present in one or more cylinders will invariably result in serious damage. 'Knockout' pots are essential for both types of engine, but it is essential that for gas-ignition reciprocating engines a small separator or knockout pot is installed close to the engine, which is designed to remove both slugs of condensate and any liquid which has condensed on the walls of the fuel gas supply pipe. High liquid level in the knockout pot should be arranged to shut down the engine.

LOAD PROFILES

The following topics need to be taken into consideration when selecting the number, type and rating of generator sets on a particular installation.

PROJECTED DEMAND

Over the life of the platform, the generation requirements may double or even triple as each new operational phase is reached. A typical demand profile is shown in the following.

Operation	Platform Power Demand (MW)
1. Drilling	3
2. Oil export	12
3. Gas compression/export	20
4. Artificial lift phase	25

VARIABILITY OF DEMAND OVER 24 HOURS

On a large oil production platform, the larger power users such as water injection pumps, main oil line pumps and gas compressors constitute the majority of the electrical demand. This will remain constant over 24 h unless some planned change of plant is necessary or a breakdown occurs.

If drilling activities are powered from the main platform system, some quite large but transient demands, possibly of the order of a few megawatts, can be expected from the rotary table or drawworks when 'difficult' mineral formations are experienced. On small installations where gas is exported without the need for compression, and where there are no 'round-the-clock' maintenance shifts, there will be a distinct profile created by the use of galley equipment and electrical water heaters in the accommodation areas.

LOW LOADING PROBLEMS

No internal combustion engine will run efficiently at loads much below 50% of its rated full power output. This is particularly a problem with reciprocating engines, as below about 30% of full power, combustion products such as soot and gum will begin to collect inside the engine in amounts sufficient to substantially reduce the power available from the engine when the load demands it. Turbochargers are particularly susceptible and will be stopped by deposits after only a few hours running at low loads. If the load profile dips dangerously low for an hour or so, followed by a period where the load is substantially higher, i.e., greater than 50% of full power, it is likely that the increased combustion will clear the engine of the build-up in these deposits.

As with a fuel gas analysis, it is important that a cyclic load profile is presented to the engine manufacturer and a statement obtained from him to the effect that the engine will continue to run without deterioration with the load profile submitted. Should such guarantees not be forthcoming, it will be necessary to reconsider some of the other engine options and/or whether the most suitable number and rating of generator sets has been selected.

CHOICE OF FUEL

The choice of fuel is usually governed by the following:

1. The quality and quantity of gas available from the field being exploited. Where the reservoir produces a preponderance of oil and there is a likelihood that there will not always be sufficient gas for fuel, then some other fuel will be required, either continuously or as a standby.
2. Availability of well gas at the times when the engine is required to run. If the engine is the prime mover for an emergency generator, it will need to run when the production process is shut down and no gas supply is available.
3. Logistical costs associated with the transport of diesel oil to the platform concerned. The use of diesel oil requires that sufficient quantities of diesel can be stored on or in the structure to allow for periods of bad weather when refuelling is not possible. As this fuel will constitute a considerable fire risk, the storage location will need to be carefully considered.
4. Comparative costs of connection to a suitable power generation complex. This may be a nearby platform with spare capacity or an onshore supply system. The costs may also need to include those for the purchase and installation of a suitable motor generator set to cater for platform supply frequency and/or to effectively reduce the supply system impedance.

MAIN GENERATION

Having established all the limitations, such as weight and dimensions, imposed by the platform structure, the supply and operating constraints can be tackled.

NUMBER REQUIRED

From the reliability point of view, three generators, each rated for the full platform load, is the optimum. This allows for one generator running, one standby and one undergoing routine maintenance. If space is in particularly short supply, it may be necessary to dispense with the third machine. If point loading on the platform structure is a problem, it may be necessary to use a greater number of smaller rated machines.

SIZE

The types of prime mover available have already been discussed in this chapter. One of the benefits of using large machines, each capable of supplying the whole system load, is that of improved motor starting capability and greater stability during power system disturbances. However, some care will have to be exercised to ensure that alternator subtransient reactances on such large machines are not so high that switchgear of sufficient fault capability cannot be obtained without going outside the normal ranges of switchgear in standard production. Few manufacturers produce standard switchgear at voltages in the 11–15 kV region above 1000 MVA. Cables capable of withstanding the associated prospective fault currents would have to be sized very much over their current ratings and would be expensive, heavy and more difficult to install. Therefore, at present, 1000 MVA is considered the practical top limit for system fault levels.

LOCATION

Generator modules should be located in an area classified as 'safe' when the release of flammable gases is considered. This is necessary to reduce the risk that flammable gas might be drawn into the engine enclosure and be ignited on hot parts of the engine. If a small quantity of gas is drawn into the engine intake, this should not cause a significant increase in engine speed as the governor should correct for the presence of this extra 'fuel'. Large concentrations of gas, however, may cause overspeed in gas turbines and, in all engines, may interfere with combustion and, if no precautions were taken, may be ignited in the engine air intake, leading to fire and/or explosion.

In general, the following precautions must be taken with gas-fuelled engine enclosures.

1. Any part of the engine, including ancillaries such as turbochargers, exhaust systems and anything else in the enclosure which may have a surface temperature in excess of 80% of the ignition temperature of the actual gas/air mixture (200°C in the case of North Sea gas), must not under any circumstances be allowed to come in contact with such an explosive gas/air mixture. This can be avoided by
   a. providing sufficient ventilation to prevent gas accumulation,
   b. enveloping the hot areas in a water-cooling jacket (note that it is fairly impractical to do this with a turbocharger),

   c. ensuring that the enclosure is always positively pressurised (with air free from gas), so that gas cannot be drawn in from outside the enclosure,
   d. keeping potential gas leak sources in the enclosure (such as fuel gas pipe flanges) to an absolute minimum.
2. A 'block and bleed system' should be included in the fuel gas supply system so that when the engine is stopped, the entire length of the fuel gas supply pipe within the enclosure is blocked by isolating valves at both ends and vented safely to the atmosphere.
3. Electrical instrumentation and controls associated with the fuel gas pipework, such as pressure and temperature transmitters, solenoid valves and throttle actuators, should all be suitable for safe use in areas where explosive gas mixtures may be present. With spark-ignition engines, the shielded type ignition system, as discussed earlier, should be used. This is because the high ignition voltages are only present within the engine with this type of system. The subject of hazardous areas is discussed in greater detail in Part 5 Chapter 4.
4. In reciprocating engines, various extra precautions need to be taken as follows:
   a. Drive belts must be of the antistatic, fire-resistant type.
   b. Cooling fan blades must be of a type which cannot cause friction sparks if they come into contact with adjacent parts.
   c. Exhaust systems must be fitted with flame traps and spark arresters.
   d. A flame trap may be required in the combustion air intake to protect against a flashback through the induction system (i.e., backfiring). This will be required even if the intake is in a safe area, if the engine may be run in emergency conditions.
   e. All diesel engines should be fitted with a 'Chalwyn' or similar air induction valve to prevent overspeeding if flammable gases are drawn in with the combustion air. This is particularly important if the engine runs on a fixed fuel-rack setting, i.e., is not fully governed.
   f. If the crankcase volume is greater than $0.5\,m^3$, relief valves must be fitted to the crankcase to prevent damage or external ignition due to crankcase explosions. The relief device must be provided with its own spark arrester/flame trap system.
   g. Care must be taken to ensure that any special design features that could cause external ignition are adequately catered for. For example, turbochargers must be water-cooled; decompression ports, if absolutely necessary, should be treated in the same way as exhaust systems.
   h. Engine governors, and the fuel injection pumps of diesel engines, should be designed so as to make reverse running of the engine impossible.

   For further information, please refer to Publication MEC-1 of the Engineering Equipment and Materials Users Association (formerly OCMA) entitled Recommendations for the Protection of Diesel Engines Operating in Hazardous Areas.

Table 2.4.1 Heat Balance Values

Energy Gain	Heat Loss	Electrical Output
Energy provided by fuel – 1500 kW		
Heat lost to engine water jacket	450 kW	
Heat radiated from engine	50 kW	
Net mechanical output from engine – 500 kW		
Losses from alternator (eff. = 90%)	50 kW	450 kW
Heat lost to turbo intercooler	50 kW	
Heat lost to engine exhaust	450 kW	
Total	1050 kW	450 kW

COOLING SYSTEMS

Although this subject and most of the following ones in this section are definitely in the realms of mechanical and other engineering disciplines, the electrical engineer needs to be vigilant for the proceedings or risks that occur when problems such as insufficient cooling arise during commissioning of the generator module. At some stage during purchase and manufacture of the generator set, the manufacturer will provide heat balance figures. Typical heat balance figures are listed in Table 2.4.1.

As with an accountant balancing his books, all the waste heat from the engine must be accounted for in the design of cooling and ventilation systems. The ratings of the engine and alternator are based on designed operating temperature bands, and if these are exceeded when the engine is running at its rated power output because of poor cooling and ventilation, the generator manufacturer will have to derate the equipment accordingly. The above-mentioned heat balance example is for a reciprocating engine, but the same principle may be applied to turbines. In generator modules where all ventilation is provided by the engine radiator fan, an allowance must be made for the temperature rise caused by heat dissipated within the module before the airflow reaches the engine radiator. Wind speed and direction will also affect the airflow through the module and when the wind is strong and blowing directly against the fan, it may stall the airflow completely, causing a rapid temperature trip. If the radiator fan is electrical, 'windmilling' of the fan should be prevented when the engine is not running, otherwise when the engine is started the fan motor may trip on overload due to the excessive acceleration time from some negative to full forward speed. If power is available from another source after the generator has stopped, it is advisable to have another smaller fan running to prevent a build up of heat in the compartment while the generator set is cooling down. Without this, the temperature in the compartment may exceed maximum allowable values for electrical equipment or insulation.

LUBE OIL SYSTEMS

On large gas turbine generator sets, lubrication is accomplished by a forced feed lube oil system, complete with tank, pumps, coolers, filters and valves. Lubricating oil is circulated to the main bearings, flexible couplings and gearboxes. A portion of the oil may be diverted to function as hydraulic oil for operation of guide vanes, etc. within the turbine. A typical arrangement is for the lube oil pumps to take their suction from the lube oil tank and the hydraulic control valves to take their suction from a bearing header. The system may contain between 500 and 2000 gallons of oil.

The electrical engineer's interest in this system is that out of, for example, six lube oil pumps on each generator set, five of them are driven by electrical motor. The following is a description of the typical function of such pumps.

1. Main pump
   The main lube oil pump is a shaft-driven positive displacement unit mounted into the inboard wall of the lower casing of the accessory gear. It is driven by a splined quill shaft from the lower drive gear and the pressure is 65 psig max. As the output pressure of this pump is engine speed dependent, with certain models of turbine, insufficient lubrication pressure is available from this pump below a certain speed, and in this case, even when it is operating satisfactorily, this pump may require to be supplemented by an electrically driven pump.
2. Auxiliary pump
   The auxiliary pump, mounted on the oil tank cover, is a submerged centrifugal type pump which provides lube pressure during startup and shutdown of the generator under normal conditions. The auxiliary pump is driven by a low-voltage 30 hp two-pole alternating current (AC) flameproof motor.
3. Emergency pump
   The emergency pump is also mounted on the tank cover and is a submerged centrifugal type which also provides lube pressure under startup or shutdown conditions. This pump is driven by a 125 V 5 hp direct current (DC) flameproof motor.

   The three remaining pumps of our typical system are listed in the following:
4. Main hydraulic supply pump and auxiliary hydraulic pump
   Failure of these pumps and the resulting low hydraulic pressure would not necessarily cause an immediate generator set failure, although the unit would eventually trip because of low lube pressure.
5. Hydraulic ratchet pump
   Loss of hydraulic ratcheting pressure or equipment would cause generator shaft bowing and also lead to excessive bearing stress. It may, however, be possible to repair the fault by, for example, replacing the motor within a few hours, i.e., before serious damage has been caused.

These three pumps would all be normally driven by low-voltage AC motors.

The basic criteria for establishing electrical supplies to generator lube oil pumps and other vital auxiliaries are as follows:

1. Although weight and space limitations usually prevent auxiliary switchboards being fed from separate generator transformers, as one would expect to find in a power station onshore, the switchboard or individual motor starters should be electrically as close to the generator as possible.
2. Each complete set of electrically driven auxiliaries should be supplied from one switchboard, so that only one supply is required to ensure that the particular set of auxiliaries is available.
3. Any standby auxiliaries should be fed from another switchboard, which obtains its supply from the generator via a different electrical route.
4. If possible, the complete set of auxiliaries should be duplicated in order that a fault on one switchboard or its incoming supply will not lead to a shutdown of the generator.
5. Loss or temporary disconnection of supplies to the emergency DC lube oil pumps should be adequately displayed on the generator control panel annunciator so that the operator is aware that loss of AC power and, consequently the failure of the AC-driven lube oil pumps, may lead to damage to the generator as it runs down.

GOVERNORS

Sophisticated electronic governors able to provide reliable service coupled with high performance have been available for some time. However, there are a few points worth considering when selecting a suitable governor. Firstly, it is advisable to obtain as many of the governor's control system parameters as possible from the governor manufacturer. These will be required if the power system is to be computer simulated before the project design stage is complete, which is usually the case. Any governor-engine-alternator system will have a finite response time. It also has a certain amount of stored energy which can be extracted from the system by sacrificing some speed, and this energy may be used while the governor is responding to a load increase to maintain the system. Apart from large power system disturbances such as large motors starting or short-circuit faults, which need to be studied by dynamic simulation, any cyclic load should be considered carefully. The following anecdote illustrates the problem. A radio transmitting station was built in an isolated location without access to any external electricity supply, depending entirely on two diesel generators for electrical power. The station was successfully commissioned with the exception that the supply frequency became unstable and the generators shut down during each broadcast of the time pips. The problem was traced to the cyclic loading imposed by the series of pips being at a frequency at or close to the natural frequency of the governor mechanical linkages. In this case the problem was overcome by substantially increasing the mass of the engine flywheel. The only cyclic loads likely to be experienced offshore are those produced by heaters which use thyristor integral cycle firing controllers. The phase angle controlled rectifiers and variable-frequency invertors used on the drilling rigs are unlikely to produce the necessary supply frequency subharmonics.

ALTERNATORS AND EXCITATION SYSTEMS

The control system parameters for the generator excitation system, like those for the governor, will be required if any computer simulation work is to be carried out.

The following aspects need particular attention in offshore systems:

1. Although, as with large onshore machines, two-pole 50 or 60 Hz designs are used because of the greater efficiency of energy transfer at the higher speeds, it should be remembered that this requires the need for complex engineering analysis and the use of high-grade materials, particularly as the generator module could be located on a 600-ft-high steel structure, 100 ft above sea level.
2. To the author's knowledge, the highest generator rated voltage offshore is 13.8 kV and alternator manufacturers have no difficulty in producing machines at this rating. Voltages up to 22 kV could be used, provided suitable switchgear is available.
3. The alternator subtransient reactance (X″d) is a useful regulator of maximum prospective fault current, and alternator manufacturers are usually prepared, within certain limits, to vary the design of the windings to enable the system designers to limit the prospective fault level to a value suitable for the switchgear available. For a machine of, for example, 30 MW, the degree of variation for X″d would be approximately 15%–21%.
4. Although, as discussed earlier, a highly reactive machine may be beneficial to limit prospective fault currents, this is accompanied by the penalty of poor motor starting performance because of the increased transient reactance (X′d). The winding reactance must therefore be optimised for the best motor starting performance allied with prospective fault capabilities within the capacity of the switchgear installed. This 'trade-off' is best accomplished using computer simulation.
5. The conventional configuration of brushless alternator with pilot and main exciter is commonly used offshore for machines of 500 kW rating and above. Machines with static (rectifier derived) excitation are acceptable, provided fault currents can be maintained for at least the full generator fault time rating. A definite time overcurrent device should be used to shut down the generator within this time, so that the machine is not left running with a fault still on the system after the voltage has collapsed.
6. The automatic voltage regulator should incorporate a means of detecting a control loop disconnection, such as that caused by open-circuit voltage transformer fuses, in order to avoid excessive voltages being developed on the machine stator if such a disconnection occurs.

NEUTRAL EARTHING

Typical low-voltage solidly earthed and medium-voltage resistance earthed systems are shown in Figs. 2.4.3 and 2.4.4. There is little difference between offshore and onshore practice with regard to generator neutral earthing. However, it is worth repeating that earth cables and earthing resistors should be adequately rated, both for

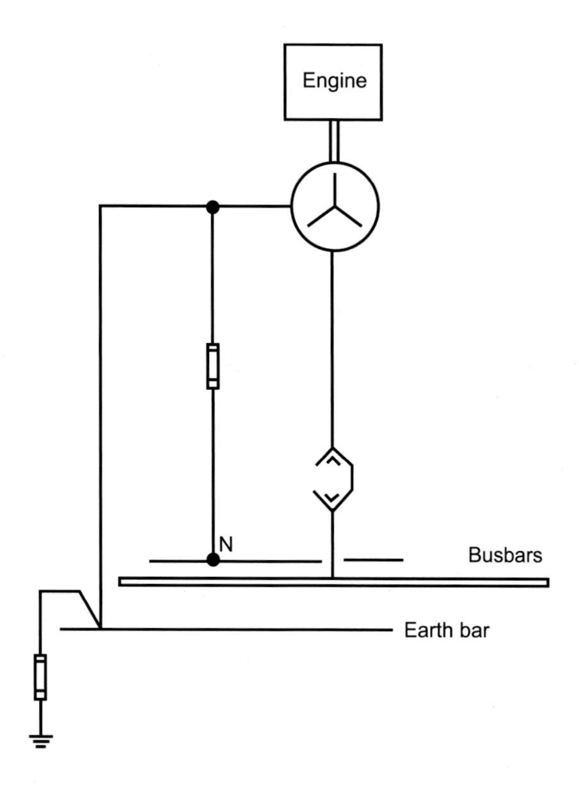

FIGURE 2.4.3

Low-voltage solidly earthed arrangement.

current magnitudes and circuit breaker tripping times. Because of the marine environment, earthing resistors tend to deteriorate quite quickly and require very regular maintenance (in the author's experience, at least once a year).

STARTING REQUIREMENTS

At some stage in the design of the platform power system, it is worthwhile carrying out a Failure Modes, Effects and Criticality Analysis (FMECA) of the system to ensure that as many operational problems can be foreseen, and catered for, in the design as possible. The FMECA methods are discussed in PART 8 Chapter 1.

Consideration of operational problems is especially important when providing for the bringing of generators into service when all or most platform services are unavailable, i.e., black-start facilities. Although written black-start procedures should be available to the operators, these should reflect the permanent facilities installed. Black starting cannot be safely or adequately catered for by describing some temporary rig in the 'Installation Standing Instructions'. For example, the installation manager would not thank the system designers if in certain conditions

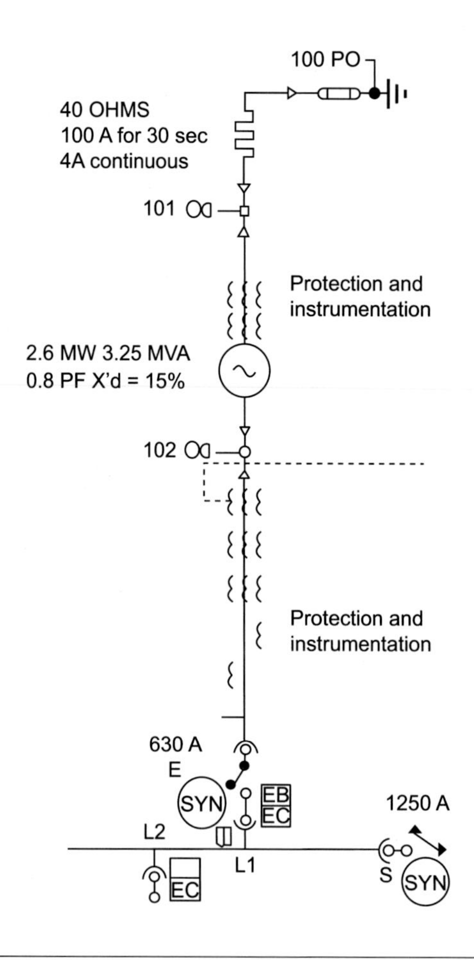

FIGURE 2.4.4

Medium-voltage resistance earthed arrangement.

it became necessary to fly a small generator set or compressor out to the platform in order to restart generators and hence continue the output of oil from the installation.

Maintenance must also be catered for in starting facilities such that whilst part of the system is being maintained, there is still at least one method of black start open to the operator. An example of this would be the need to allow for planned or unplanned outages in the emergency switchboard whilst still providing electrical supplies to main generator facilities, such as engine cranking motors and lube oil auxiliaries; a provision which has often been overlooked in the past.

KEY SERVICES GENERATION

For intermediate operating conditions, when, for example, certain process or service machinery is required, but production is shut down or the level of production does not require the running of main generators, key services or submain generators may be run. These would allow all the utilities, such as accommodation ventilation and cooling systems, plus the statutory services, to be operated without the need to run the main generators at inefficient power levels. Key service generators can also be used to provide peak load power, particularly during a planned outage of a main generator. It is best, however, to avoid running machines of dissimilar size in parallel whenever possible, as the shorter time constants usually associated with the smaller machine tend to cause it to react more quickly to step increases in load and this leads to system stability problems.

Emergency generation is covered in PART 2 Chapter 3.

CHAPTER

Generation and Distribution Switchgear and Transformers

5

SWITCHGEAR – GENERAL REQUIREMENTS

Switchboards are the nodal points in any electrical system. They must be designed to provide a safe, reliable means of carrying and directing current flows where they are required and must be able to withstand the thermal and magnetic stresses involved in clearing faults. This fault capability must include:

1. With the circuit breaker initially connecting two healthy sections of the system, the ability to interrupt a fault immediately after it occurs in a fast and reliable manner. This is known as the fault breaking capability.
2. The ability to connect a live section of the system to a faulty section and then to immediately interrupt the resulting fault, in a fast and reliable manner. This is known as the fault making capability.

For an in-depth discussion on the operation of circuit breakers, the reader is recommended to refer to the *J&P Switchgear Book* (See Bibliography). The following is a brief introduction to the principles and should assist the engineer in selecting switchgear of adequate performance for the application.

Switchboards on vessels and platforms are normally classified in one of the following categories:

- Main switchboards, to which the generators are connected.
- Primary distribution switchboards, supplied either directly or via transformers from the main switchboard.
- Emergency switchboards, fed normally from the main switchboard, but with a direct supply from the emergency generator for emergencies. On ships emergency generators have traditionally not been provided with paralleling facilities, as it was considered that the emergency generator should be reserved for emergencies. (Note: On platforms, emergency generators are increasingly being fitted with paralleling facilities. However, the benefits of easy load testing of the emergency generator and flexibility of supply have to be weighed against cost of upgrading the emergency switchboard fault rating to allow this.)
- Main and emergency distribution boards fed from the appropriate supply and located close to the supply users.

Offshore Electrical Engineering Manual. https://doi.org/10.1016/B978-0-12-385499-5.00007-8

THE MECHANISM OF SHORT CIRCUIT CURRENT INTERRUPTION

Fig. 2.5.1 shows the dc or asymmetrical component that will exist in the current flowing in at least one of the phases unless the short circuit appears when circuit voltage is at a maximum. This voltage decays after a few cycles and the resulting decrement curve can be seen. On an offshore platform, where power is more than likely to be locally generated, there will also be an ac or symmetrical decrement associated with

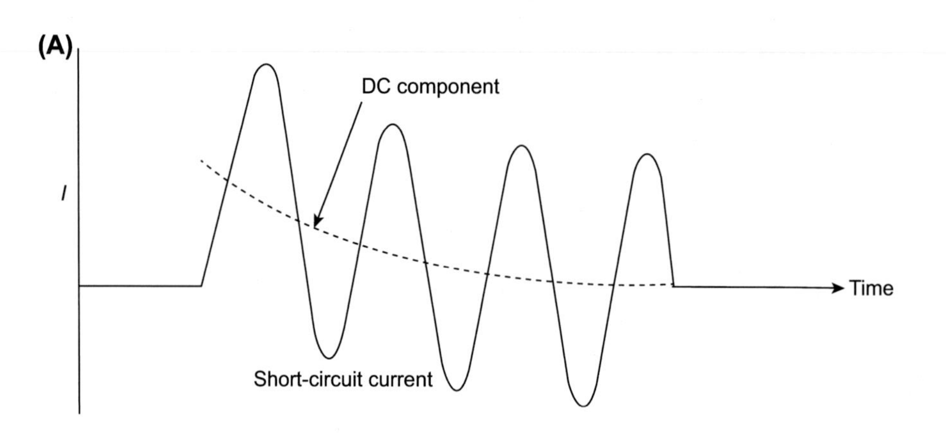

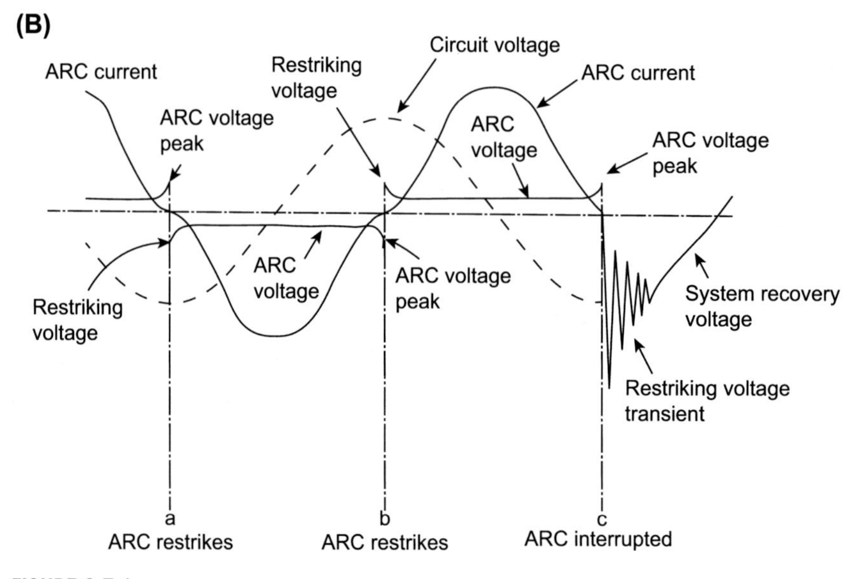

FIGURE 2.5.1

Interruption of asymmetrical short circuit current: (A) short circuit current (B) circuit breaker current interruption.

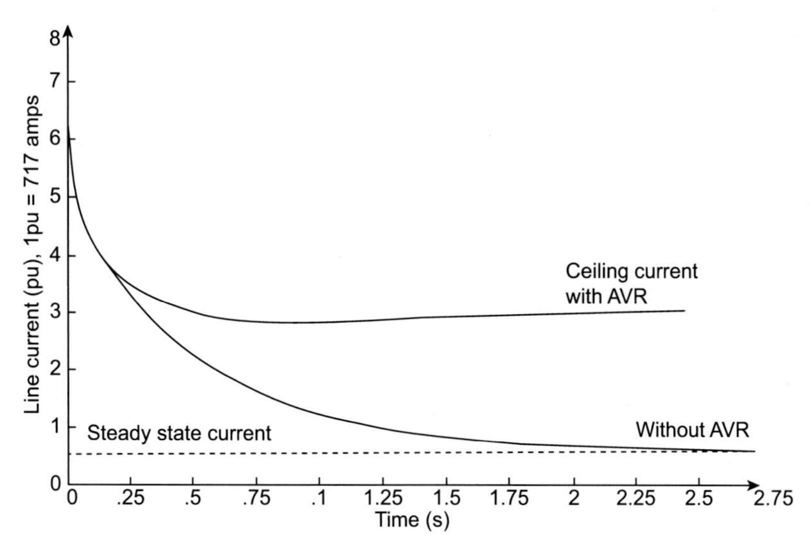

FIGURE 2.5.2

Typical generator decrement curve: $I_{SC} = I_{FU}/X_d$, where X_d is generator subtransient reactance.

the dynamic reactances of the generators. A typical generator decrement curve is shown in Fig. 2.5.2.

Fault making and breaking currents for particular operating configurations must be calculated in a standard manner if the performance of switchgear from different manufacturers is to be compared with any degree of consistency.

Fig. 2.5.3 shows waveforms for two fault situations, (A) at system voltage peak, and (B) at system voltage zero. Situation (A) shows no dc component and the peak current is $\sqrt{2}$ times the symmetrical current. In situation (B), the maximum dc component is present and the peak current is $2\sqrt{2}$ times the symmetrical current. The power factor of the fault current depends on the R/X ratio of the circuit, where R is the effective resistance of the circuit and X is the effective reactance. If the total impedance of the circuit was only resistive, the fault current which flows will be symmetrical, irrespective of the point on the current wave at which the fault occurs. Therefore the asymmetry and hence the ratio of rms symmetrical current to peak current varies with power factor. This is shown by the graph in Fig. 2.5.3.

BREAKING CURRENT

Referring to Fig. 2.5.1, that when the fault occurs with the circuit breaker initially closed, the fault current decays in line with the generator impedance (Fig. 2.5.1A), and therefore the switchgear operating time allows the fault current

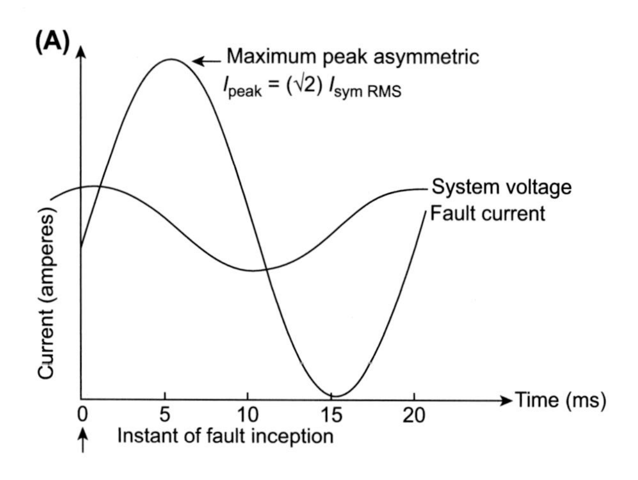

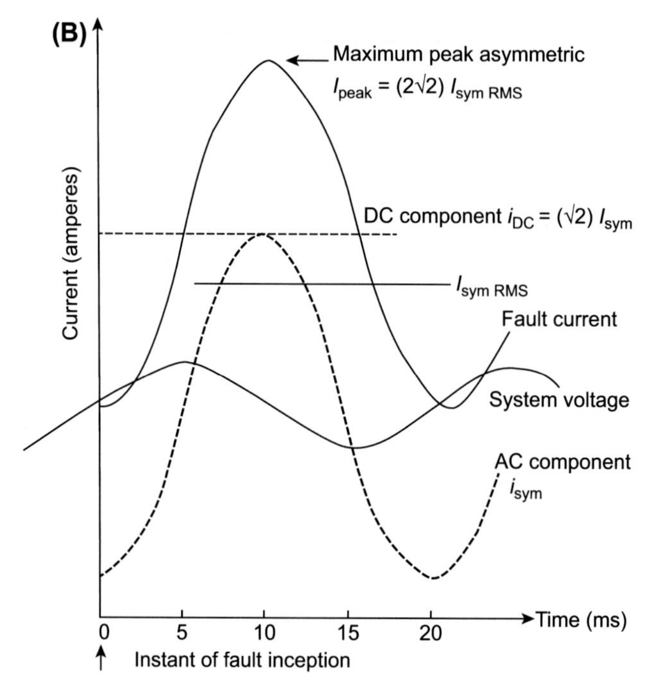

FIGURE 2.5.3

Fault inception at different points on the waveform. (A) System voltage peak. (B) System voltage zero.

to decay before the contacts have to deal with the arc. The magnitude and duration of the arcing currents shown in Fig. 2.5.1B will vary depending on the type of interrupting device (fuse, circuit breaker etc.) and the dielectric it uses (e.g., air, vacuum, SF6 etc.).

MAKING CURRENT

The most onerous fault for a circuit breaker to handle is where a short circuit fault exists on a part of the system which is isolated from the generator(s), and the intervening circuit breaker is closed. This would mean that the full system voltage is applied with no decrement. If the fault is close to the circuit breaker, then the intervening impedance is small, giving the highest instantaneous current flow.

TYPES OF INTERRUPTER

HRC CARTRIDGE FUSES

1. A fuse is basically a device with a central conductor that is designed to melt under fault conditions. Interruption is achieved by spacing the two ends sufficiently far apart for the arc to be naturally extinguished. However, to obtain a consistent performance characteristic, fuse elements are carefully designed for particular voltage and current ratings. The ceramic cartridge tube is filled with powdered quartz and sealed. The silver element is not a continuous strip of silver, but is necked in short sections to reduce pre-arcing time. It may also have sections of low melting point, 'M'-effect material, to improve performance at low fault levels (see Fig. 2.5.4).
2. The physical design of the element such as the length and shape of the necks will depend on the application of the fuse and its operating voltage. If the maximum arc voltage is exceeded, the pre-arcing time will be reduced and the fuse will not operate according to its standard characteristic. Since a fuse is purely a means of protection against overload and fault currents, isolators or contactors are installed in series to carry out normal switching operations.
3. The advantage in using fuses is that the resulting switch-fuse device is often less expensive, smaller and lighter than the equivalent circuit breaker, particularly in the higher power ranges. Exceptions to this for low voltage distribution equipment are some of the higher performance current limiting miniature circuit breakers. These will be discussed later in the chapter.
4. The disadvantages with fusegear are:-
   a. HRC fuses must be replaced, once operated and no matter how reliable the circuit is, some holding of spare fuses will be necessary and replacement of the larger bolted fuses is time-consuming. Permanent power fuses are available, which use the thermodynamic characteristics of liquid sodium to interrupt fault currents for a few milliseconds during which time a suitable switching device is operated to isolate the faulty circuit. This type of fuse is not (to the author's knowledge) used offshore.
   b. The standard ranges of fuses available are limited in current and fault capacity below the maximum ratings available in switchgear ranges.

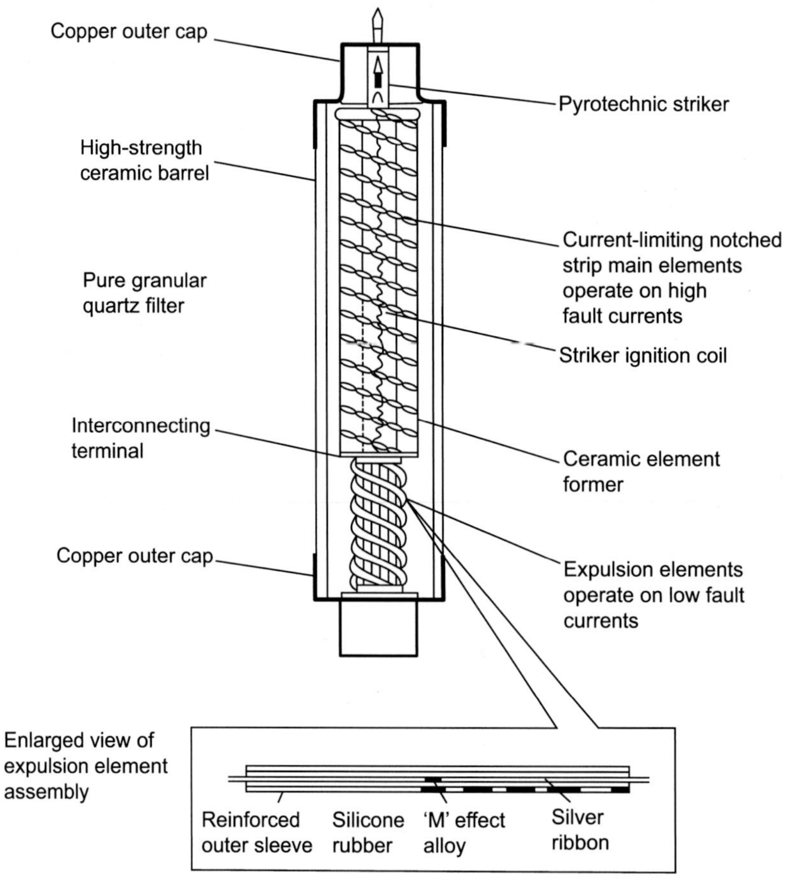

FIGURE 2.5.4

Diagram showing the interior of typical HRC fuse.

Courtesy GE Electrical Distribution & Control.

AIR CIRCUIT BREAKERS

An air circuit breaker (ACB) is a device where the circuit is made or interrupted by moving contacts located in atmospheric air. The ACB relies on moving the contacts sufficiently far apart to extinguish the arc under short circuit conditions. To assist in this operation, the design of ACBs has been greatly improved since the original invention, notably, in the following ways:

1. The use of 'trip-free' mechanisms, i.e., where the geometry of the mechanism is such that tripping can occur even while a closing operation is under way.
2. Separate arc and current carrying contacts for better thermal rating.
3. The use of blowout coils and arc-chutes. A blowout coil uses the fault current to produce a strong magnetic field which pulls the arc away from the contacts

into the arc-chute. The arc-chute is a series of parallel insulators designed to extinguish the arc by splitting and cooling it. Because of the need for blowout coils, performance will be partially related to the fault current magnitude. This effect will appear as a 'critical current' below which the arc is not drawn into the chute and arc contact wear will be accelerated. Thus the risk that fault clearance will not be achieved will be increased. This type of switchgear also tends to be bulky, mechanically complex and hence more costly compared with other forms now available. However, this equipment is well proven and its use well established offshore, since it has the ability to interrupt the high magnitudes of fault current found in offshore systems.

BULK OIL CIRCUIT BREAKERS

The bulk oil circuit breaker is a device in which the moving contacts are totally immersed in a container of mineral oil. As oil is a better insulant and has a higher specific heat than air, contact gaps may be reduced and the better heat dissipation means that the overall cubicle size may be reduced compared with those for an ACB of the same rating. Arc extinguishing is assisted by convection currents in the oil, produced by the heating effects of the arc. No weight saving is likely, however, because of the weight of oil required. Arcing within the oil causes hydrogen gassing, which must be vented from the container, and although with modern designs the risk is minimal, open flames or even explosions may occur under severe fault clearing duty or normal fault clearing with badly contaminated oil. The insulating oil can deteriorate over a number of operations or after a long period of time, and requires regular sampling and testing. Sufficient replacement oil must be stocked offshore or at least made available offshore prior to testing.

Although this form of circuit breaker is still in common use onshore and has been installed for main switchboards offshore, it is now generally considered undesirable in an offshore environment and its use may be questioned by a number of certifying authorities and underwriters.

LIMITED OIL VOLUME CIRCUIT BREAKERS

If oil is injected at high velocity between the contacts as they open, an efficient means of arc extinguishing may be obtained. The amount of oil required is much less than with a bulk oil circuit breaker, and therefore the fire risk is reduced. However, as with the air break and bulk oil, maintenance requirements are usually heavy compared with vacuum and SF6.

VACUUM CIRCUIT BREAKERS AND CONTACTORS

At the heart of the vacuum circuit breaker is a device called the vacuum interrupter, one of which is required for each pole. The interrupter (see Fig. 2.5.5) consists of a ceramic tube with metal seals at both ends. The fixed contact is mounted on one metal

(A)

(B)

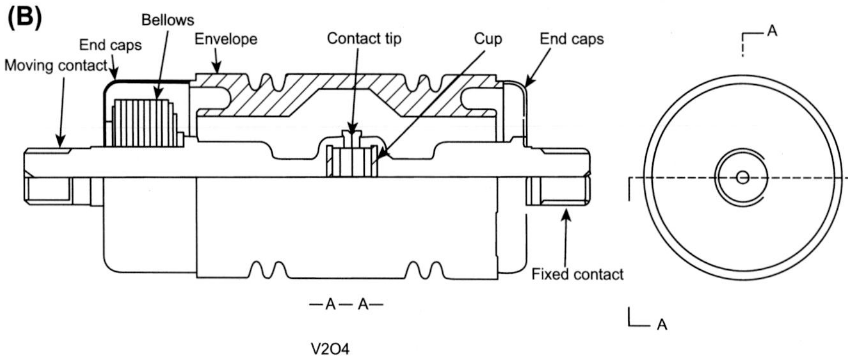

FIGURE 2.5.5

Interior of typical vacuum interrupter. (A) Photograph of sectioned interrupter. (B) Sectional drawing.

Courtesy GEC Alsthom Vacuum Equipment Ltd.

seal while the moving contact at the other end of the tube is free to move inside a metal bellows which maintains the vacuum seal. The ceramic tube is usually in two parts, to allow the insertion of a sputter shield designed to prevent contact metal condensing on the ceramic tube and providing a conducting path between poles. The vacuum is much harder at 10^{-8} to 10^{-5} mbar, than in a fluorescent lamp and should only allow a few free molecules. Thus the majority of ionised particles required to support an arc are provided by molecules from contact metal, the metallurgy of which is vital to the satisfactory operation of the interrupter. 'Hard' contact metals will not provide sufficient molecules and arcing will be extinguished prematurely, leading to current chopping and high voltage transients. However, if the contact metal is too soft, contact wear will be accelerated and contact welding may occur. The difference in contact metal is one of the essential differences between the interrupter and the contactor 'bottle', the contactor having a softer contact metal. Vacuum contactors still have a significant fault rating, around 7 kA, but, as with other forms of contactor, require to be protected by fuses for currents of greater magnitude if contact welding is to be avoided. The circuit breaker resulting from this technology is compact and light, with a much reduced maintenance requirement. Foundation shock due to operation is very much reduced due to the lightness of contacts and the small distances they have to move apart for fault clearance. Interrupters and contactors can be checked for adequate vacuum by applying a test voltage across the contact when in the open position. 25 kV ac is required for testing a 12 kV interrupter. Due to the permeability of the various ceramic and metal construction materials, the operational life is limited due to a slow loss of vacuum, but by using a 'getter' it is normally in excess of 20 years.

Short circuit ratings of up to about 40 kA are available for circuit breakers operating up to 12 kV. This may not be sufficient on some of the larger installations where prospective fault currents and operating voltages may be higher. Because of the size weight and space advantages of vacuum interrupters and contactors, and the reduced fire hazard, this type of equipment is recommended for offshore use.

SULPHUR HEXAFLUORIDE (SF6) CIRCUIT BREAKERS

The SF6 medium voltage circuit breaker competes with the vacuum device for low weight and bulk and can be considered as an alternative in most cases. Sulphur hexafluoride has a dielectric strength several times that of air and good heat dissipation properties. Low pressures are required to be maintained (usually less than 2 bar) and it is unusual if topping up is required more than once every 2 years. Suitably high fault ratings are available for most offshore situations.

SWITCHBOARD CONSTRUCTION

GENERAL

The following paragraphs list some important recommendations for switchgear procurement specifications. The configuration of switchboards is discussed in PART 1 Chapter 2 and reliability aspects in PART 8 Chapter 1. In all cases, offshore switchboards must be highly resistant to the salt-laden corrosive atmosphere, and sufficiently moisture protecting to prevent the ingress of water, particularly from above (IP54 minimum). The enclosure must also physically protect operators from any arcing, flames or flying fragments due to mal-operation.

Major changes to switchboards, once installed offshore, are extremely expensive, and depending on the planned operational life expectancy of the offshore installation, as much spare equipped capacity should be incorporated into the switchboard as possible. The spare auxiliary contacts of circuit breakers, contactors, switches and relays should all be wired out to terminals as a matter of course by the manufacturer. Whether racked or single mounting methods are used for protection relays, space for additional relays or spare relay rack spaces should be provided.

Facilities for adding extra switchboard cubicles at either end should also be provided.

It is also important that facilities for cabling are as flexible as possible, and that sufficient space is available for installation of the largest cables without exceeding bending radii limits. Neutral connection arrangements for outgoing supplies with neutrals must be provided. Facilities need to be provided for the earthing of cable armour. If cable entry is at the top, there should be no risk of moisture entering via the cable entry even if the cables do not have drip loops.

The space requirements for circuit breaker handling trucks must not be forgotten. Functional test facilities should be built in to the switchboard using a test panel cubicle with bus wiring to each unit to avoid dangerous trailing leads. The test facility must be interlocked to ensure that live operation of field equipment is impossible using the test panel. Every unit on the switchboard will require suitable permanent labeling and interlocking, shuttering and maintenance padlocking facilities. In fact, all the usual requirements for onshore substations such as safety rubber matting, earthing facilities and safety testing equipment will be required.

Fault calculations are covered in PART 4 Chapter 6.

Protection relay schemes are covered in PART 4 Chapter 7.

MAIN SWITCHBOARDS 6.6–13.8 KV

The overriding consideration for any main switchboard must be the short circuit capability of the circuit breakers, because of the fault current capability and proximity of the installed generation. Generator operational configurations which produce prospective fault MVAs of more than 1000 should be avoided as downstream equipment and cables would require to be of special nonstandard manufacture with all the expensive development and testing this would entail, in order to obtain sufficient rating.

The following are typical examples of switchgear presently available:

Manufacturer	Type	Maximum Fault Capacity (kA)	Max. Operating Voltage (kV)
ABB	Unigear (air insulated)	31.5	46
ABB	ZX (gas insulated)	40	40
ABB	Is Limiter[a]	210	40.5
GE (USA)	SecoGear (vacuum)	40	17.5
Powell	PowlVac (vacuum)	63	15
Mitsubishi Electric	MS-E	40	15
Schneider Electric	F400 (vacuum/SF6)	40	40.5
Siemens	8DA10/8DB10 (vacuum/SF6)	40	42

[a] *Recommended for upgrade projects only, where original fault ratings are exceeded.*

As discussed earlier in the chapter, a check will have to be made to ensure that the asymmetrical breaking capacity is adequate, allowing for the decrement in the value of this between fault inception and contact opening. Generator ac decrement must also be taken into account. Providing the switchgear fault make rating is adequate; problems with fault break ratings may be overcome by delaying the circuit breaker opening until the fault current has decayed to a value within the rating. The use of bus bar reactors is not recommended due to offshore weight and space limitations. Because of the high prospective fault currents, it is likely that any large motors supplied directly from this switchboard will require circuit breaker rather than fused contactor switching. To avoid shutting down generators or other vital equipment to carry out maintenance on the switchboard, a duplicate busbar switchboard may be considered. This is not often specified, however, because of the extra complexity, cost, weight and space involved. If it is likely that further generators will be required to be installed, due to a later operational phase such as artificial lift, then the switchboard will require to be rated for this future load and fault rated for the future prospective fault current capacity of the expanded system. Sufficient spare equipped circuit breakers should be provided for the expansion.

LARGE DRIVE SWITCHBOARDS 3.3–6.6 KV

Development of motor controlgear at up to 6.6 kV has resulted in very compact units where relatively low load currents are switched by vacuum contactors protected from short circuit faults by suitable HRC fuses. For incoming and outgoing distribution, circuit breaker cubicles are provided, the whole forming a composite switchboard of low weight and compact dimensions. The prospective fault level on this switchboard can be regulated to some extent by adjusting the reactance of the supply transformer windings. Therefore, motor control is usually by fused contactor rather than circuit breaker.

UTILITY SERVICES AND PRODUCTION SWITCHBOARDS

Because of the interdependence of various systems on an offshore installation, as can be seen by the examples in PART 1 Chapter 2, the low voltage switchboards must be considered as just as vital as their medium voltage neighbours. Maintenance of circuits for such supplies as machinery auxiliaries and hazardous area ventilation must be given high priority. An example of a generator lube oil auxiliary system is given in PART 2 Chapter 4.

EMERGENCY SWITCHBOARDS

The function of the emergency switchboard is described in PART 1 Chapter 1.

It is beneficial to provide synchronising facilities for the switchboard's associated emergency generator. The generator has automatic start facilities which will initiate a start following a main generation failure, provided the start signal is not inhibited by one of the safety systems. The synchronising facility gives a convenient means of routine load testing for the generator, and allows for changing over to main generation after a shutdown incident, without a break in the supply. The switchboard should also include facilities to prevent the generator from starting when a fault exists on the switchboard. Interlocking must be provided between the emergency generator incomer and the incomer from the rest of the platform power system if synchronising facilities are not available. As the emergency switchboard usually feeds all the ac and dc secure supply battery chargers and other vital equipment, it is important that planned switchboard maintenance outages are catered for in the design. It is not usual to go to the expense of a duplicate bus switchboard, but certain battery chargers and other vital equipment are usually fed from an alternative switchboard via a changeover switch. These supplies should also include those necessary for starting other generators and for safe area ventilation, the basic philosophy being to allow continued safe oil production whilst the switchboard is being serviced.

DRILLING SUPPLIES

The drilling electrical system is usually independent of the installation system, with its own diesel generation. The reason for this is partly to do with organisation, since the drilling may be carried out by a different company who provide the complete drilling package, including generation and switchgear.

If the drilling system is of the same operating frequency as the rest of the platform, then an interconnector of some kind between the systems is mutually beneficial, provided there is no equipment in the main system that is particularly sensitive to the harmonics generated by the silicon controlled rectifier (SCR) equipment. The reliability and maintainability of the drilling electrical system is vital, as failures at particular times in the drilling operation may increase the risk of blowouts or cause the abandonment of a producing well. The switchgear used for drilling distribution is of the conventional motor control centre type, with the exception of the SCR cubicles. The SCR cubicles cannot be isolated individually due to the permanent interconnection arrangement between each of the variable speed drives. An 'assignment' switch on the driller's console allows the connection of any SCR cubicle to any dc drive motor in the system. This arrangement allows for dc drive motors to be reassigned to another SCR cubicle if a fault develops in the first cubicle. To allow plenty of ventilation, the SCR cubicles are of a much more open design than other offshore switchboards. The assignment contactors are usually arranged in the upper section of the panel, while the SCR assemblies are in the lower section. Fig. 2.5.6 shows typical SCR cubicle schematic diagram.

To avoid obstruction of cooling air, there are usually no insulating barriers between the interconnecting busbars, contactors and equipment within the cubicle. However, it is sometimes necessary if drilling is to continue, for the rig electrician to change SCRs with the cubicle only isolated by the assignment contactor at the top. This problem is usually overcome by using a removable insulating barrier which can be carefully located below the contactor before working on the SCR assembly.

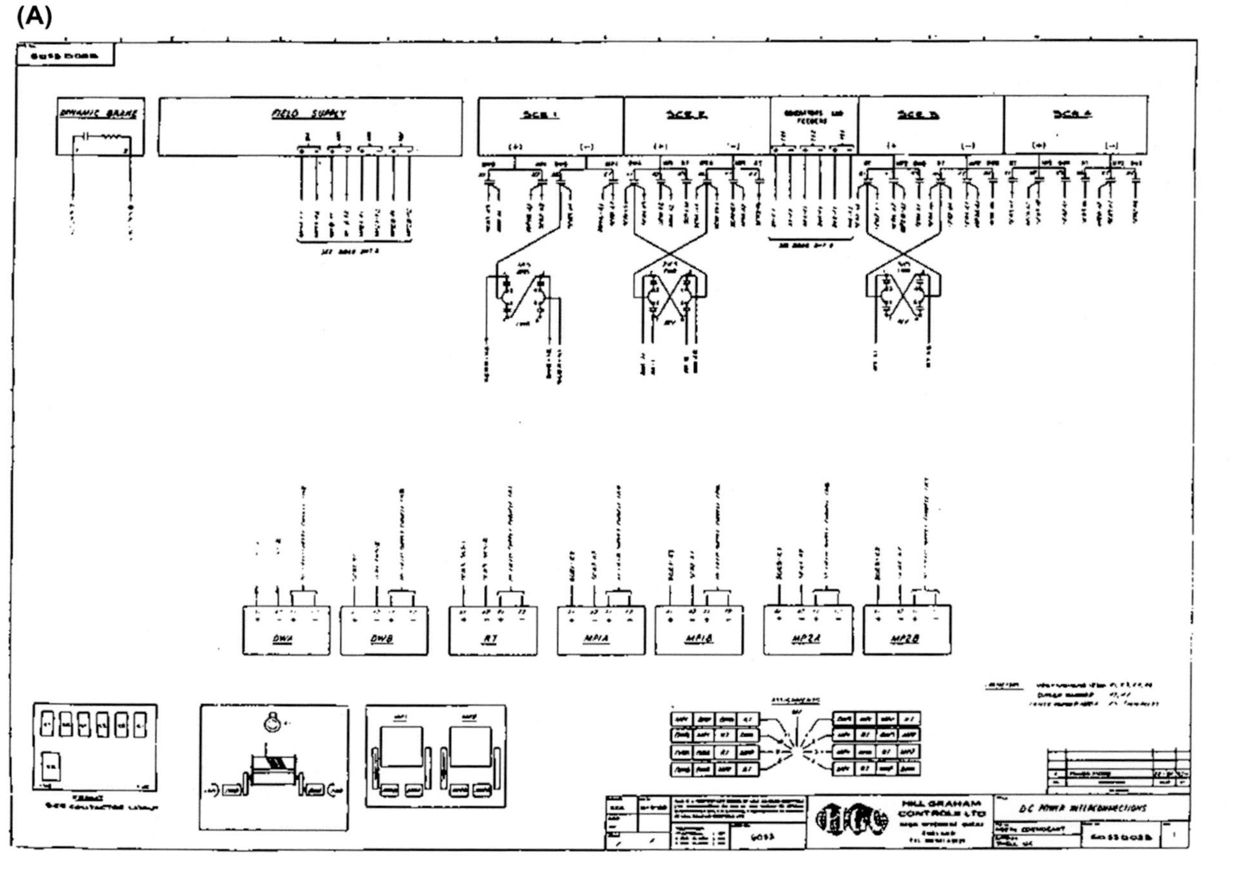

FIGURE 2.5.6

Typical SCR cubicle schematic diagram. (A) Block diagram of typical drilling silicon controlled rectifier (SCR) cubicle. (B) Schematic diagram of typical drilling SCR cubicle.

Courtesy Hill Graham Controls Ltd.

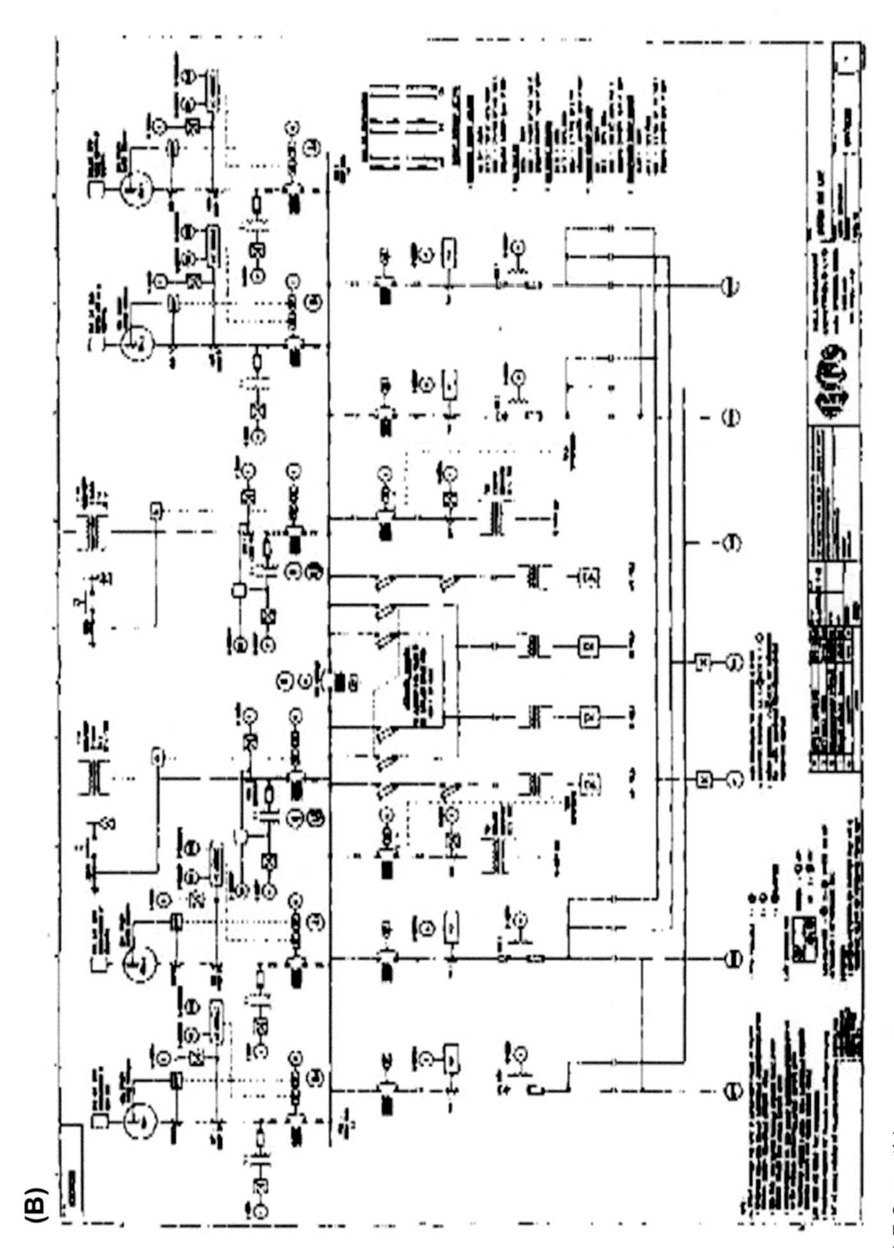

FIGURE 2.5.6, cont'd

LIVING QUARTERS SUPPLIES

Electrical supplies for living accommodation are important for the well-being of the offshore staff. Failure will not only bring discomfort, but will present a serious health hazard, when frozen food begins to thaw, toilets cannot be flushed etc.

As mentioned in PART 1 Chapter 2, domestic pipework should be kept out of switchrooms where possible. Where pipes do have to pass through switchrooms, possible sources of leaks such as flanges, valves etc., should be kept well away from switchboards or avoided altogether.

Most offshore accommodation is sealed and positively pressurised to prevent ingress of gas, should any process leaks occur, and so the operation of air conditioning is also vital. Should some serious incident result in a gas cloud developing, leading to ingestion of gas by the accommodation module HVAC, all generation is usually shut down and doors and fire dampers closed as automatic actions of the emergency shutdown (ESD) system and fire and gas systems. The actions taken however will depend on many design and operational factors associated with the particular installation. When the incident is over, certified hazardous area fans fed from another source may be used to purge the remaining gas from the module. Some operators prefer to allow natural ventilation to clear the gas, as purging fan systems have been known to ignite the gas and cause an explosion.

A serious fire outside the accommodation module may lead to its envelopment in dense smoke. Smoke detectors located at the ventilation air intakes should signal the fire and gas system to close down the ventilation system automatically, shutting fire dampers and doors to seal the module.

PROCESS AREA DISTRIBUTION

By far the largest power consumers on any offshore oil installation are the process modules. Apart from the large process drives, power is required for level control and circulating pumps, agitators, centrifuges, compressors and ventilation fans. Lighting and instrumentation power must also be provided, and for maintenance, sockets for welding transformers and other temporary equipment. The production switchboards located in safe area switchrooms provide the majority of controlgear for this smaller type of equipment. Welding sockets and portable tool sockets are equipped with isolators. A popular type of socket unit is certified Ex 'e' but an Explosion-proof cassette containing the isolator contacts is housed within the Ex 'e' enclosure. All such sockets are fed from shutdown contactor feeders in the production switchboards, so that should gas be detected, all portable equipment in the area may be immediately isolated.

Lighting distribution is discussed in PART 2 Chapter 14.

TRANSFORMERS

Offshore distribution transformers are usually of either sealed silicon oil filled or encapsulated resin types. Standard mineral oil filled types are too great a fire hazard and askarel-based insulants are regarded as a health hazard due to the presence of PCBs and the use of all such material was prohibited in most parts of the world in the early 1990s.

Air cored types are not recommended offshore in external locations because the salt-laden environment tends to lead to insulation problems even with good enclosure IP ratings. The sealed silicon-filled type have the advantages that a faster repair can normally be obtained and Buchholz type pressure sensing for winding faults can be fitted. Being heavy devices, transformer weights, in particular, need to be checked against switchroom maximum floor loadings. If a transformer is used to interconnect the main system with the drilling system, the heating effect of SCR harmonics may need to be considered.

High voltage transmission transformers in offshore wind farm substations are covered in PART 6 Chapter 6.

CHAPTER

Direct Current Generators

6

INTRODUCTION

The author is not aware of any direct current (DC) generators being used on offshore installations, but there are some older ships with DC systems.

The DC output current is generated in the same way as in the alternating current (AC) machine by conductors in the armature cutting lines of magnetic flux. However, the AC generated is mechanically switched to a unidirectional current by means of the commutator, a drum made up of copper segments that are held together by clamping rings and keyed to the armature shaft. DC generators for supplying ship services are usually compound wound; i.e., there are field winding connected to the output terminals in both parallel and series to provide a suitable generator load/voltage output characteristic.

Although compounding of field windings should maintain reasonably constant generator output voltages at a constant ambient temperature, DC generator output voltage is fairly sensitive to ambient temperature changes, and field rheostats (variable resistor field regulators) are used to compensate for changes in shunt field resistance due to these variations in temperature. When tested during manufacturing in temperate ambient conditions (e.g., around 15°C), the voltage across the field regulator should not be less than 14% of the generator rated terminal voltage to achieve rated output voltage with cooling air at tropical ambient temperatures.

COMPOUND-WOUND GENERATORS

These have both series and shunt field windings (Fig. 2.6.1) and their terminal voltage is a combination of both series and shunt generator characteristics. Standard practice is to connect the series field to the negative pole; both the generator and switchgear manufacturers need to consult, as uniformity is essential.

Fig. 2.6.2 shows the typical voltage characteristics, in which curve A represents the voltage due to the series field and curve B represents the voltage due to the shunt field. The series field will produce additional ampere turns proportional to the load and thus offsets the drooping characteristic of a shunt machine. The resultant (compound) output voltage is approximately the sum of the ordinates of the two separate curves, with any difference being because of the fact that, as a compound machine, the excitation voltage of the shunt winding is almost nearly constant.

Offshore Electrical Engineering Manual. https://doi.org/10.1016/B978-0-12-385499-5.00008-X

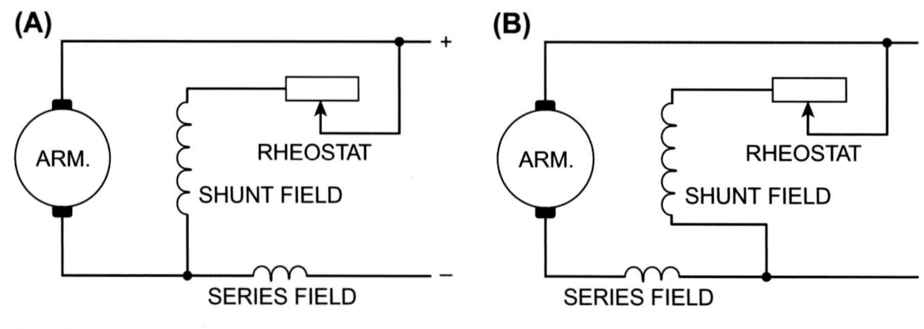

FIGURE 2.6.1

Compound-wound direct current generator: (A) short shunt connected and (B) long shunt connected.

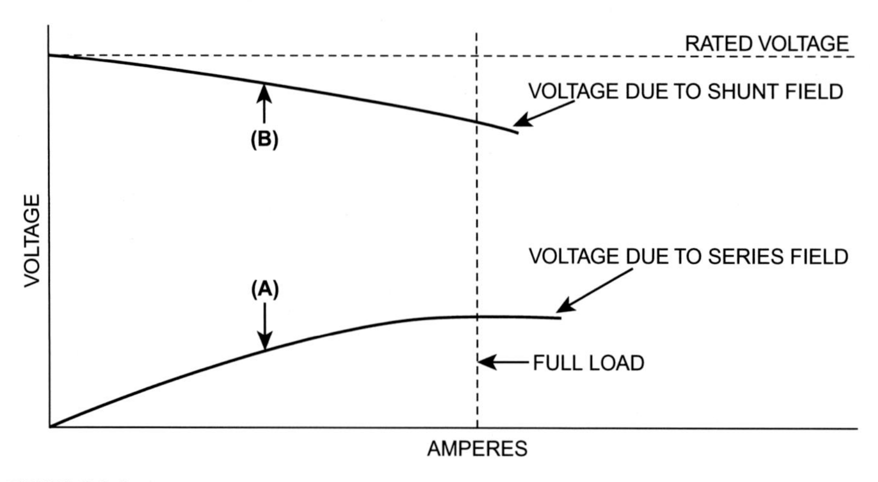

FIGURE 2.6.2

Characteristics of a compound-wound generator: curve A, as a series generator and curve B, as a shunt generator.

Variations in characteristics can be obtained by altering the strength of the series winding, such as undercompounding, flat compounding or overcompounding (Fig. 2.6.3), although the arrangement usually adopted is flat compounding.

It can be observed that the curves in Fig. 2.6.3 are not completely linear because of the saturation effects, and so even with flat compounding, a hump of 2%–3% is normal, but in small generators may be as much as 6%–7%.

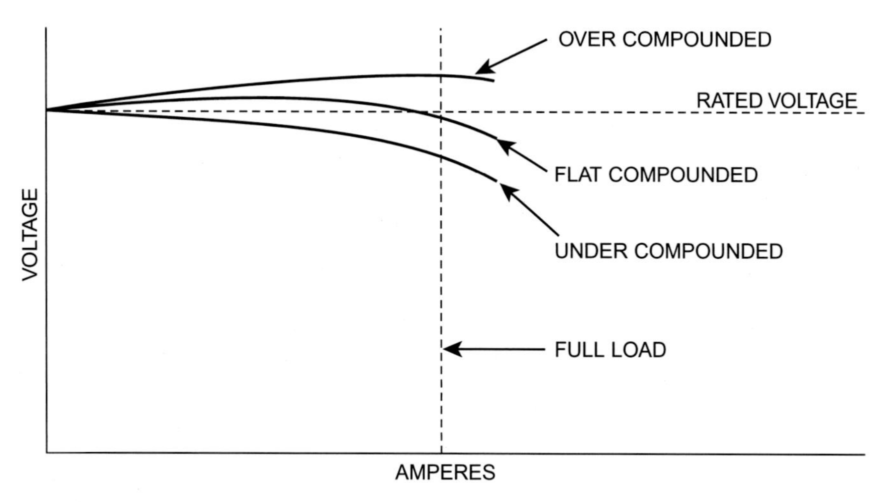

FIGURE 2.6.3

Compounding characteristics.

ADJUSTMENT OF COMPOUNDING

Voltage characteristics may be modified by adjusting the airgap during manufacturing by inserting or removing shims between the poles and the yoke. Other methods are

- the use of a diverter resistance connected in parallel with the series winding,
- adjustment of the commutator brush position.

The shunt field may be connected in 'short shunt' (Fig. 2.6.1A) or 'long shunt' (Fig. 2.6.1B). The effect on the characteristics is usually minimal, but machines should always be operated as originally built. In small generators, the effects of armature reaction are reduced by actually mechanically shifting the position of the brushes. The practice of shifting the brush position for each current variation is not practiced except in small generators. In larger generators, other means are taken to eliminate armature reaction. Compensating windings or interpoles are used for this purpose (see Fig. 2.6.4).

PARALLEL OPERATION

The general rule is for the use of machines with flat-compounded characteristics when running in parallel. When two DC machines are operating in parallel, a change in load will cause change in speed and therefore in terminal voltage. Therefore, 'equaliser' connections will be required to avoid the machine with higher revolutions

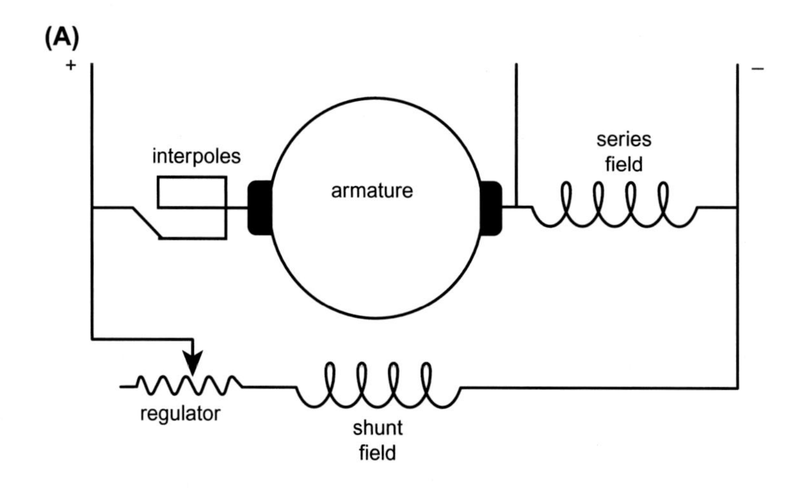

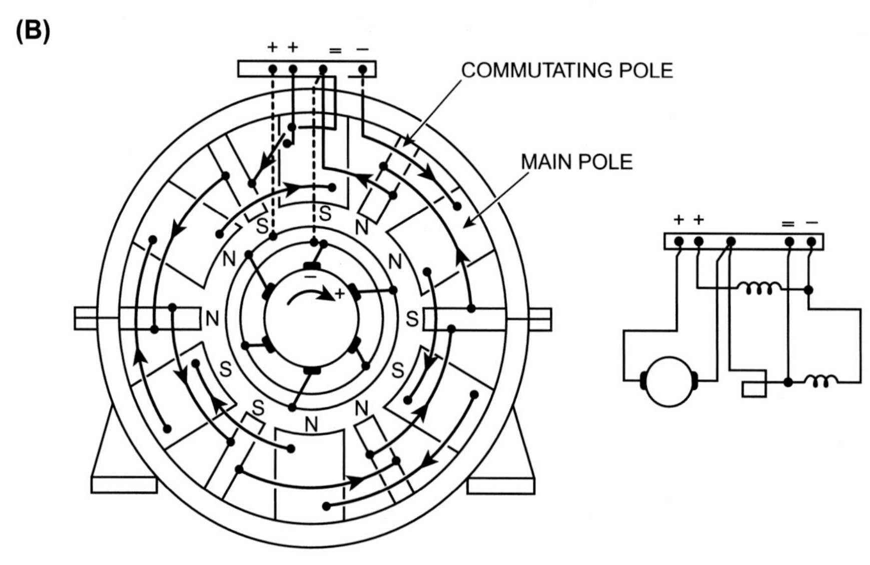

FIGURE 2.6.4

Interpoles. (A) A schematic and (B) series and commutating pole connections of a compound-wound direct current generator configured to avoid magnetisation of the rotor shaft (the shunt field winding is omitted). The series field is connected for clockwise current flow, and the commutating poles for anticlockwise flow.

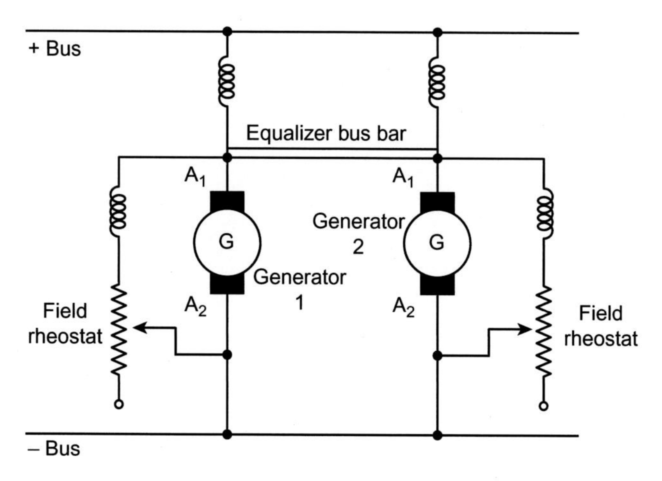

FIGURE 2.6.5

Schematic for two direct current generators operating in parallel.

per minute from grabbing the entire load. The only correct way to run compound machines in parallel is to connect the series fields in parallel, the paralleling connections being known as 'equalisers' (see Fig. 2.6.5 and Bibliography).

From Fig. 2.6.5, if the load sharing is disturbed and generator 1 starts taking more than its proper load share, its series field current will increase, and this increased field current flow will be split between the two generator series field windings via the equaliser busbar. Thus the two generators are affected in a similar way, preventing generator 1 from taking extra load.

In order to have a proper division of load from no load to full load, it is required that the regulation of each generator must be same. The series field resistances should be inversely proportional to the generator ratings.

CHAPTER

Direct Current Switchgear

7

INTRODUCTION

New variable drive equipment is likely to be via variable-frequency inverters, and therefore this chapter applies mainly to existing equipment found on offshore installations.

The direct current (DC) switchgear used offshore is unlikely to be of very high rating, but there are a number of areas where it is likely to be used as follows:

Drilling equipment[1]

- Rotary table motor drives
- Drawworks drives
- Top drill drives
- Mud pump drives
- Cement pump drives

DC-secure power supplies

- Level zero isolation for batteries

Floating installations

- Thruster drives
- Anchor winches

This list is typical, not comprehensive, as every installation will be different.

SWITCHING DIRECT CURRENTS

Switching in DC circuits is more onerous because unlike alternating currents, there is no natural current zero at any time, and it is invariably necessary to break the full voltage by increasing the arc resistance using suitably rated arc chutes to cool and quench the arc. In other words, the DC must be forced to zero by the circuit breaker.

In most designs the fault current is used in blowout coils to magnetically pull the arc into the arc chutes, so that the higher the fault current the stronger the magnetic field produced. In early designs, there tended to be a minimum 'critical current' required to pull the arc into the arc chute, and so 'low' fault currents could leave the

[1]As mentioned earlier, in more modern rigs, inverter-fed variable-frequency induction motor drives will be used.

Offshore Electrical Engineering Manual. https://doi.org/10.1016/B978-0-12-385499-5.00009-1

dangerous situation of the arc remaining unquenched, outside the chute. Modern circuit breakers are designed to avoid this condition.

Typical DC loads and fault currents are highly inductive, and circuit breakers must be capable of dissipating all the energy stored in the circuit until arc extinction. Such highly inductive loads will maintain the arc for longer whilst the stored energy in the load circuit decays. This will place more stress on the contacts, so two-stage contacts are now used, where the magnetic coils ensure that the *arcing* contacts take the onerous arcing duty whilst the *main* contacts will carry the load current when the circuit breaker is in the closed position.

All circuit breaker mechanisms (including alternating current) should be 'trip free', i.e., if a fault occurs during closure (making fault), the tripping action must still be available and the circuit breaker does not need to close first. Obviously, once the circuit breaker is closed and latched, all necessary protection device trips must still be operationally live. Trip circuit supervision should be provided, where the trip circuit continuity is monitored and provided with an alarm facility in a manned area. Fig. 2.7.1 shows a typical modern DC circuit breaker.

Other arc quenching methods may be used.

- 'Arc runners' will assist with avoidance of critical current by leading the arc away from contacts, also driven by electromagnetic forces from blowout coils.
- A 'puffer' stream of air may be used to assist moving the arc into the arc chute.

SPECIFICATION

Modern DC circuit breakers should be specified for

- rated voltage and impulse voltage,
- maintaining dielectric strength,
- rated switching current for load and overload,
- trip time performance, i.e., high speed, and current limiting,
- current sensing – directional or bidirectional,
- thermal performance for continuous rated current,
- fault performance, i.e., arc containment of pressure/safe venting of gasses and heat dissipation.

Variable-frequency inverter drives are covered in PART 2 Chapter 11.

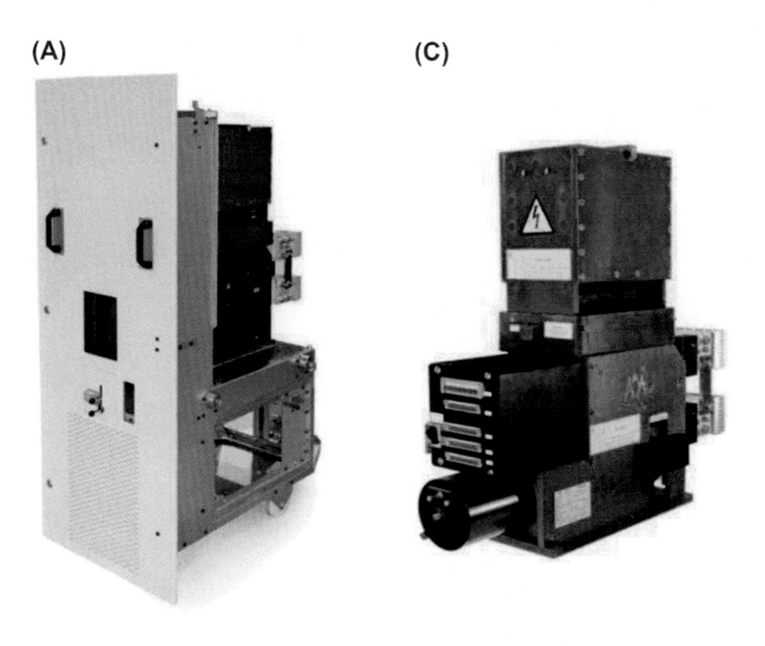

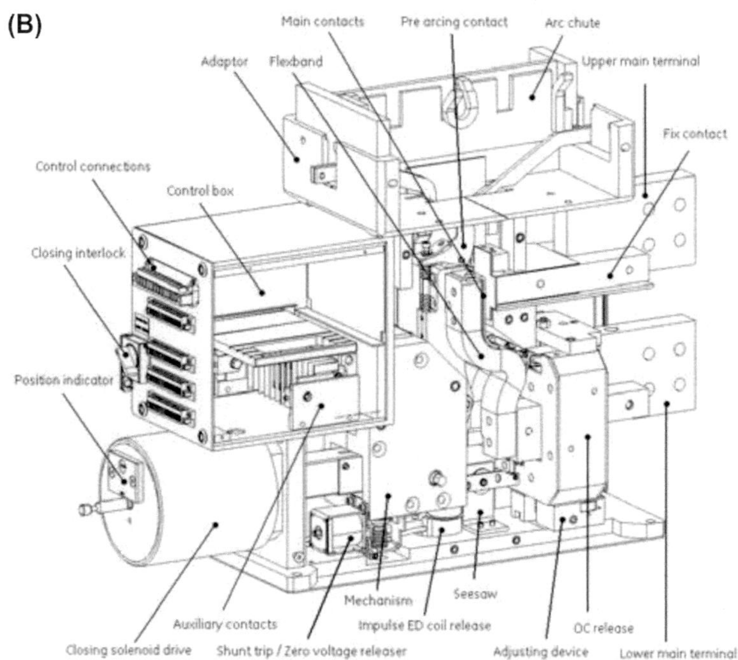

FIGURE 2.7.1

A typical direct current circuit breaker: (A) truck, (B) three-dimensional drawing and (C) circuit breaker (removed from truck). *ED*, electrodynamic; *OC*, overcurrent.

Courtesy GE Gerapid.

CHAPTER

Electrical Cables

8

SELECTION

Cables used on any offshore installation will require the following attributes:

1. They should have stranded copper conductors for smaller cross-section and better flexibility.
2. They should be voltage and fault rated for the system in which they are operating.
3. They should be rated for the normal maximum current flow in the circuit without exceeding the maximum conductor temperature or the temperature class limit if passing through a hazardous area.
4. If they are involved in a fire, smoke and acid fume emission should be low. The insulation compound used should have an oxygen index of at least 30.
5. They should normally be armoured for mechanical protection. Braided or wire armouring may be used; the braided type is slightly more flexible but is often more difficult to gland. These criteria cannot be rigidly applied, however, as special cables will be needed for instrumentation, diving umbilicals, downhole pumps, etc.
6. The cables designed for the most onerous duty are those for electrical fire pump supplies. These fire survival cables are designed to continue to supply current after an hour at a temperature of 1000°C, followed by hosing down with high-pressure water jets whilst being hit with a hammer. The difficulty lies in finding a suitable support system that will survive the same treatment.
7. The most common cable type in use is the EPR/CSP type similar to that found in merchant shipping, which meets the high oxygen index and fire-retardant requirements of BS 6883 and IEC 93-3.
8. Cables must be sized to allow for circuit current and maximum voltage drop as with any onshore system. AC cables can be sized using a chart generated by spreadsheet, as shown in Fig. 2.8.1. Care will still need to be exercised to ensure that close excess protection is still applied and that cable fault ratings are adequate.
9. In Europe, IEC 61892 is the prime standard for electrical installations on mobile and fixed offshore units and should be used for guidance in the sizing of conductors (BS IEC 61892 is expected to replace the IEE Recommendations for the Electrical and Electronic Equipment of Mobile and Fixed Offshore Installations in the United Kingdom, assuming that this is not affected by Brexit).

Offshore Electrical Engineering Manual. https://doi.org/10.1016/B978-0-12-385499-5.00010-8

CABLE SELECTION FOR 415 VOLT MOTORS
STARTING CURRENT = 6#FLA

Frequency = 60Hz
CABLE SWA/EPR/CSP

KW	FLA AMP	CABLE X-SECTION	CABLE AMP	DISTANCE METRES	VOLT DROP PER METRE	DERATE FACTOR	POWER FACTOR	SIN@	RESIST-ANCE PER METRE	REACTANCE PER METRE	VOLTAGE KILOVOLTS	react-ance @ 50Hz
5.5	9.002	1.500	22.00	75.084	1.105	0.409	0.850	0.527	0.0138	0.000164	0.415	0.000136
5.5	8.897	2.500	33.00	136.454	0.608	0.270	0.860	0.510	0.0076	0.000151	0.415	0.000125
7.5	12.133	1.500	22.00	55.079	1.507	0.551	0.860	0.510	0.0138	0.000164	0.415	0.000136
7.5	12.133	2.500	33.00	100.066	0.829	0.368	0.860	0.510	0.0076	0.000151	0.415	0.000125
11.0	17.794	2.500	33.00	68.227	1.217	0.539	0.860	0.510	0.0076	0.000151	0.415	0.000125
11.0	17.794	4.000	44.00	108.986	0.762	0.404	0.860	0.510	0.0047	0.000149	0.415	0.000124
15.0	24.265	2.500	33.00	50.033	1.659	0.735	0.860	0.510	0.0076	0.000151	0.415	0.000125
15.0	24.265	4.000	44.00	79.923	1.039	0.551	0.860	0.510	0.0047	0.000149	0.415	0.000124
18.5	29.927	6.000	58.00	97.145	0.854	0.516	0.860	0.510	0.0031	0.000142	0.415	0.000118
15.0	24.265	10.000	77.00	199.031	0.417	0.315	0.860	0.510	0.0018	0.000140	0.415	0.000116
22.0	35.180	6.000	58.00	81.787	1.015	0.607	0.870	0.493	0.0031	0.000142	0.415	0.000118
22.0	35.180	10.000	77.00	135.967	0.610	0.457	0.870	0.493	0.0018	0.000140	0.415	0.000116
30.0	47.973	10.000	77.00	99.709	0.832	0.623	0.870	0.493	0.0018	0.000140	0.415	0.000116
30.0	47.973	16.000	103.00	155.021	0.535	0.466	0.870	0.493	0.0012	0.000131	0.415	0.000109
37.0	59.166	16.000	103.00	125.693	0.660	0.574	0.870	0.493	0.0012	0.000131	0.415	0.000109
37.0	59.166	25.000	135.00	192.905	0.430	0.438	0.870	0.493	0.0007	0.000124	0.415	0.000103
45.0	71.141	16.000	103.00	103.644	0.801	0.691	0.880	0.475	0.0012	0.000131	0.415	0.000109
45.0	71.141	25.000	135.00	159.272	0.521	0.527	0.880	0.475	0.0007	0.000124	0.415	0.000103
55.0	86.950	25.000	135.00	130.313	0.637	0.644	0.880	0.475	0.0007	0.000124	0.415	0.000103
55.0	86.950	35.000	165.00	175.520	0.473	0.527	0.880	0.475	0.0005	0.000122	0.415	0.000101
75.0	118.569	35.000	165.00	128.715	0.645	0.719	0.880	0.475	0.0005	0.000122	0.415	0.000101
75.0	118.569	50.000	200.00	167.849	0.494	0.593	0.880	0.475	0.0004	0.000120	0.415	0.0001
90.0	142.282	50.000	200.00	139.874	0.593	0.711	0.880	0.475	0.0004	0.000120	0.415	0.0001
90.0	142.282	70.000	250.00	191.506	0.433	0.569	0.880	0.475	0.0003	0.000117	0.415	0.000097
110.0	173.901	70.000	250.00	156.686	0.530	0.696	0.880	0.475	0.0003	0.000117	0.415	0.000097
110.0	173.901	95.000	300.00	198.604	0.418	0.580	0.880	0.475	0.0002	0.000114	0.415	0.000095
132.0	208.681	95.000	300.00	165.503	0.502	0.696	0.880	0.475	0.0002	0.000114	0.415	0.000095
132.0	208.681	120.000	360.00	202.774	0.409	0.580	0.880	0.475	0.0002	0.000112	0.415	0.000093
160.0	252.946	150.000	400.00	192.407	0.431	0.632	0.880	0.475	0.0001	0.000112	0.415	0.000093
160.0	252.946	185.000	460.00	223.580	0.371	0.550	0.880	0.475	0.0001	0.000112	0.415	0.000093
200.0	316.183	185.000	460.00	178.864	0.464	0.687	0.880	0.475	0.0001	0.000112	0.415	0.000093
200.0	316.183	240.000	550.00	211.013	0.393	0.575	0.880	0.475	0.0001	0.000111	0.415	0.000092
250.0	395.229	300.000	630.00	191.553	0.433	0.627	0.880	0.475	0.0001	0.000110	0.415	0.000091

FIGURE 2.8.1

Spreadsheet-generated cable sizing chart. For basis see PART 4 Chapter 5. *CSP*, chlorosulphonated polyethylene; *EPR*, ethylene propylene rubber; *SWA*, steel with armoured.

INSTALLATION

1. Cables should be routed in such a way as to facilitate maintenance and the installation of additional cables, with minimum need for expensive scaffolding. For example, a cable route located under a deck, so that there is no immediate access and the sea is directly below it, is not recommended. The cost of a small cabling modification in such a location would be overshadowed by the enormous scaffolding bill.
2. Cable routes should avoid known fire risks where possible. Cables to main and standby machinery should be run on separate routes.
3. Cables should be laid parallel on cable ladders and tray in a neat and orderly fashion.
4. Where heavy three-phase currents are carried, cables are usually single core. Single-core cables of the same conductor cross-section usually have a higher fault rating than their three-phase equivalents and are easier to install, having a lower weight per metre and a lower bending radius. To avoid eddy currents being induced in local steelwork, such cables must be run in a trefoil configuration. In some situations the cables must be run in a flat profile because of space limitations, in which case a balanced arrangement must be adopted which still avoids the promotion of eddy currents. Support or other steelwork must not pass between phases, as the steel will heat up owing to hysteresis loss. Where two or more trefoil cable groups run on the same route, they should be at the same horizontal level with a clear space between the groups of at least one cable diameter.
5. Cable bends should not be tighter than the minimum bending radius specified by the manufacturer. Drip loops should be provided at external cable terminations. These will also be useful if the cable has to be reterminated at some later date. In any case, straining of cables at cable glands should be avoided, and cables should be perpendicular to gland plates for a minimum of 100 mm before entering the gland.
6. Expansion loops will also be required across the expansion gaps necessary between module walls and support steelwork.
7. Some cable insulation materials tend to become brittle with lower temperatures. This is particularly the case with polyvinyl chloride (PVC) compounds. In low-temperature conditions, care needs to be taken when installing most cable types. It is desirable that cables with PVC insulation should not be installed directly at ambient temperatures of 5°C and below. In such conditions the cable should be stored for at least the previous 24 h at a temperature of 20°C or more.
8. Cables should be clamped or cleated to a rack or tray with suitable ties. The usual arrangement is to use nylon 12 ties (not nylon 66, as it becomes brittle in the offshore environment), supplemented at suitable intervals (depending on the weight) by ethylene-vinyl acetate-coated stainless steel banding or cable cleats of the correct diameter. Ties should be nonmagnetic for single-core cables cleated in flat formation. The recommended cleating/banding intervals are shown in Table 2.8.1.

Table 2.8.1 Recommended Stainless Steel Cleating/Banding Intervals

	Maximum Spacing of Stainless Steel Banding (mm)	
Overall Diameter and Type of Cable	**Horizontal Run (to 60 degrees From Horizontal)**	**Vertical Run (to 30 degrees from Vertical)**
Single-core cables in trefoil formation	2000	1500
Multicore cables (above 35 mm outer diameter)	1500	1000

9. Where trefoil cleats are used for single-core cables, stainless steel banding should also be fitted for fault current bracing at 1 m intervals, with a band always located close to each cleat.
10. The positive and negative cables forming a direct current circuit should be run side by side as far as possible to reduce magnetic effects.
11. Spare cores in multicore cables should be coiled back for future use, not cut off at the gland.

TRANSITS, GLANDS AND CONNECTORS

Wherever cables have to penetrate a wall or enclosure which must remain gastight or has a fire protecting function, it will need to be installed using a transit, gland or connector.

TRANSITS

Transits consist of a steel frame which is welded or bolted on to the wall or floor where a suitable slot has been cut. Some transit frames are split so that they can be placed around existing cables without the need for the cables to be reinstalled. Where the cables pass through the frame, they are individually enclosed in shaped flexible packing blocks as shown in Fig. 2.8.2. Each layer of cables and surrounding blocks is anchored into the frame by a stayplate. To fill up any unused space, blank blocks are inserted and the whole assembly is compressed down by a compression plate and threaded bolt. The space left at the top is then filled by an end packing which is fitted with compression bolts to expand sandwiched flexible blocks and complete the sealing process. The resulting arrangement provides an effective and reusable method of penetrating a steel bulkhead or deck and can withstand gas pressures of up to 4 bar. Good fire tightness properties can be obtained when the transits are used in conjunction with mineral wool or compounds such as Mandolite 25 which cover the entire fire wall.

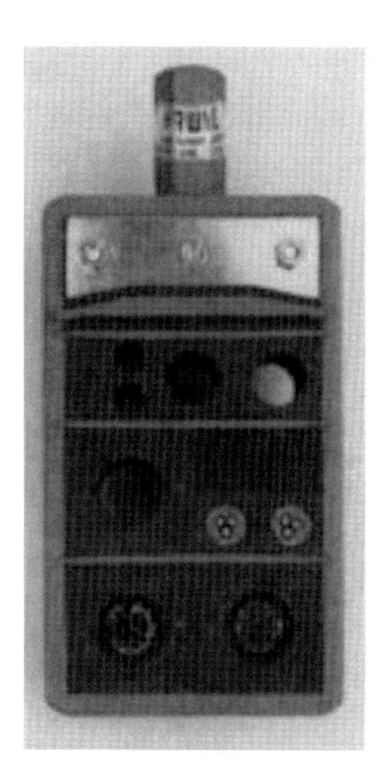

FIGURE 2.8.2

Illustration of a typical cable transit.

Courtesy Hawke Cable Glands Ltd.

GLANDS

Glands are normally of the flameproof type with inner and outer seals, as shown in Fig. 2.8.3. Flameproof barrier glands may be used where the inner seal is replaced by a setting compound which forms a seal around individual cores. The benefit of being able to use more economical cable constructions through the use of barrier glands is usually outweighed by the disadvantage that installation, particularly compound filling, is more difficult, and the cable can only be freed from the gland by cutting it off at the entry point. In either case the gland must be correctly assembled and installed (without cutting or tampering with its components) if certification is to remain valid.

CONNECTORS

Connectors of various types may be used, particularly for diving and subsea applications. Some connectors are designed to be connected with circuits live underwater. Some may be hazardous area certified, although in this case the circuits would need to be intrinsically safe or provided with some means of preventing live disconnection.

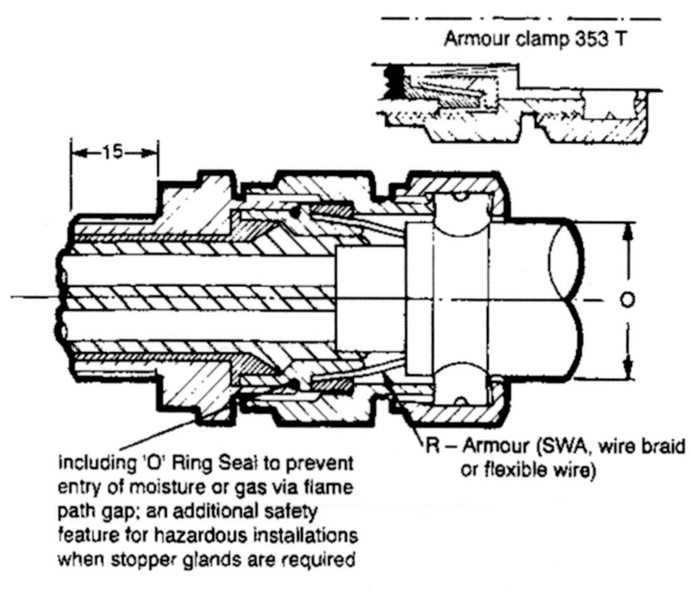
Armour clamp 353 T
15
O
including 'O' Ring Seal to prevent entry of moisture or gas via flame path gap; an additional safety feature for hazardous installations when stopper glands are required
R – Armour (SWA, wire braid or flexible wire)

FIGURE 2.8.3

Sectional diagram of an explosion-proof cable gland. *SWA*, steel with armoured.

Courtesy Hawke Cable Glands Ltd.

BUS DUCTING

The transmission of heavy electrical currents over short distances and in congested areas using cables presents two main difficulties. First, cables of cross-sectional area up to 630 mm^2 are available from manufacturers' standard ranges. However, to cope with loadings of up to several thousand amperes, three or more such large cables per phase may be required. The bunching of such a number of large cables in a congested space requires that their current-carrying capacities need to be heavily derated. Second, the bending radii of such cables are large leading to difficulties in installation and termination.

In order to overcome such problems, bus ducting may be used. This has the benefits of high current-carrying capacity and the inherent facility for the fabrication of right angle or other bends to order. The bus ducting used must be dust and moisture protective to at least the same standard as the switchgear to which it is connected, or Ingress Protection (IP)66 standard if it is necessary to run the duct outside the switchgear module. Accurate measurement is required to ensure a close fit and to avoid stressing of the duct or terminating equipment. Unless certaint that it is not possible for external ducting to be live during a gas release incident, it is recommended that such ducting is hazardous area certified.

CABLES FOR INTRINSICALLY SAFE CIRCUITS

1. Intrinsically safe (IS) cables should have a light-blue outer sheath and be screened (shielded).
2. Requirements for armouring vary between countries and standards, but mechanical protection should be provided, particularly in cables containing multiple IS loops. Such multicore IS cables are acceptable, provided each IS loop pair is individually screened. If the IS cables are subject to frequent flexing, a separate cable for each loop, incorporated into a purpose-designed umbilical with hydraulic or pneumatic hoses (if any), is recommended.
3. It must be possible to complete a valid loop safety calculation for every loop, so it is essential that comprehensive cable electrical parameters, including inductances and reactances, are available for the cable used.
4. All IS cables should be segregated from non-IS cables by using separate cable-racking or earthed-metal partitions.
5. In the United Kingdom, cable screens are earthed at the nonhazardous end, but earth or ground loops must always be avoided.
6. IS circuits should also be separated from non-IS circuits in enclosures (junction boxes, etc.). At bare live connections (terminals) a minimum distance of 50 mm should be allowed between IS and non-IS circuits. Between IS circuits, the distance should be a minimum of 6 mm. IS and non-IS wiring may be contained in the same enclosure, provided the same segregation is maintained. IS and non-IS earth bars should be located at opposite ends of the enclosure. Good labelling and strict adherence to light-blue sheathing for IS cables/wiring is essential.

Please refer to

PART 4 Chapter 5 — Cable Sizing Calculations

PART 6 — Any Variations for Floating Installations and Ships

CHAPTER

Motors

9

INTRODUCTION

Machinery drives for offshore installations range in size from fractional horsepower cabin ventilation fans to the 15 MW or more machines required for gas injection compressors. Some concern has been expressed over the practicality of installing very large electric motors offshore owing to the stresses imposed on the structure, especially during starting. A figure of around 25 MW depending on the application (e.g., less for reciprocating pump drivers) should be considered as a ceiling figure for offshore electric motor ratings.

VOLTAGE LEVELS

As with onshore industrial plant, motor operating voltage levels correspond to ranges of motor power ratings in order to keep current and voltage drop magnitudes within practical limits. For weight saving purposes, the step points at which higher operating voltages are selected tend to be lower offshore as in the following table.

Voltage Level	Rating
415–460 V	Up to 150 kW
3300–6600 V	Up to 1000 kW
11–13.8 kV	Over 1000 kW

Higher voltages such as 22 kV could be utilised provided the economics are favourable and proven ranges of machine at these higher voltages are available.

STARTING

Where an individual motor load represents a substantial part of the power system capacity, it is desirable to obtain designs of squirrel-cage motors with the lowest starting current characteristics compatible with the driven equipment. Both reduced voltage starting of cage motors and the use of slip ring motors will add considerable weight to the drive package, and therefore direct-on-line low-starting-current machines are preferred unless technical considerations for selection are overridden by those of cost and delivery. Motors with starting currents ranging from 3.5 to 4.5, instead of the normal 6–10 times full load, are usually available from the larger manufacturers.

Offshore Electrical Engineering Manual. https://doi.org/10.1016/B978-0-12-385499-5.00011-X

SPEED

The shaft speed of a pump, compressor or fan is critical to its performance.

It is therefore necessary to obtain the required speed or speed range either by installing a gearbox between the motor and the driven equipment or by selecting a suitable system frequency and pole configuration for the motor. In the 1930s the US Navy changed from direct current to alternating current systems and selected 440V 60Hz as the operating parameters. Since then the whole of the NATO fleet and the majority of commercial ship designs have standardised on these parameters. This has greatly assisted in improving the availability of 60Hz options on the standard equipment ranges of most European manufacturers. With small low-voltage motors the increased efficiency as a result of higher pump speeds at 60Hz is marginal, but with motors having ratings of the order of several megawatts the weight and power savings can be substantial.

With smaller installations, however, operating at 60Hz may be a disadvantage, where it is decided to select a reciprocating engine rather than a gas turbine main generator prime mover. The problem is that the optimum engine speed of around 1500rpm is better suited to generating at 50Hz. Reciprocating engines running at 1200rpm tend to have too low a power-to-weight ratio, and operating at 1800rpm leads to short cylinder life or even piston speeds which would be beyond the design limitations of the engine. The higher synchronous speeds obtainable at 60Hz also lead to higher inherent noise levels, although this can be deadened with better module sound insulation.

From a machinery standpoint, a major disadvantage in adopting 60Hz for an offshore installation is related to testing the equipment before installation on the platform, as full-load tests cannot be carried out using the British and European national supply networks. Until recently, tests were mainly carried out at 50Hz and the results extrapolated to give projected machine characteristics at the design operating conditions. However, test facilities are now available in the United Kingdom for motors of up to 6MW at 60Hz.

For larger machines, where capital investment is high and full-load tests are considered essential, it is usually possible to arrange full-load tests in conjunction with testing of the main generators to be installed on the offshore installation. Although this procedure is usually expensive, the costs should be more than offset by the benefits of adopting the higher frequency.

Once the system frequency has been established, it becomes increasingly expensive to change and hence may no longer be considered a variable after that point in the system design. It is therefore important to consider the number, rating and purpose of the larger drives on the installation at an early stage in the power system design, before the frequency is selected.

However, in many cases the shaft speed available from a motor, even from a 60Hz two-pole machine (i.e., 3600rpm), is lower than the required shaft speed, and it will still be necessary to install a gearbox in the drive string. Once the requirement for a gearbox has been established, changes in the drive ratio have only minor effects, and the motor speed may be chosen to give the optimum motor design in terms of dimensions, weight, reliability, noise emission and so on.

POLE CONFIGURATION

At a given frequency, motor synchronous speed is determined by the number of poles incorporated in the motor stator, and this governs the maximum operating speed of the machine. The fastest possible speed of both induction and synchronous motors is with a two-pole configuration, which gives synchronous speeds of 3000 rpm at 50 Hz and 3600 rpm at 60 Hz, respectively.

As size and weight penalties are usually incurred by increasing the number of poles, only two- and four-pole machines are normally used offshore. An occasional exception is large reciprocating compressor drivers, where eight-pole (or more) synchronous machines may be used. These machines, apart from giving the required lower speed, also reduce the current fluctuations caused by the cyclic torque variations associated with reciprocating machinery.

The advantages and disadvantages of two-pole against four-pole machines are listed in the following.

1. Advantages of two-pole machines
   - **a.** Energy conversion within the higher-speed two-pole machine is usually more efficient, giving some reduction in size and weight for a given output. However, because of the higher rotor speeds, especially at 60 Hz, mechanical forces on the rotor cage, known as hoop stresses, become significant and limit the maximum dimensions of rotor that may be manufactured using conventional materials. Metallurgically more exotic materials may be used to extend this limit with, of course, the accompanying large increases in the cost of the machine.
   - **b.** Manufacturers differ in the application of a practical maximum rating limit for a two-pole motor, but as a general rule, this is between 3 and 5 MW. Therefore motors above 5 MW should not be considered as a feasible alternative unless the cost of using exotic metals in the rotor is outweighed by the savings accrued by, for example, the elimination of a gearbox.
   - **c.** Below this limit, and within manufacturers' normal product ranges, the use of a two-pole machine should, in comparison with an equivalent four-pole machine, provide dimension and weight savings roughly proportional to power rating; however, for small machines of only a few kilowatts, there is little benefit.
2. Disadvantages of two-pole machines
   - **a.** The starting torque of two-pole machines is less, requiring the driven equipment to have a lower moment of inertia. It may also prove more difficult to accelerate the machine up to operating speed, where driven machinery cannot be run up to speed unloaded. The speed/torque characteristics of pumps in particular should be carefully studied to avoid any problems. Starting currents are also likely to be higher, and with large machines, this may lead to unacceptable voltage dips.
   - **b.** Irregularities in the core stampings, which are inevitable unless very high levels of quality control are applied, generate more magnetic noise in two-pole machines. A characteristic low-frequency (twice slip frequency) growl can be heard from these motors.

c. Rotor imbalance is more likely and can cause more vibration on two-pole machines.
d. The higher rotor speeds associated with two-pole machines will shorten bearing life or require a more expensive higher-performance bearing.
e. Poorer heat dissipation within the rotor necessitates increased cooling airflow rates in two-pole machines. The effect of higher fan speeds and increased airflow rates is to increase noise emission from the machine.
f. From the above-mentioned points, it can be seen that the two-pole machine generally scores on weight and dimensions but suffers from the design limitations associated with higher speeds. Certain applications such as direct drive ventilation fans and axial compressors can often take considerable advantage of the higher-speed machine. Nevertheless, if no significant benefit is obtained from the higher speed then the four-pole machine should be used.

COOLING AND INGRESS PROTECTION

As discussed in earlier chapters, only fully enclosed types of motor are normally considered suitable for offshore installations.

Three typical enclosure types are to be found offshore.

1. Totally enclosed fan ventilated
   a. In this motor type, the stator windings are enclosed within a finned motor casing. Cooling is achieved by the mounting of a fan on the nondrive end of the rotor shaft, external to the stator end plate, in order to blow air over the external cooling fins. The fan fits inside a cowling which deflects the air over the fins. An ingress protection (IP) rating of IP55 is typically achieved with this design, which provides a high level of physical protection. The majority of low-voltage machines, especially in hazardous areas, are of this type.
   b. The chief disadvantage with this type of motor is the inherently high noise level produced by the external fan. Methods are, however, available to reduce this noise, such as the use of acoustically treated fan cowls or the use of a machine oversized for the application but fitted with an undersized fan.
   c. This design is unsuitable for the larger medium-voltage machines but can be used advantageously with the smaller 3.3 or 4.16 kV machines.
2. Closed air circuit, air cooled (CACA)
   a. As motor sizes increase, it becomes less and less likely that heat generated in the rotor and stator windings can be dissipated to the machine casing at a fast-enough rate to prevent design insulation temperature rises being exceeded. It then becomes necessary to pass a cooling medium (usually air)

through the inside of the motor to remove surplus heat. With the majority of CACA types found offshore, air is usually forced through the windings by a rotor-mounted fan within the motor casing. As the machine is still required to be enclosed, this cooling air is recirculated in a closed circuit through the machine, and heat is extracted by an air-to-air heat exchanger mounted on top of the motor. A second rotor-driven fan is often required to force the external air over the heat exchanger.

   b. The CACA design suffers from the same noise and weight problems as the totally enclosed fan cooled (TEFC) type, and the air-to-air heat exchanger only adds to weight and bulk. Nevertheless, it does provide an adequate and simple method of cooling the larger motors and, important from the reliability point of view, requires no external services in order to continue operating.

3. Closed air circuit, water cooled (CACW)

   a. If the air-to-air heat exchanger on the CACA machine is replaced by an air-to-water unit, we then have a CACW machine. The machine is dependent on an adequate supply of cooling water for continued operation. The benefits of this arrangement are that the bulk of the heat exchanger is very much reduced, and there is no requirement for the secondary cooling fan and consequently there is a substantial reduction in noise.

   b. The disadvantages are the machine depends on the cooling water supply and hence there is a reduction in reliability and the presence of water under pressure around the machine is a hazard.

To summarise, the following is recommended for enclosure selection:

Cooling Type	Application
TEFC	Smaller low-voltage machines
CACW	Where the TEFC design is not practicable, i.e., with larger 3.3/4.16 kV and all higher-voltage machines
CACA	Larger machines where the cooling water supply is uneconomic or the machine must operate during a cooling system outage

PARTICULAR APPLICATIONS

These are covered in PART 2 Chapter 12.

Hazardous area topics are discussed in PART 5 Chapter 4 and are not presented here. Readers who are likely to be specifying motors for hazardous areas would be advised to read PART 5 Chapter 4.

CHAPTER

Motor Control Gear

10

INTRODUCTION

This chapter deals with the design and operation of motor control gear but assumes some familiarity with low-voltage switchgear and motor control centres (MCCs).

Motor protection is covered in PART 4 Chapter 3 and requirements for large process drives are covered in PART 2 Chapter 12.

LOW-VOLTAGE SWITCHGEAR AND MOTOR CONTROL CENTRES FOR OFFSHORE USE

It is recommended that low-voltage switchgear and control gear for use offshore should be continuously rated for operation at the maximum ambience with module ventilation failed. This may require allowing a little more space in the switchroom but will provide operators with more flexibility during upset conditions. In highly ambient climates, this may not be possible, in which case some form of backup ventilation for critical switchrooms is recommended.

Other design issues which will help operators are as follows:

- Contactor thermal ratings should *not* include any 'heat sink' effects obtained by using oversized cables.
- Ensure (by testing if necessary) that contactor coils will maintain closure of contactors down to at least 80% of nominal voltage.
- Provide reacceleration timers with adjustment facilities so that the timers may be set based on the results of fault clearance simulation studies.
- In low-voltage switchgear, arc faults may be prolonged and damaging. Therefore arc fault detection is recommended, such as ABB's REA or Siemens' Siprotec systems.

Star delta starters are frequently used offshore to reduce transient currents, in the same way as they are used onshore. However, because of the additional switching involved, they are less reliable and may require more maintenance.

Auto transformer starting are used offshore but should be used sparingly because of the additional weight and space they take up. Using an electronic soft starter may be a better option.

Offshore Electrical Engineering Manual. https://doi.org/10.1016/B978-0-12-385499-5.00012-1

INTELLIGENT MOTOR CONTROL CENTRES

In a fully integrated intelligent MCC, all nonpower functions are digital and software controlled.

Ideally, all the units should also have input points to monitor devices such as the disconnect switch, contactor, overload relay and a hand-off-auto selector switch. A network scanner module or network linking device must also be provided to collect and distribute the device data in the MCC. An integrated intelligent MCC should have at least the following components available.

1. *Digital overload relays*: Each motor starter should have
   a. built-in network communication;
   b. input points (for monitoring disconnect or selector switch);
   c. output points (for controlling contactor);
   d. light-emitting diodes for status indication;
   e. protective functions – thermal overload, underload, jam, current imbalance, stall, phase loss, zero sequence ground fault and positive temperature coefficient thermistor input;
   f. programmable parameters for the protective functions – trip level, warning level, reacceleration time delay and inhibit window, with the facility to enable programming via a Laptop;
   g. warning alarms to alert users of a potential trip and allow actions to avert impending downtime. Time delays and inhibit windows allow recognition of abnormal current loads (e.g., extended starting times with high currents), without nuisance tripping;
   h. current monitoring – phase, average, full load, ground fault, imbalance percent and percent thermal capacity used are important monitoring features;
   i. diagnostics – device, warning and trip status; time to overload trip; history of last five trips and time to reset.
2. *Small input/output (I/O) module for 'nonintelligent' units (e.g., isolators)*: The preferable solution is an I/O module within the unit – small enough so that the MCC unit size is not altered – to link the device and the network. The I/O module should have an adequate number of inputs and outputs, according to the unit functions (for a starter, four inputs and two outputs).
3. *Network communication interface unit with input points*: Intelligent devices often require an external communication module. Ideally this module should contain input points (again, to eliminate wiring to a distant I/O chassis) (five inputs each).

MOTOR CONTROL CENTRE SOFTWARE

Integrated intelligent MCCs require dedicated software that provides the operator with essential data on MCC functions and is integrated into the operator's displays. MCC software should eliminate the need for creating customised MCC screens within operator interface software.

The following checklist identifies key benefits available from integrated intelligent MCC software.

1. *Can be accessed at any network level*: The user should be able to view the MCC by plugging into any network level, such as DeviceNet, ControlNet or Ethernet. This feature gives the user flexibility to locate the software on a maintenance laptop, in a control room or at an engineer's desk.
2. *Operates in a familiar environment*: The software will be easiest to use if it behaves according to known operating environments, e.g., Windows.
3. *Initiates network communication*: Establishing devices as recognised entities on a network can be the most time-consuming step. In the optimal situation, the MCC manufacturer downloads user-specific information, such as node addresses (per user specification or a standard scheme) and baud rate, and then tests the entire system for accurate function and communication.
4. *Display of preconfigured screens showing most common parameters*: Intelligent MCC software can be arranged to access the user's specific data files and may facilitate the generation of corresponding screen display windows.
5. *Includes unique MCC documentation to initialise screens*: The application program, on installation, should access specific information to generate screens containing data pertinent to that particular MCC.
6. *Event logging*: Alarms, faults and warnings should be logged as required.
7. *Documentation database*: Current wiring diagrams and schematics, user manuals and spare parts lists, available on-screen can be very useful for diagnosing problems. However, it can be difficult to keep all this material up to date.

Please also refer to PART 2 Chapter 15.

MEDIUM-VOLTAGE STARTERS

DIRECT ON LINE STARTING

Vacuum contactors provide a reliable form of motor control for medium-voltage motors. They are designed for routine starting of the motor and have insufficient fault rating for fault clearance on most systems. This is because of the 'softer' metallurgy of the contacts, designed for more frequent operation than the equivalent circuit breaker, and also to avoid current chopping transients which would damage the motor insulation. Hence, they are protected from the higher prospective fault currents with appropriately rated fuses. The fuses will also reduce the risk of contact welding.

ELECTRONIC STARTERS

Electronic starters are often called 'soft' starters and use electronics to control voltage and current during the motor starting. They are limited in use because of the effect on motor starting torque and must be carefully matched with the torque requirements of the driven machinery and the torque characteristics of the motor.

For the large drives, such as water injection, gas export compressors and main oil line pumps, computer simulation will be necessary to ensure that the available main generation on the platform can start the drive successfully, without the platform electrical system collapsing.

See also PART 2 Chapter 11 and Chapter 12.

CHAPTER

Power Electronics (Semiconductor Equipment)

11

INTRODUCTION

Since the first edition of this book, solid-state devices have improved significantly in their capability to handle higher powers and voltages.

Although thyristors are still in use for such applications as integral cycle fired heater controllers, new variable-frequency converters operating alternating current (AC) induction motors are now more likely to be utilised for drilling motor drives. The heart of these drives is usually made up of a set of high-voltage insulated gate bipolar transistors, using deionised water cooling.

Power MOSFETs (metal oxide semiconductor field-effect transistors) are now commonly used in switched-mode power supplies for computers and small control panels. Secure supply systems now have more intelligent monitoring, with better diagnostics, so that, for example, the need for replacement of electrolytic capacitors can be spotted earlier.

ENVIRONMENTAL CONDITIONS

These devices are able to operate satisfactorily in air-conditioned pressurised modules, using the same conditions as would be provided for controlgear and switchgear.

Where full convertor systems are used, harmonics can be regulated and kept within industry-accepted limits.

UNINTERRUPTIBLE AND SECURE POWER SUPPLIES

Because of the need to maintain control of the process, and to provide electrical supplies for statutory safety equipment such as navaids and marine and helicopter radios, whether generation is available or not, a number of battery-fed secure supply systems are required on every offshore installation.

The grouping of these supplies so that they are fed from a common battery/charger system should be considered carefully to avoid common mode failures, i.e., situations where several vital supplies are lost owing to the failure of a common component. Emergency lighting fed from a central battery is particularly prone to

Offshore Electrical Engineering Manual. https://doi.org/10.1016/B978-0-12-385499-5.00013-3

this form of failure, as battery or distribution board failure will extinguish *all* the lamps on the system.

Although luminaires with integral emergency batteries require slightly more maintenance than luminaires without batteries, the benefit, in terms of increased system reliability by avoiding the risk of common mode failure, is very considerable.

DIRECT CURRENT SUPPLIES

The majority of secure supply systems are direct current (DC) and are required for the following.

Engine-starting batteries: The normal criterion for generator starting is that at least six consecutive start attempts each of 10 s cranking are possible from a fully charged battery without further recharge. In the case of fire pump start batteries, this is raised to 12 attempts from each of the two banks of batteries. This would be further reinforced by a second means of starting, such as a hydraulic pump device or compressed air motor.

Class A equipment: Fire and gas panels, emergency radios and public address (PA) systems are usually called category A or class A equipment. They are provided with duplicate chargers and batteries, each battery/charger being capable of supplying the rated load for the specified discharge time with mains failed. It is important to ensure that the manufacturer's design provides a fully duplicated system. There should be no common components in the rectifier or voltage regulation system, and the connection point for the two supplies, i.e., the distribution board, should be protected by blocking diodes and fuses from failure of one of the battery/charger units.

It is also normal practice to obtain the AC supplies for the two chargers from different points in the supply system in order to increase overall reliability.

Class B equipment: All other DC supplies, with the possible exception of some subsea equipment (see Part 2 Chapter 11), are supplied by single battery/charger units. However, for improved reliability a compromise of two 100% rated chargers with one battery may be used. As most of the components of the system are in the charger, an improvement in reliability can be obtained without the extra weight or bulk of a second battery.

Switchgear tripping supplies: Offshore practice is similar to that for onshore substations. For reasons of reliability, each supply should be dedicated to one switchboard. Batteries should have sufficient capacity for the tripping and closing of all circuit breakers on the switchboard in succession, i.e., twice, followed by the simultaneous tripping of all the circuit breakers. There should also be sufficient capacity to supply the continuous drain imposed by any control relays and indication lamps for a reasonable period (e.g., 8 h).

Secure DC and AC (battery-backed) power supplies are required to maintain power for the following:

A1 *An installation alarm system* – an acoustic alarm that is audible throughout the installation, to warn of danger, signal assembly at muster stations or to prepare for evacuation.

A2 *A PA system* – to enable those managing the emergency to broadcast messages to all personnel onboard, to warn of hazards, to issue instructions and to inform of incident status.

A3 *Muster station communications* – to provide communications between muster stations and the emergency control centre (ECC)/emergency control point (ECP).

A4 *Emergency response/support teams communications* – to provide communications between members of emergency response teams and with the ECC/ECP from all areas of the installation.

A5 *Other communications* – the provision of other communications for the purpose of emergency response, between persons on the installation, as identified by the prevention of fire and explosion, and emergency response (PFEER) regulation 5 assessment.

Objectives for external communications include the following:

A6 *Ongoing operations* – to enable the ECC/ECP to communicate with any ongoing associated operations, e.g., pipeline, heavy lifts, supply or diving operations.

A7 *Onshore emergency services and support* – to enable the ECC/ECP to inform or call for external assistance or support.

A8 *Nearby offshore installations* – to enable the ECC/ECP to communicate with nearby offshore installations that may be able to provide assistance or to advise of changes in pipeline status that could have an effect on the safety of their operations.

A9 *Vessels in vicinity* – to enable the ECC/ECP to alert and communicate with any ships in the vicinity, including standby boats or any other vessels, that may be able to provide assistance during an emergency.

A10 *Aircraft communications* – to enable the ECC/ECP and Helicopter Landing Officer to communicate with helicopters involved in evacuation and rescue operations. Note that battery-backed helideck lighting and a directional navigation beacon will also be required (see CAP437 in Bibliography).

A11 *TEMPSC (totally enclosed motor propelled survival craft) communications* – to enable persons in survival craft to communicate with ships and helicopters in the vicinity.

ALTERNATING CURRENT SUPPLIES

The uninterruptible power supply units are required where process controls, instrumentation, telemetry and telecommunications equipment require a secure source of AC power to avoid disruption of remote control, blank sections in logging records or loss of voice communication with the shore base.

Provided the platform generated frequency and voltage are reasonably stable and within a few percent of the nominal values, the inverter will normally remain in synchronism with the generated supply. During generated voltage or frequency excursions, the inverter will isolate itself from the main supply and continue to feed the load by itself. A mains bypass supply is also available and, if the inverter fails, it is automatically connected to the load. However, in some designs, if the bypass supply is outside the inverter voltage or frequency tolerance, it will not be made available to the load. This can be a problem if the generated supply is being obtained from the emergency generator, which will not normally have such good voltage or frequency holding capabilities as the larger machines.

It should be remembered that the inverter itself is generating a sine wave from a power oscillator and therefore cannot produce a fault current much greater than its rated current. For example, a 20-A inverter with the bypass supply unavailable is able, at best, to blow a 6A fuse in a reasonable time. If the distribution system requires larger fuses, then an inverter of higher rating will normally be required. It will not be necessary to provide a larger battery, however.

SELECTION OF VOLTAGE TOLERANCES

The general rule for establishing voltage tolerances is to optimise these for the minimum output cable cross-sectional area and battery size. Having been provided with or having calculated the diversified load current, discharge duration and voltage tolerances required for the equipment being supplied, the engineer can establish the minimum battery size from the manufacturer's tables.

However, this takes no account of the voltage drop in the DC output cable. This voltage drop must be catered for in the supply system to ensure that the lower tolerance of the supplied equipment is not exceeded, as shown in Table 2.11.1, for a nominal 24V DC system. The fire and gas panel supply shown in the table requires a 150-mm^2 cable because the voltage drop has been restricted to 0.5V DC. If the minimum voltage at the charger is raised to 23.1V DC, it would be possible to reduce the cable size to 70 mm^2. However, to raise this voltage, it would be necessary to increase the battery capacity, assuming the required discharge time remains constant.

Table 2.11.1 Voltage Tolerances

			Volts at Charger		Volts at Equipment	
Equipment	**Load (A)**	**Cable (mm^2/ length)**	**Min.**	**Max.**	**Min.**	**Max.**
Fire/gas panel	40	150/50m	22.5	26.4	22.0	26.4
ESD panel	12.5	35/45m	23.1	27.0	22.5	26.9
PA system	25	95/100m	22.6	26.4	21.6	26.4

ESD, emergency shutdown; PA, public address.

BATTERIES

TYPES OF BATTERIES

The following three types of batteries are used commonly offshore.

Lead–Acid Planté Cells

This type of cell is appropriate for most applications where a suitable ventilated battery room can be provided. It should not be used in a hot vibrating environment such as adjacent to engines. This is because the electrolyte will tend to evaporate and require frequent topping up owing to the heat, and the vibration will tend to cause the oxide coating to fall off the plates. With careful maintenance, these batteries may last in excess of 20 years.

Lead–Acid Recombination Cells

These cells are a relatively recent development and provide several advantages over the Planté cells. The cell contains no free electrolyte, the entire electrolyte being contained in an absorbent blotting-paper-like material.

Except for a small safety vent, the cells are sealed for life; unlike the Planté cells, they normally emit only molecular quantities of hydrogen. Because of their recent availability, no operational life expectancy figure is available but from present experience it is known to be in excess of 8 years.

Excessive charging of recombination cells will permanently damage them by electrolysing the small amount of electrolyte they contain. Once this has been vented, it cannot be replaced. Therefore it is vital that, with the chargers used, there is a very low probability that a fault will occur which allows charging voltages greater than the maximum recommended by the manufacturer. For the same reason, a boost charging facility must not be fitted and engine-starting batteries of this type must not be connected to an engine-driven charging alternator, as this will more than likely exceed the permitted charging voltage.

If, on discharge, the cell voltage falls below a certain minimum value, it will no longer be possible to recharge the cells affected, making it necessary to replace them. Therefore, it is important that during the installation and commissioning period, cells are fully recharged at least every 6 months.

Despite these considerations, these batteries are becoming very popular, mainly because of the saving that can be obtained from the reduction in maintenance required, as well as because reliable electronic chargers are available with good voltage regulation and current limiting facilities.

Nickel–Cadmium Cells

These cells are generally more capable of withstanding heat, shock and vibration than both of the lead–acid types. A nickel–cadmium battery is 35% lighter than the equivalent lead–acid one. However, nickel–cadmium batteries are considerably more expensive, especially in comparison with recombination cell batteries of the same capacity and voltage.

Although able to endure longer periods of inaction without charge, they may require more skilled maintenance than lead–acid batteries for the following reasons:

1. There is a memory effect associated with repetitive cycling of the battery which may cause the battery voltage to be depressed below the value expected from its discharge characteristic. The effect is usually temporary and can be reversed by reconditioning cycles.
2. Sustained boost charging may have the effect of depressing the discharge voltage slightly. As mentioned earlier, this can be reversed by reconditioning cycles.

BOOST CHARGING FACILITIES

It is common to find a boost charge mode available on battery chargers, the intention being to enable the operator to reduce the battery recharge time.

With the electronic chargers now available, battery recharge durations are short enough for most purposes without having to recourse to a manually initiated boost charge mode. Boost charge facilities are not recommended on any battery systems because of the excessive gassing they can cause. If they must be fitted, some form of timer should be incorporated to prevent sustained overcharging. They should not be fitted to the chargers of sealed lead–acid cells to avoid permanent cell damage.

VENTILATION AND HOUSING OF VENTED BATTERIES

Special provisions must be made for the housing of batteries offshore to minimise the risks due to electrical ignition hazard, evolved hydrogen and electrolyte spillage.

The battery room provided should be lined on the walls and floor with a rubber-based or similar electrolyte (either sulphuric acid or potassium hydroxide)-resistant material. Owing to the risk involved in any contact between the two forms of electrolyte, it is not advisable to house a mixture of lead–acid and nickel–cadmium cells in the same room. Inadvertent topping up with the wrong type of electrolyte would also be extremely hazardous for the personnel concerned.

Both types of cell evolve hydrogen, and this must not be allowed to build up to explosive levels. A ventilation system must be provided that is sufficient to dilute the hydrogen well below this level at all battery states. A hydrogen detector is usually provided, and any increase in hydrogen concentration would be indicated at a manned control room for the installation. Cells should be racked in tiers so that each cell is easily accessible for maintenance or replacement.

To cater for major gas leak conditions, an explosion-proof isolator must be provided for each battery. These double-pole isolators would be remotely operated through the emergency shutdown system should such a serious gas leak occur.

SEALED CELLS

Although very little hydrogen is emitted from lead–acid recombination cells, it is still advisable for batteries of significant size to be located in a dedicated purpose-built room. Failure of the battery charger to limit charge voltage would lead to a short-lived evolution of hydrogen, and the room ventilation would need to cater for this. A dedicated room also prevents the access of unauthorised personnel who may not be aware of the dangers associated with stored electrochemical energy; tools or other metal objects inadvertently placed across battery terminals will cause heavy short-circuit currents to flow.

SOLID-STATE CONTROLLERS

BURST-FIRED (INTEGRAL CYCLE FIRED) PROCESS HEATER CONTROLLERS

Burst firing consists of supplying a series of whole mains cycles to the load. This method of control is used for controlling heaters, but not motors, for obvious reasons. It has the benefit of avoiding the production of nonsinusoidal waveforms (Fig. 2.11.1). Unfortunately, it can produce flicker in lighting if the heater it is controlling is large (Fig. 2.11.2).

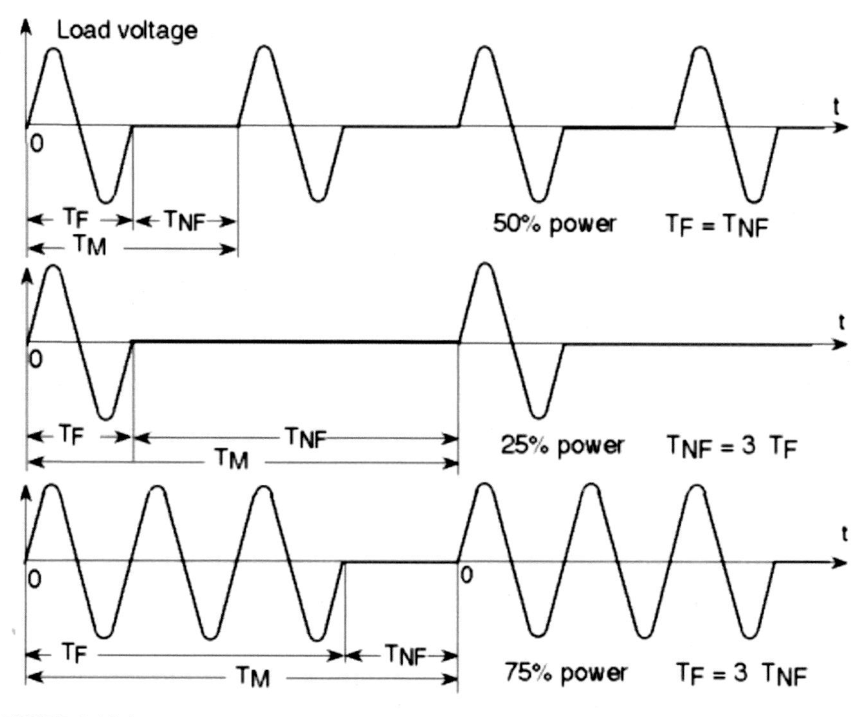

FIGURE 2.11.1

Waveforms of a burst-fired controller. T_F, firing time; T_M, modulation time; T_{NF}, nonfiring time.

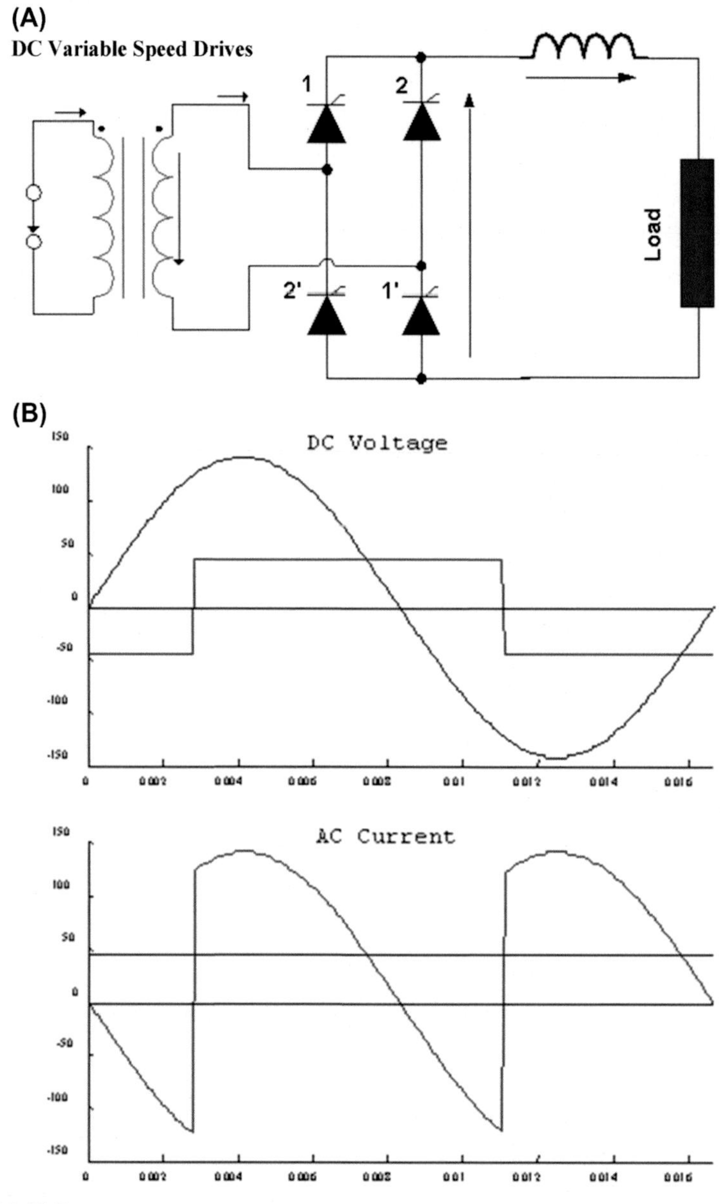

FIGURE 2.11.2

(A) A typical thyristor phase-controlled bridge. (B) Direct current (DC) voltage and alternating current (AC) under continuous conduction for a large value of smoothing reactance.

DIRECT CURRENT VARIABLE-SPEED DRIVES

Thyristors or Silicon controlled rectifiers (SCRs) have been used for some years for the simple speed control of DC motors, particularly for offshore drilling equipment. SCRs known commonly as thyristors are three-junction PNPN semiconductor devices which can be regarded as two inter-connected transistors that can be used in the switching of heavy electrical loads. They can be latched-"ON" by a single pulse of positive current applied to their gate terminal and will remain "ON" indefinitely until the anode to cathode current falls below their minimum latching level SCRs have largely been replaced by water-cooled insulated gate bipolar transistors (IGBTs).

ALTERNATING CURRENT VARIABLE-FREQUENCY DRIVES

A pulse width modulation drive consists of four stages:

1. A rectifier which converts the fixed (mains) frequency power into DC.
2. A DC link circuit which provides inputs to the inverter and includes a filter which reduces harmonics reflected back into the fixed frequency system. The filter unit may be passive or active. Passive filters consist of a smoothing network of inductors and capacitors which is switched in and out by the control circuitry, whereas active filters consist of a full inverter circuit capable of injecting a variable corrective current to cancel out the major harmonics.
3. The inverter.
4. The induction motor designed to operate with the variable-speed drive package above.

The rectifier circuit of a pulse width modulated drive normally consists of a three-phase diode bridge rectifier and capacitor filter. The rectifier converts the three-phase AC voltage into DC voltage with a slight ripple. This ripple is removed by using a capacitor filter. (Note: the average DC voltage is higher than the root mean square (rms) value of incoming voltage by AC (rms) $\times$ 1.35 = VDC).

The control section of the adjustable-frequency drive accepts external inputs which are used to determine the inverter output. The inputs are used in conjunction with the installed software package and a microprocessor. The control board sends signals to the driver circuit which is used to fire the inverter.

The driver circuit sends low-level signals to the base of the transistors to tell them when to turn on. The output signal is a series of pulses (Fig. 2.11.3), in both the positive and negative directions, that vary in duration. However, the amplitude of the pulses is the same. The sign wave is created as the average voltage of each pulse, and the duration of each set of pulses dictates the frequency.

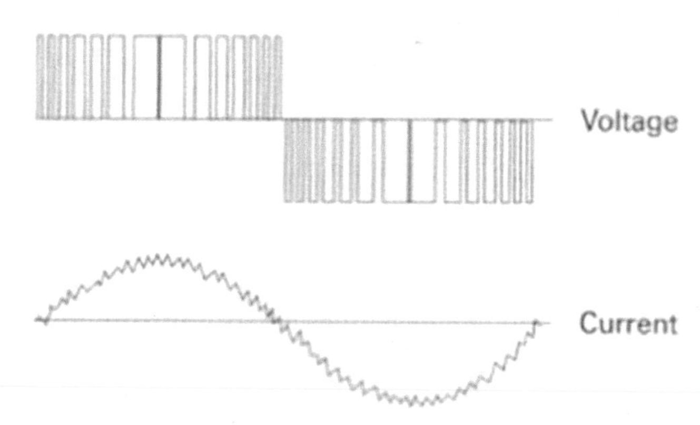

FIGURE 2.11.3

Variable-frequency drive output.

By adjusting the frequency and voltage of the power entering the motor, the speed and torque may be controlled. The actual speed of the motor, as previously indicated, is determined by

$$N_s = [(120 \times f)/P] \times (1 - S)$$

where N, motor speed; f, frequency (Hz); P, number of poles; S, slip.

VARIABLE-SPEED DRIVE OR VARIABLE-FREQUENCY DRIVE

Both systems have benefits, but technical variable-frequency drives (VFDs) are better because of the higher torques available and less maintenance. However, because of the more complex electronics, quicker obsolescence is likely to be a problem with VFDs.

CHAPTER

Process Drives and Starting Requirements

12

INTRODUCTION

This chapter is a continuation of PART 2 Chapter 9.

Machinery drives for offshore installations range in size from fractional horsepower cabin ventilation fans to the 15 MW or more machines required for gas injection compressors. Some concern has been expressed over the practicality of installing very large electric motors offshore because of the stresses imposed on the structure, especially during starting. A figure of around 25 MW, depending on the application (e.g., less for reciprocating pump drivers), should be considered as the normal ceiling figure for offshore electric motor ratings. Where the design is outside general engineering experience/practice, special studies will be required to ensure that structural resonance, shock, noise and vibration, metal fatigue, electric system dynamics, etc. are not 'showstoppers' and this will add significant cost to the project.

Summarising from PART 2 Chapter 9.

VOLTAGE LEVELS

Voltage Level	Rating
1. 415–460 V	Up to 150 kW
2. 3300–6600 V	Up to 1000 kW
3. 11–13.8 kV	Over 1000 kW

Higher voltages such as 22 kV could be utilised, provided the economics is favourable and proven ranges of machine at these higher voltages are available.

STARTING

Where an individual motor load represents a substantial part of the power system capacity, it is desirable to obtain designs of squirrel-cage motors with the lowest starting current characteristics compatible with the driven equipment. Both reduced-voltage starting of cage motors and the use of slip ring motors will add considerable

Offshore Electrical Engineering Manual. https://doi.org/10.1016/B978-0-12-385499-5.00014-5

weight to the drive package, and therefore direct-on-line, low-starting-current machines are preferred, unless technical considerations for selection are overridden by those of cost and delivery. Motors with starting currents ranging from 3.5 to 4.5, instead of the normal 6–10 times full load, are usually available from the larger manufacturers.

SPEED

Higher pump speeds at 60 Hz is marginal, and with motors having ratings in the order of several megawatts, weight and power savings can be substantial.

With smaller installations, however, operating at 60 Hz may be a disadvantage where it is decided to select a reciprocating engine rather than a gas turbine main generator prime mover. The problem is that the optimum engine speed of around 1500 rpm is better suited to generation at 50 Hz. Reciprocating engines running at 1200 rpm tend to have too low power-to-weight ratio, and operating at 1800 rpm leads to short cylinder life or even piston speeds which would be beyond the design limitations of the engine.

The higher synchronous speeds obtainable at 60 Hz also lead to higher inherent noise levels, although this can be deadened with better module insulation.

From a machinery standpoint, a major disadvantage in adopting 60 Hz for a European offshore installation is related to testing the equipment before installation on the platform, as full-load tests cannot be carried out using the British and European national supply networks. Until recently, tests were mainly carried out at 50 Hz and the results extrapolated to give projected machine characteristics at the design conditions. However, currently, test facilities are available in the United Kingdom for motors of up to 6 MW at 60 Hz. For larger machines, where capital investment is high and full-load tests are considered essential, it is usually possible to arrange full-load tests in conjunction with testing of the main generators to be installed on the offshore installation. Although this procedure is usually expensive, the costs should be more than offset by the benefits of adopting the higher frequency.

Once the system frequency has been established, it becomes increasingly expensive to change and hence may no longer be considered a variable after that point in the system design. It is therefore important to consider the number, rating and purpose of the larger drives on the installation at an early stage in the power system design, before the frequency is selected. However, in many cases the shaft speeds available from a motor, even from a 60 Hz two-pole machine (i.e., 3600 rpm) is lower than the required shaft speed, and it will still be necessary to install a gearbox in the drive string. Once the requirement for a gearbox has been established, changes in the drive ratio have only minor effects and the motor speed may be chosen to give the optimum motor design in terms of dimensions, weight, reliability, noise emission, etc.

POLE CONFIGURATION

The fastest possible speed of both induction and synchronous motors is with a two-pole configuration which gives synchronous speeds of 3000 rpm at 50 Hz and 3600 rpm at 60 Hz, respectively. As size and weight penalties are usually incurred by increasing the number of poles, only two- and four-pole machines are normally used offshore. The only occasional exception being large reciprocating compressor drivers, where eight-pole (or more) synchronous machines may be used. These machines, apart from giving the required lower speed, also reduce the current fluctuations caused by the cyclic torque variations associated with reciprocating machinery. The advantages and disadvantages of four-pole against two-pole machines are listed in the following.

Advantages

a. Energy conversion within the higher speed, two-pole machine is usually more efficient, giving some reduction in size and weight for a given output. However, because of the higher rotor speeds, especially at 60 Hz, mechanical forces on the rotor cage, known as hoop stresses, become significant and limit the maximum dimensions of rotor that may be manufactured using conventional materials. Metallurgically more exotic materials may be used to extend this limit with, of course, the accompanying large increases in the cost of the machine.
b. Manufacturers differ in the application of a practical maximum rating limit for a two-pole motor, but as a general rule, this is between 3 and 5 MW. Therefore motors above 5 MW should not be considered as a feasible alternative unless the cost of using exotic metals in the rotor is outweighed by the savings accrued by, for example, the elimination of a gearbox.
c. Below this limit, and within the manufacturer's normal product ranges, the use of two-pole machines should, in comparison with an equivalent four-pole machine, provide dimension and weight savings roughly proportional to power rating; however, for small machines of only a few kilowatts, there is little benefit.

Disadvantages

a. Less starting torque is available from two-pole machines, requiring the driven equipment to have a lower moment of inertia. It may also prove more difficult to accelerate the machine up to operating speed, where driven machinery cannot be run up to speed unloaded. The speed/torque characteristics of pumps in particular should be carefully studied to avoid any problems. Starting currents are also likely to be higher, and with large machines, this may lead to unacceptable voltage dips.
b. Irregularities in the core stampings, which are inevitable unless very high levels of quality control are applied, generate more magnetic noise in two-pole machines. A characteristic low-frequency (twice slip frequency) growl can be heard from these motors.

c. Rotor imbalance is more likely and can cause more vibration on two-pole machines.
d. The higher rotor speeds associated with two-pole machines will shorten bearing life or require a more expensive higher-performance bearing.
e. Poorer heat dissipation within the rotor necessitates increased cooling airflow rates in two-pole machines. The effect of higher fan speeds and increased airflow rates is to increase noise emission from the machine.

COOLING AND INGRESS PROTECTION

As discussed in earlier chapters, only fully enclosed types of motor are normally considered suitable for offshore installations.

To summarise from PART 2 Chapter 9, the following is recommended:

Machine	Design
Smaller low-voltage machines	TEFC type
Where the TEFC design is not practicable, CACW type, i.e., with larger 3.3/4.16kV and all higher-voltage machines	CACA type
Larger machines where cooling water supply is uneconomic or machine must operate during a cooling system outage	CACA type

CACA, *closed air circuit, air cooled;* CACW, *closed air circuit, water cooled;* TEFC, *totally enclosed fan cooled.*

SPECIAL APPLICATIONS

The following section provides some advice on the selection of motors for particular applications.

Hazardous area topics are discussed in PART 5 Chapter 4 and are not discussed in detail here. However, readers who are likely to be specifying motors for hazardous areas are advised to read PART 5 Chapter 4 (see also Fig. 2.12.1).

RECIPROCATING PUMPS AND COMPRESSORS

If an induction motor is used to drive a large reciprocating pump or compressor, the heavy cyclic torque fluctuations demanded from the motor will in turn demand heavy current fluctuations from the supply. When the motor load is a significant part of the installation generating capacity, instability of voltage and power may result.

An alternative is to use a synchronous motor with a squirrel-cage damping winding imbedded in the rotor. If a steady torque is being developed by the machine, the

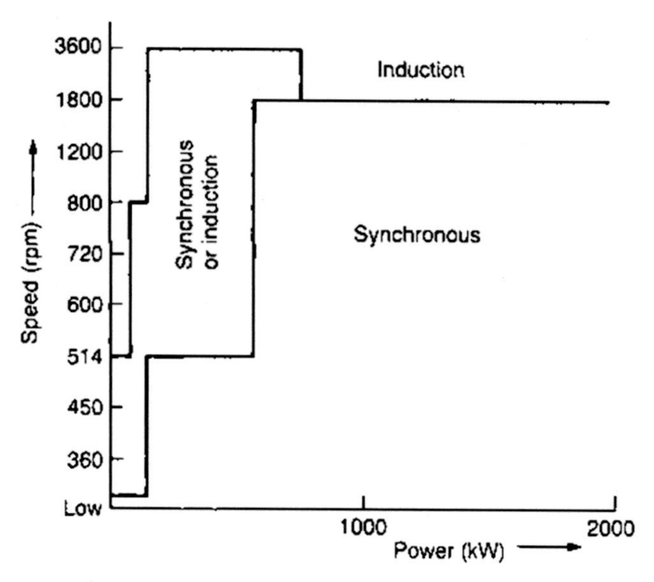

FIGURE 2.12.1

Ratings of induction and synchronous motors for compressor drivers.

load angle would remain in an equilibrium position. As the rotor of a synchronous motor running in synchronism with the supply experiences a torque proportional to its angular displacement from the equilibrium position and also possesses rotational inertia, it constitutes an oscillatory system similar to the balance wheel of a clock. If J is the moment of inertia of the rotor in kilogram metre squared, then it can be shown that natural frequency of the rotor will be

$$f=\frac{1}{2\pi}\sqrt{\mathrm{Ts/J}\times(\text{No. of pole pairs})}\ \mathrm{Hz}$$

where Ts = 3VI/cos θ nm; V, the system voltage; I, current produced by the field induced voltage; θ, the load angle.

Synchronous motors driving reciprocating machinery receive torque impulses of a definite frequency and for satisfactory operation the natural frequency of the rotor must be at least 20% higher or lower than the frequency of the torque impulses (Fig. 2.12.2).

The imbedded squirrel-cage damping windings, used for starting, will produce some corresponding current fluctuation with torque, but this is not excessive as can be the case with an equivalent induction motor. Such windings produce damping torques proportional to the angular velocity of any rotor oscillation and hence reduce the synchronous motor's tendency to hunt because of the alternating currents induced in the other windings and current paths of the rotor, giving rise to destabilising torques.

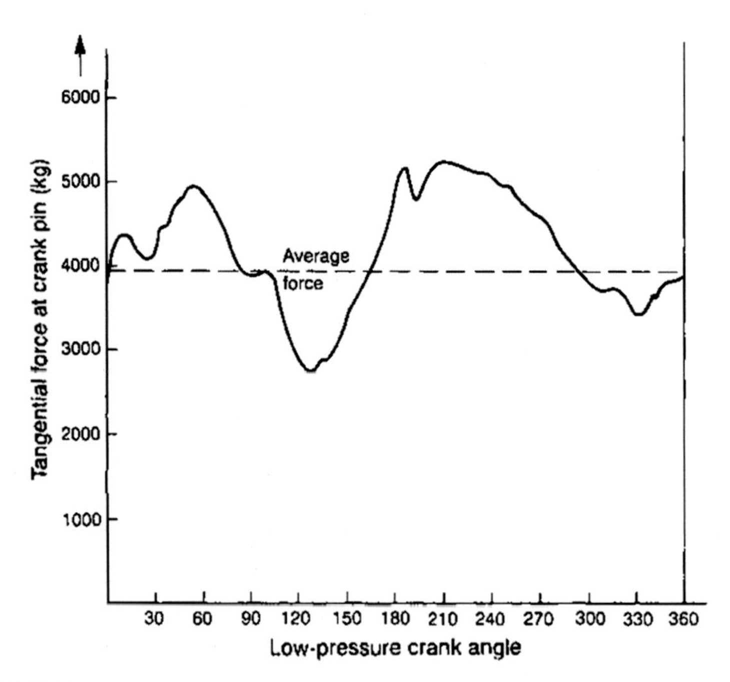

FIGURE 2.12.2

Typical torque requirements over one cycle of reciprocating compressor operation, full load.

GAS COMPRESSORS

It is worth remembering that large gas compressors, whether using induction or synchronous motors, are highly dependent on their auxiliaries for reliable and safe operation. The following is a list of basic requirements:

1. hazardous area ventilation fans,
2. main and standby lube oil pumps,
3. main and standby seal oil pumps,
4. cooling water pumps,
5. motor ventilation fans.

This list is not exhaustive, as it does not include the various installation utilities and safety systems which have to provide continuous permissive signals to allow starting or continued running of the compressor package.

DIRECT CURRENT DRILLING MOTORS

The conventional arrangement in a drilling rig is to utilise 750V direct current (DC) machines run from phase angle-controlled thyristor units.

Typically, the machines must be capable of accepting a voltage variation of 0–750 V DC and continuous load currents of 1600 A. For mud pump duty, two motors will run in parallel on one silicon controlled rectifier (SCR) bridge, whilst for the drawworks and rotary table, one SCR bridge will be assigned to each motor with appropriate current limiting devices in operation. The motors must also be capable of producing around 600 kW continuously and 750 kW intermittently at 1100 rpm. These machines are shunt-wound machines with class H insulation, derived from railway locomotive designs. However, as they operate in hazardous areas, the construction is closed air circuit, water cooled to restrict surface temperatures, with the enclosures being pressurised to prevent ingress of explosive gas mixtures.

This type of motor is often provided without a terminal box, the winding tails passing through a sealed and insulated gland to a separate flameproof terminal box. It is recommended, however, that motor-mounted terminal boxes should be used, as the exposed winding tails are difficult to protect mechanically.

POWER SWIVELS

On some drilling rigs the drill string is powered by the swivel instead of the rotary table. The system consists of the swivel powered by a hydraulic motor fed by hoses from a hydraulic power pack which is located in a pressurised room. Power pack consists of a swash plate pump driven by a medium-voltage motor. The motor is a conventional squirrel-cage type, as described earlier.

SEAWATER LIFT PUMPS

In floating installations or those with hollow concrete legs, conventional pumps may be used to obtain source water for cooling because the pump may be located at or near sea level.

However, on steel jacket construction platforms, it is necessary to draw seawater for cooling and firefighting up to the topsides using seawater lift pumps.

There are three basic types of drives for seawater lift pumps:

a. submersible electric
b. submersible hydraulic
c. electric shaft driven

The last two types use standard forms of electric motor as the power source, in which the power is transmitted mechanically or hydraulically to the location of the pump (see Fig. 2.12.3).

Therefore, only the first type will be discussed as a special application. Submersible pumps for this duty are of small diameter and usually very long, consisting essentially of a series of small induction motors all mounted on the same shaft. The motor-pump string, having been connected to a special flexible cable, is

FIGURE 2.12.3

Typical electric shaft-driven fire water pump.

Courtesy SPP Offshore, a division of SPP Ltd.

lowered down the suction or 'stilling' tube. Alternatively, the stilling tube may be sectionalised and the motor fixed in the lowest section of the tube. The tube is then lowered down a platform riser and jointed section by section until the motor suction is 10 or more metres below the sea surface at the height of the lowest expected tide. For cooling purposes, the motor must be below the pump, and therefore, difficulty is often experienced in avoiding damage to cables which have to pass between the pump and the stilling tube in order to reach the motor terminals. Bites need to be taken out of the pump retaining flanges to allow the cables past, and the cable overall diameter must allow a loose fit through these to avoid damage. A photograph of this type of pump set is shown in Fig. 2.12.4.

DIESEL–ELECTRIC FIRE PUMPS

Statutorily, every offshore installation has to be provided with at least two (depending on the capacity) serviceable fire pumps, each of which must be powered independent of the other. A third pump must be provided to cater for unavailability during servicing. Pumps must also be physically segregated and located geographically well away from each other to minimise the risk of both pump systems being damaged by the same fire or explosion. Further details on capacities of pumps etc. are obtainable from the Department of Energy Guidance Notes on Offshore Firefighting Equipment (see Appendix A).

FIGURE 2.12.4

(A) Diesel–electric fire pump. The pump (foreground) is installed in a vertical submerged stilling tube. (B) Rear view.

Courtesy SPP Offshore, a division of SPP Ltd.

A typical fire pump arrangement consists of one motor-driven pump powered by the installation power system and two pumps driven by dedicated direct diesel or diesel generator sets. If a diesel generator set is to be dedicated to the supply of a particular submersible pump, weight and cost may be reduced by dispensing with any switching device and, in the manner of a diesel–electric locomotive, by cabling the generator directly to the pump. The usual arrangement is to use a fire-survival cable from the generator to a terminal box at the top of the stilling tube where it is connected to the flexible stilling tube cable. The advantage of this arrangement is that the motor starts with the generator, effectively providing a reduced voltage start characteristic. If the motor were started from a switching device, then a generator of lower rating than that required can be used. Another benefit of this arrangement is that a generator voltage can then be specified which minimises the stilling tube cable diameter, i.e., copper and insulation cross sections can be traded off. The fire pump diesel generator module is designed to be as independent as possible from other platform systems, and if using cooling water bled from the fire pump system itself, it only requires to obtain combustion air from its surroundings. The photograph in Fig. 2.12.5 shows a typical diesel generator of the current design.

FIRE PUMP DIESEL ENGINE STARTING REQUIREMENTS

The starting requirements for fire pump diesel engines are laid down in the (American) National Fire Protection Association (NFPA) Specification No. 20. This requires that the engine be provided with two batteries each capable of 12 start attempts, specified both for minimum cranking duration and interval between each cranking. The engine must also have an independent means of starting, such as a hydraulic or compressed air device complete with some form of accumulator.

FIGURE 2.12.5

A typical fire pump/emergency generator module designed for fire survivability and independent operation.

Courtesy SPP Offshore, a division of SPP Ltd.

DOWNHOLE PUMPS

Downhole pumps are used in the latter stages of well life to provide 'artificial lift' when wellhead pressure or crude oil flow rates need to be improved. Downhole pumps are extremely rugged devices, having to work in a 'hole' less than 0.25 m in diameter, thousands of feet below the sea bed and at pressures and temperatures near the design limits. The pump–motor string although of very small diameter, is often in excess of 10 m long. The pumps are expensive proprietary devices, with their manufacturers' closely guarding the design secrets. The installation of a downhole pump may cost in excess of £4 million and as usually a minimal warranty is provided once the pump is installed; however, that sum may need to be spent again a few days after installation if nothing happens when the start button is pressed.

However, much can be done at the surface to improve the reliability of the pump. Notably, the use of variable-frequency convertors to provide a very 'soft' start for the motor has been proved to be very successful in the past few years. Downhole pump motor nominal surface voltages are usually around 2 kV to allow for voltage drop in the cable and to trade off copper and insulation cross sections to minimise motor dimensions. As the pump may be capable of lifting in excess of 20,000 barrels per day of oil and water from the well, the frequency convertor and transformer units tend to be large.

MAIN OIL LINE AND WATER INJECTION PUMPS

Apart from the hazardous area requirements, the motor drivers for these pumps will be very similar to those used onshore. Because of the pressures involved (typically 200–400 bar g), care needs to be taken in the design of controls to ensure that no undue stresses are put on the pump or pipework, particularly during start-up. It should not be possible to start the motor if the pump motor set is running backwards because wrong valves are being inadvertently opened. As well as causing shock to the pipework, such maloperation may draw starting currents of excessive magnitude from the system, possibly causing damage to the motor windings and/or power system instability.

CHAPTER

Subsea Supplies and Cathodic Protection

13

INTRODUCTION

To those new to the subject of offshore electrical engineering, using electrical power on equipment which is totally immersed in highly conducting salt water may seem somewhat foolhardy. In actual fact, electrical power has been used in submersible pumps and for underwater arc welding since around the beginning of the last century. Providing certain precautions are taken, using electricity under the sea can be comparatively safe.

SUBSEA POWER SUPPLIES

Permanent arrangements will be required for power to operate valves, switches and sensors on underwater wellhead manifolds and the new generation of subsea production wells which are serviced periodically by special vessels. Where power and control is required at the manifold from a platform some distance away, an unearthed supply will be required at a suitable voltage to compensate for the impedance of the subsea cable. It is usual, however, for power for operation of valves etc. to be transmitted hydraulically with electrical actuation from a secure (UPS) ac supply on the adjacent platform.

It is recommended that all ac supplies used are three phase and isolated via the delta connected, unearthed secondary windings of transformers. This provides the best scope for incorporating protection devices and single earth faults can be arranged to operate alarms whilst the circuit remains in operation.

DIVER'S LIFE SUPPORT EQUIPMENT

The main requirement for the North Sea diver is heat, both for body and respiratory gas heaters.

The design of respiratory gas heating, if fitted to the diver's suit must be such that there is no possibility of electrocution through the mouth. If the diver's suit is heated electrically then the elements of heated undergarments should be screened such that there is no possibility of the heating elements touching the diver's skin, or the elements shorting internally and causing localised heating. Both respiratory gas heaters and electric body heaters should use a low voltage dc supply no greater than

Offshore Electrical Engineering Manual. https://doi.org/10.1016/B978-0-12-385499-5.00015-7

24 Vdc. Heater elements, including supply cables, should be screened with the screen continuity continuously monitored.

A line insulation monitor (LIM) designed on fail-safe principles should be incorporated into the supply circuit to detect any deterioration in the supply cable or heater element insulation resistance. The sensitivity of the LIM should be such as to ensure a tripping response in a few milliseconds to a significant fall in resistance.

By far the most common form of diver's suit heating in the North Sea, however, is the 'hot water machine' from which heated seawater is pumped from the surface. The heated water is directly in contact with the diver's skin and escapes from wrist and ankle bands in the suit. There is a remote possibility that an earth leakage current could travel down the heated water via the hose and be earthed via the diver, but if the heater element tank in the hot water machine is adequately earthed, there should be no danger, especially as the incoming cold water also provides a conducting path.

If the diver is to use any handheld electrical equipment, sensitive earth leakage current (RCD) protection will be required. RCDs with a trip current of 4 mA are recommended (see Codes of Practice in Bibliography).

SUBSEA COMPLETION MODULES

With the development of remotely operated vehicles (ROV's) and with oil and gas exploration in deeper waters, subsea completions are now both more robust and also far more complex, due to the well monitoring and control telemetry facilities available. Fig. 2.13.1A–H is a series of CAD 3D pictures of various modules and connectors. Some connectors have special keys/implements to allow mating of connectors using ROV tools.

FIGURE 2.13.1

(A) Subsea Christmas Tree (XMT) with Twinlock subsea control module (SCM) integrated in the foreground. (B) 199098_POD-01 – Twinlock SCM with top cover removed for clarity. (C) 199098_POD-02 – Twinlock SCM with top cover shown as built. (D) Module mounting base. (E) Older generation Monolock SCM with cover shown as built. (F) Oil filled EFL (Tronic) (2) – typical remotely operated vehicle (ROV) mateable Tronic electrical flying lead (EFL). (G) Older generation Monolock SCM hydraulic/electrical connection detail for SCMMB interfacing. Optical iConSEM_with new logo_original_1035 × 1057 (2) – iCon subsea electronics module (SEM) as built with optical and electrical connection interfaces. (H) Optical iConSEM, with new logo original_1035 × 1057 (2) – iCon SEM as built with optical and electrical connection interfaces.

(A) Courtesy Aker Solutions, Dyce, Aberdeen. (B) Courtesy Aker Solutions, Dyce, Aberdeen. (C) Courtesy Aker Solutions, Dyce, Aberdeen. (D) Courtesy Aker Solutions, Dyce, Aberdeen. (E) Courtesy Aker Solutions, Dyce, Aberdeen. (F) Courtesy Aker Solutions, Dyce, Aberdeen. (G) Courtesy Aker Solutions, Dyce, Aberdeen. (H) Courtesy Aker Solutions, Dyce, Aberdeen.

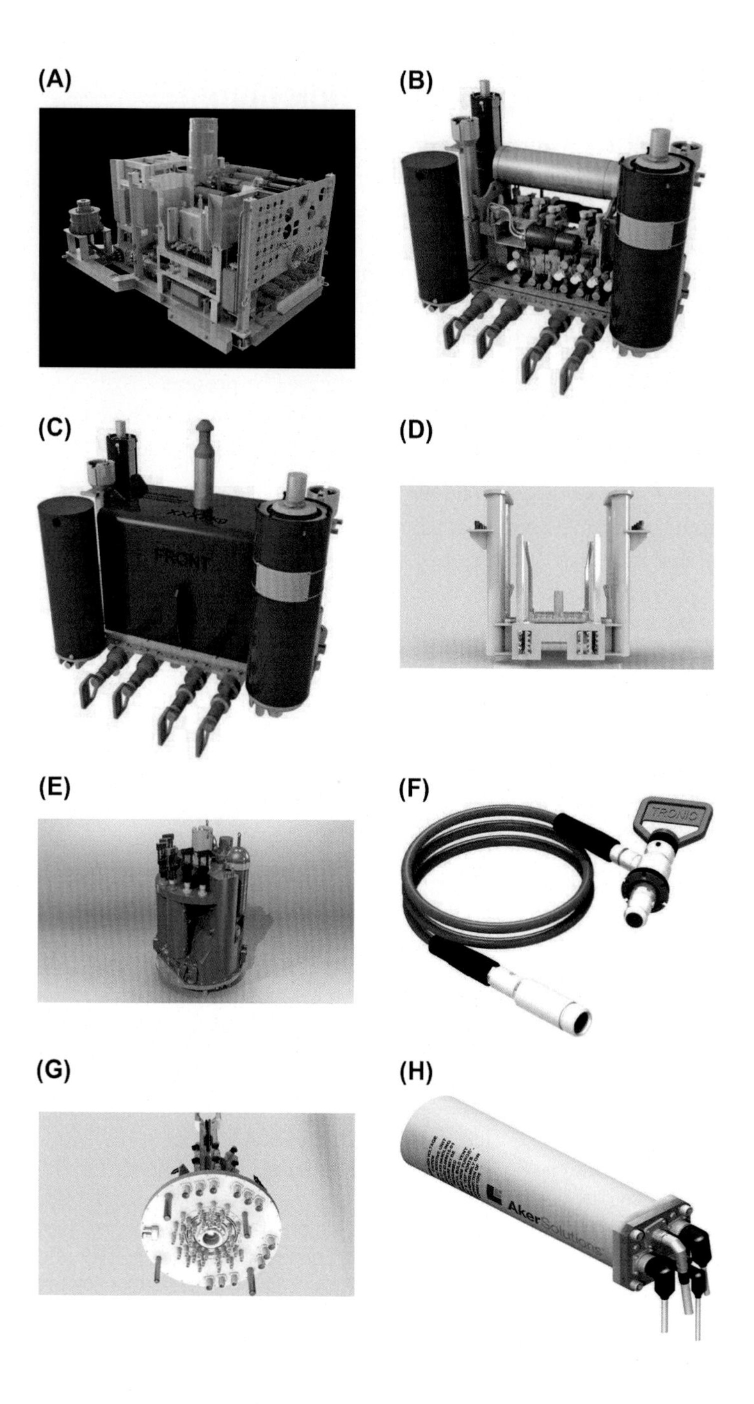
(A)
(B)
(C)
FRONT
(D)
(E)
(F)
TRONIC
(G)
(H)
AkerSolutions

DIVING CHAMBERS FOR SATURATION DIVING

As the lives of divers may depend on the electrical supply being available at the chamber, particularly if deep diving involving compression chambers is being undertaken, the electrical supply systems of the associated support vessels must be of the highest integrity and reliability.

The equipment used inside diving chambers is, where possible, to be designed using the principles of hazardous area 'intrinsic safety'.

This is for the following reasons:

1. To prevent ignition should flammable well gases be present
2. It will help to prevent fires in a chamber with an oxygen enriched atmosphere
3. It will ensure earth fault currents are limited to very low values.

Table 2.13.1 shows the effects on the human body of various magnitudes of ac current and from this it can be seen that values beyond 18 mA are dangerous. In salt water resistances tend to be low and very low voltages are required to achieve this threshold value. All electrical circuits within the chamber will need to be protected against earth faults by using current operated earth leakage circuit breakers set to trip at not more than 15 mA in 30 ms. The power supply within the chamber should be low voltage and

Table 2.13.1 Physiological Effects of Electricity on Humans

Current at 50 to 60 Hz rms mA	Duration of Shock	Physiological Effects on Humans
0–1	Not critical	Range up to threshold of perception, electric current not felt.
1–18	Not critical	Independent release of hands from objects gripped no longer possible. Possibly powerful and sometimes painful effect on the muscles of fingers and arms.
18–30	Minutes	Cramp-like contraction of arms. Difficulty in breathing. Rise in blood pressure. Limit of tolerability.
30–50	Seconds to minutes	Heart irregularities. Rise in blood pressure. Powerful cramp effect. Unconsciousness. Ventricular fibrillation if long shock at upper limit of range.
	Less than cardiac cycle	No ventricular fibrillation, heavy shock.
50 to a few hundred	Above one cardiac cycle	Ventricular fibrillation. Beginning of electrocution in relation to heart phase not important. (Disturbance of stimulus conducting system?) Unconsciousness. Current marks.
	Less than cardiac cycle	Ventricular fibrillation. Beginning of electrocution in relation to heart phase important. Initiation of fibrillation only in the sensitive phase. (Direct stimulatory effect on heart muscle?) Unconsciousness. Current marks.
Above a few hundred	Over one cardiac cycle	Reversible cardiac arrest. Range of electrical defibrillation. Unconsciousness. Current marks. Burns.

either direct current (e.g., 24 Vdc) or possibly at a frequency greater than 20 kHz which, apparently, the human responds to in a similar way to direct current.

Any sockets used should be the underwater type that can be connected while immersed in seawater without hazard. Unused sockets should be provided with insulating caps. If no suitable intrinsically safe equipment can be obtained, equipment incorporating other types of protection may be used.

Any electric motors in the chamber should be of the flameproof or increased safety type, and designed to withstand stalling currents for long periods without detriment. It is more than likely that surface Ex 'd' certification will prove invalid for equipment to be used in underwater chambers, because of the higher 'breathing atmosphere' pressures required for long-term compression diving or the oxygen enrichment. The equipment enclosure will require pressure testing to 1.5 × the normal peak explosion pressure × working pressure of chamber (bar. A).

Intrinsically safe equipment safety margins are also reduced by increase in pressure, and such systems will also require reassessment when used in this environment.

Other types of certified equipment may be affected by the higher humidity or by water ingress because of the higher pressures likely to be experienced by underwater equipment.

INDUCTIVE COUPLERS

If the core laminations of a transformer are cut in half, the primary coil is wound on one half and secondary is wound on the other; then, when the two halves are mated together, a rather inefficient transformer is formed. Although this principle is used in the inductive couplers used throughout the North Sea and elsewhere, the design is less common now than the underwater mateable 'Tronic' type of connector. Fig. 2.13.1 shows a photograph of the mating side of an inductive coupler. The core and windings are encapsulated so that the seawater at North Sea seabed pressures cannot penetrate into the live parts of the device. Only the surface of the core, which is ground flat, is exposed.

Because of the extra losses produced by the position of the windings and the air gap in the core, more heat is generated than in a conventional transformer and this must be removed by a core heat sink connected to the stainless steel housing. Inductive couplers are seldom used these days, since electrical connectors have been developed now, which can be mated live underwater (Fig. 2.13.2).

SUBSEA UMBILICALS AND POWER CABLES

An 'umbilical' is the term used for obvious reasons to describe a bundle of hydraulic hoses, signal, control and power cables bundled together, armoured and sheathed to form a single entity and used to provide services from a surface installation or vessel to equipment below the surface, often on the seabed. Individual cables within the umbilical may also be screened or armoured, or both, and the whole umbilical may be strengthened against tension damage during deployment by the provision of an imbedded steel tension cable.

FIGURE 2.13.2

Inductive coupling units.

Courtesy Aker Solutions Ltd.

A subsea cable is usually of simpler construction containing the three main power cores operating at usually between 11 and 33 kV, with telecommunications and protection pilot wires often located in the centre of the cross-section. The cable will be armoured and sheathed overall and may also contain a tension wire. Cables are brought on to a platform via a 'J'- tube, a length of steel guide pipe fixed at various points to the installation jacket or hull and curling out at the bottom to receive the cable from the sea bed.

Whether a subsea link is required between two or more platform electrical power systems or between a surface installation and seabed equipment, the need for both a very high level of manufacturing quality control strict adherence to the planned delivery dates is vital.

Several very expensive mistakes have occurred in recent years, where umbilicals 10 or more kilometres long have had to be scrapped because of the discovery after completion that a vital cable within the umbilical is faulty. Apart from the financial penalty involved, the time delay involved in reconstructing the umbilical threatened to delay the laying of the umbilical beyond the North Sea summer weather window.

The thorough quality control procedures adopted during construction should include the following:

1. Discussions with the manufacturers in order to arrive at the optimum cable construction for the particular application. This will include:
   a. Length of umbilical.
   b. Depth of water and hence pressure.
   c. Type of cable laying equipment used, handling conditions and tension likely to be applied.
   d. Electrical parameters required to ensure satisfactory operation of the equipment at either end. These should include as a minimum, resistance, reactance, capacitance and attenuation over an agreed frequency range.
   e. Storage conditions.

2. Analysis of samples of all materials used for the minimum performance required. Insulation thicknesses should be more generous than those for cables used on land. If screen drain-wires are required, ensure that these do not have a 'cheese-wire' effect on insulation when the umbilical is in tension.
3. Close control of conductor and insulation diameter and thickness of each cable during manufacture.
4. Samples of each cable should be taken at the start and finish of each stage of the cable manufacturing run for analysis, and results of analysis known before continuing with the next stage.
5. The final testing of each cable should include insulation resistance, high voltage testing and measurement of total cable resistance, reactance, capacitance and attenuation over a given frequency range. If dc switching or digital signal are being transmitted, low attenuation at MHz frequencies may be important in order to avoid degradation of the signal over the length of the cable. Such readings may also detect faults, if attenuation readings for known healthy cables are available for comparison.
6. A sample of the completed cable should undergo high voltage testing in water at the pressure equivalent to at least the maximum depth of water likely in the installed location. The test is carried out using a pressure vessel with a cable gland at each end. The cable sample is passed between the two glands which are then sealed. The vessel is pressurised and the high voltage can then be applied at the cable ends protruding from the glands. Another high voltage test often conducted on a sample is to submerge the sample in a tray of salt water with one exposed conductor in air at either end. A high voltage is then applied between a submerged conductor (or the tray) and the exposed conductor in air. Both of the above tests are carried out to the maximum voltage available or until a voltage breakdown occurs in the sample.
7. Once all the cables have successfully completed their testing, they may be incorporated into the umbilical. However, if there is a delay between cable and umbilical manufacture or the cable has to be transported to a different, further insulation testing will be required immediately prior to umbilical manufacture to ensure that transit or coiling damage has not occurred. It should also be remembered that faults due to poor curing of the (plastic) insulation may appear several weeks after manufacture.

CATHODIC PROTECTION

INTRODUCTION

Cathodic protection provides an effective method of mitigating corrosion damage to subsea metal structures, whether their surfaces are coated or not.

Impressed current cathodic protection is preferred for long-term protection of both platform structures and subsea pipelines whereas galvanic or sacrificial anode systems are normally only used where power supplies are unavailable, or for temporary protection during construction, tow-out and commissioning. A galvanic anode system may also be used to complement an impressed current system in order to achieve rapid polarization and as part of an overall design to protect the structure. Although there are guides and codes of practice (such as the one listed in the Bibliography) to assist the engineer, the design of a good system depends to a large extent on past experience with similar structures and subsea environments.

TYPES OF SYSTEM

Impressed Current Systems

In this type of system, an external source of power, usually a transformer-rectifier unit provides the driving voltage between the anode and the structure to be protected. The negative terminal of the dc source is electrically connected to the structure and the positive terminal, similarly, connected to the anode. The output voltage of the dc source is adjustable so that the protection current may be varied either by a trimming potentiometer or automatically through a control signal loop to provide the required voltage at the reference half-cell. Transformer-rectifiers for this use are usually rated in the range of 20–500A and from 30 to 120V. Impressed current anodes, unlike galvanic ones, must be well insulated from the protected structure if they are to be mounted on it. Where it is found necessary to place impressed current anodes very close to the protected structure, dielectric shielding between anode and structure must be provided.

As it has a uniformly low electrical resistivity, sea water is an excellent medium for the application of cathodic protection and facilitates an even current distribution over the surface to be protected. Bare steel submerged in seawater is easily polarised if an adequate current density is maintained.

Impressed current anodes may be of silicon cast iron, lead alloy or graphite. Lead silver alloy or platinised anodes, although more expensive at the installation stage, may be used to provide a longer life and hence maintenance cost savings. Impressed current anode systems should be designed for a life of at least 10 years. High silicon cast iron corrodes relatively slowly and in seawater, where chlorine is produced at the anode surface, chromium may be added to further improve longevity. The resistance of a single anode in free flowing seawater is less than that of the same anode in mud or silt, and it may be necessary to install groups of anodes, sometimes mounted on a wood or concrete framework in order to reduce silting up and maintain effective contact with the seawater. Care must be taken to avoid incidental rubbing or scraping of the active anode surface by suspension or stabilising rods or cables, since the effect of this is to accelerate dissolution at that point. For obvious reasons, suspending the anode by the conducting cable is not advised. The number of anodes is determined by the required anode life, the allowable current density and the circuit resistance of the system. Anode life will be affected by the operating current density and the total current magnitude. In free flowing seawater, a maximum output of about $10\,A/m^2$ is normal practice, giving an anode consumption rate in the region of 0.4 kg/A year.

Platinised anodes are generally constructed from solid or copper-cored rod about 12 mm in diameter and 1500 mm long. A layer of platinum approximately 0.005 mm thick is deposited on this. A similar thickness of platinum may be used on anodes of other shapes and other materials, such as platinised titanium, platinised niobium or lead silver alloy. Graphite is not generally suitable for use offshore. For reasons of economy, platinised titanium anodes should not be used where silt or mud may accumulate, above the lowest tide level, or where fluctuating voltages, single (i.e., low frequency) ac ripple or anode voltages higher than 7 V are present, since the expected long life will not materialise in these conditions. The system design should be such that these anodes are not expected to carry current densities of more than 700 A/m^2. The 0.005-mm thick platinising dissolves at the rate of about 10 mg/A year, and although the abrasive action of water-borne sand may reduce this to some degree, a life expectancy of 15 years may be realistically hoped for.

The anodes are usually mounted on supports cantilevered out, but electrically insulated from the protected structure. In shallow waters, seabed nonmetallic frames may be used to support the anodes above the mud or silt level. To protect any offshore platform, distributed multiple anode arrangements will be required to be mounted over the members of the complex subsea structure to provide a uniform current density. The size and design of the cathodic protection scheme will depend on the current density required to bring the structure up to the level of protection required. Sea water velocity and oxygen content will affect chemical activity and hence the current density.

Statistical data will be required from the platform site, if an accurate figure for current density is to be determined. However, data may be available for similar environments as a guide. Some figures are given below as an indication of the magnitude required:

Location	Environment	Typical Current Required (mA/M^2)
Persian Gulf	Seawater	20
	Below mud level	50
North Sea	Seawater	50
	Below mud level	90
Mexican Gulf	Seawater	30
	Below mud level	60

It should be remembered that there will be little or no protection afforded by these systems above the low tide level and additional forms of protection will be required above this level.

Impressed Current Systems on Submerged Pipelines

Offshore submerged pipelines generally require a minimum pipe-to-water potential of −0.9 (ON) volts with reference to a silver/silver chloride half-cell. This corresponds to a current density of 2.5 mA/m^2.

GALVANIC ANODE SYSTEMS

In the galvanic or sacrificial anode system, the driving voltage between the structure to be protected and the anodes is developed directly by the electrolytic potential between the two (different) metals involved. The anodes are usually of aluminium alloy or magnesium, types which are less affected by insulation from oil wetting such as 'Galvanum III' being often preferred. Galvanic anodes may be placed very close to, or in contact with, the structure to be protected. Galvanic anode systems should be designed for a life of 25 years. Galvanum III or similar anodes are suitable for use on both subsea pipelines and offshore structures and is particularly suited for use in saline muds or silted-over conditions. On subsea pipelines it is used in the form of 'bracelets'. On subsea structures, bar type anodes are mounted at suitable intervals.

Although used extensively on ships, zinc and magnesium anodes are not suitable for offshore use mainly due to their faster rate of consumption and hence costs involved in replacement. Supplementary magnesium anodes may be used in order to provide a temporary boost in polarising current, however. As cathodic protection starts to operate, a layer of alkaline material is formed on the protected structure by cathodic electrode reaction. Provided this material is not dislodged, its presence reduces the current density required to maintain protection. On offshore structures the polarisation-boost anode may take the form of a 10 mm × 20 mm ribbon of magnesium attached to the structure supplementing the permanent system.

CATHODIC PROTECTION CALCULATIONS

This section is for guidance only and results should be heavily weighted by previous experience and measured data from previous structures and tests carried out for the new structure to be protected.

Impressed current circuits basically consist of a dc power source driving a current through the anode, the electrolyte (i.e., seawater) and back through the structure. The empirical rule used for calculating the current is:

$$\text{Total Current (Amps)} = \frac{\text{Surface Area } \left(\text{m}^2\right) \times \text{mA/m}^2}{1000}$$

The surface area used is that of all metal surfaces submerged in the electrolyte at mean high water.

Having determined the total current, the component resistances

Cable resistance (both poles) R_C
Anode to electrolyte resistance R_A
Cathode (structure) to electrolyte resistance R_E
Linear resistance of the structure R_S

are adjusted in order to match the total dc output of the available power source. The linear resistance of an offshore platform may be neglected in most cases, but where length is extensive in comparison with cross-section as in subsea pipelines, R_S should be evaluated.

A similar calculation is required for galvanic anodes as follows:

$$I_a = \frac{E_d}{R_a} = \frac{E_a - E_{cu}}{R_a} A$$

where I_a, Anode current output (A); R_a, Anode to electrolyte resistance (Ohms); E_a, Open circuit potential between anode and electrolyte, measured by reference electrode (V); E_{cu}, Structure to electrolyte potential when fully protected. (V); E_d, Driving voltage between anode and protected structure.

$$(V) = E_a - E_{cu} = V_s + V_a$$

where V_s, structure to electrolyte potential and V_a, anode to electrolyte potential.

Typical values of E_a are:

Galvomag	1.7V
Magnesium	1.5V
Zinc	1.05V
Galvanum III	1.05V

Calculation of Anode Consumption: the expected life of an anode in years is given by:

$$Y = \frac{CR \times W}{8760 \times A}$$

where Y, Life (years); CR, Consumption rate in Ah/kg; W, Anode weight (kg); A, Anode output (A); A typical value of CR for magnesium is 1230Ah/kg.

CHAPTER

Offshore Lighting

14

INTRODUCTION

If those readers unfamiliar with offshore conditions imagine being woken up in the top bunk of a pitch black cabin by a strange alarm sounding, possibly accompanied by the sound and shock of explosions, they will appreciate that every aid to orientation and escape is vital.

Particularly since the Alexander Kielland disaster some years ago, when an accommodation semi-submersible capsized at night with heavy loss of life, offshore engineers have been aware of the importance of good escape lighting. Visual information from our surroundings is of vital importance in any environment if comfort and safety are to be maintained. In the offshore environment, bad lighting can very easily lead to accidents and injury and should be considered as an essential topic of any design safety audit. Since the time of writing of the first revision of this book, a few years after the Piper Alpha disaster (1988), the **Prevention of Fire, Explosion and Emergency Response Regulations** have come into force in UK waters. These regulations, among other issues, require platform alarms and emergency lighting to meet minimum standards for safe evacuation. The regulations not only cover the requirements for illumination of escape routes, but also illumination of any safety equipment required to be operated during an emergency.

LIGHTING CALCULATIONS

POINT SOURCES

In the visible wavelength range, radiant flux or electromagnetic radiation is considered to have a luminous flux associated with it. This luminous flux is a measure of human visual response, and a point source of light emitting a uniform intensity of 1 cd in all directions emits a total flux of 4π lumen (lm).

The illumination effect of a point source of light is shown in Fig. 2.14.1. A point source S, emitting luminous flux in all directions, illuminates a plane surface P. The flux $d\Phi$ intercepted by an element of area dA on P is the flux emitted within the solid angle $d\omega$ subtended at the source by the element dA, assuming no absorption of light takes place in the space between the source and the surface. The term $d\Phi/d\omega$ is called the *luminous intensity* I of the source of the direction being considered. The accepted

Offshore Electrical Engineering Manual. https://doi.org/10.1016/B978-0-12-385499-5.00016-9

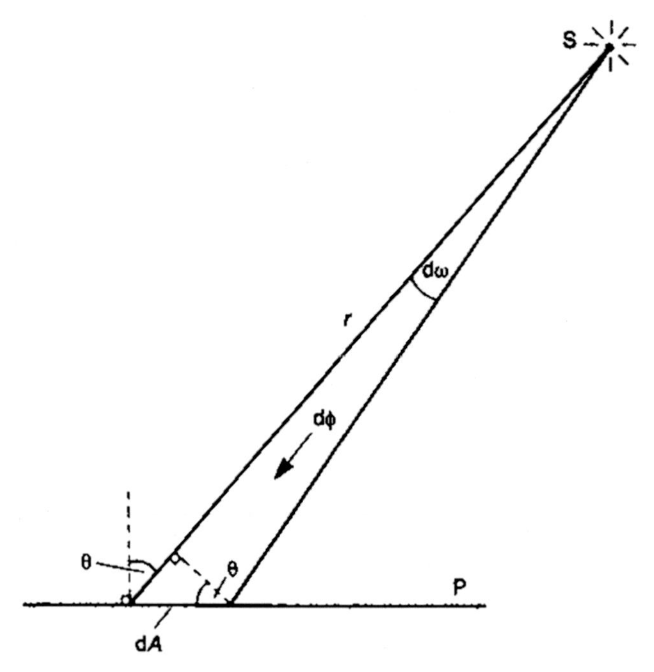

FIGURE 2.14.1

Illumination effect of a point source of light.

unit for luminous intensity I is the candela (cd) or lumen per steradian; the imperial unit used prior to this was the candlepower.

The *illuminance*, denoted by E, is the luminous flux falling on a surface. The unit of illuminance is the lumen per square metre, such that:

$$E = \frac{I \cos \theta}{r^2} \quad (2.14.1)$$

The inverse square law is only strictly applicable to point sources of light, but can be applied for practical purposes to large diffusing sources provided that r is more than five times the largest dimension of the source.

A typical point source calculation sheet is shown in Fig. 2.14.2.

LINEAR SOURCES

Large fluorescent lamps are obviously not point sources and any calculation must take this into account. In Fig. 2.14.3 a linear source of length l is represented by AB. The illuminance is required on P in the plane CD which is parallel to the source. APEB is a plane which passes through the axis of the source, is at an angle θ to the vertical and makes an angle ϕ with the normal PN to the plane CD.

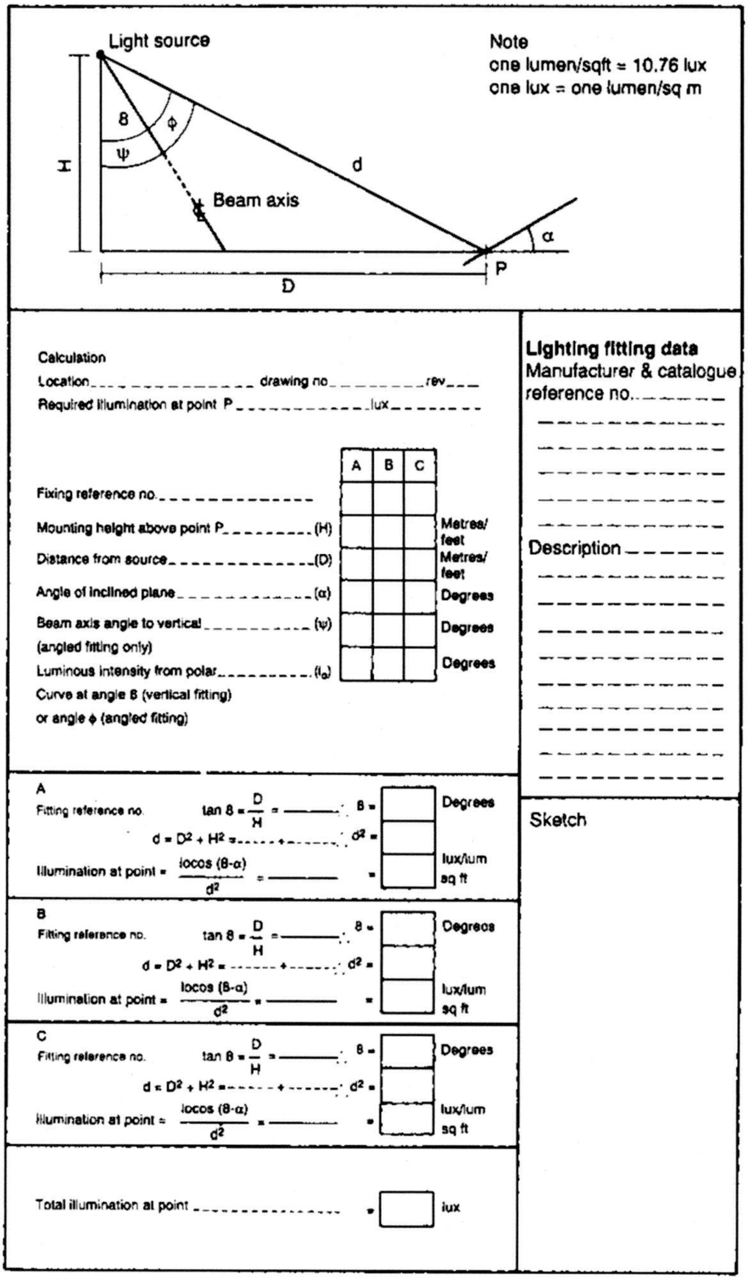

Light source

Note
one lumen/sqft = 10.76 lux
one lux = one lumen/sq m

θ φ ψ H d Beam axis α P D

Calculation
Location __________ drawing no __________ rev ___
Required illumination at point P __________ lux __________

	A	B	C	
Fixing reference no. __________				
Mounting height above point P __________ (H)				Metres/feet
Distance from source __________ (D)				Metres/feet
Angle of inclined plane __________ (α)				Degrees
Beam axis angle to vertical __________ (ψ) (angled fitting only)				Degrees
Luminous intensity from polar __________ (I_o) Curve at angle θ (vertical fitting) or angle φ (angled fitting)				Degrees

A
Fitting reference no. $\tan\theta = \frac{D}{H} = $ ______ ∴ θ = ☐ Degrees
$d = D^2 + H^2 = $ ______ + ______ ∴ $d^2 = $ ☐
Illumination at point = $\frac{I_o \cos(\theta - \alpha)}{d^2} = $ ______ = ☐ lux/lum sq ft

B
Fitting reference no. $\tan\theta = \frac{D}{H} = $ ______ ∴ θ = ☐ Degrees
$d = D^2 + H^2 = $ ______ + ______ ∴ $d^2 = $ ☐
Illumination at point = $\frac{I_o \cos(\theta - \alpha)}{d^2} = $ ______ = ☐ lux/lum sq ft

C
Fitting reference no. $\tan\theta = \frac{D}{H} = $ ______ ∴ θ = ☐ Degrees
$d = D^2 + H^2 = $ ______ + ______ ∴ $d^2 = $ ☐
Illumination at point = $\frac{I_o \cos(\theta - \alpha)}{d^2} = $ ______ = ☐ lux/lum sq ft

Total illumination at point __________ = ☐ lux

Lighting fitting data
Manufacturer & catalogue reference no. __________

Description __________

Sketch

FIGURE 2.14.2

Typical point source calculation sheet.

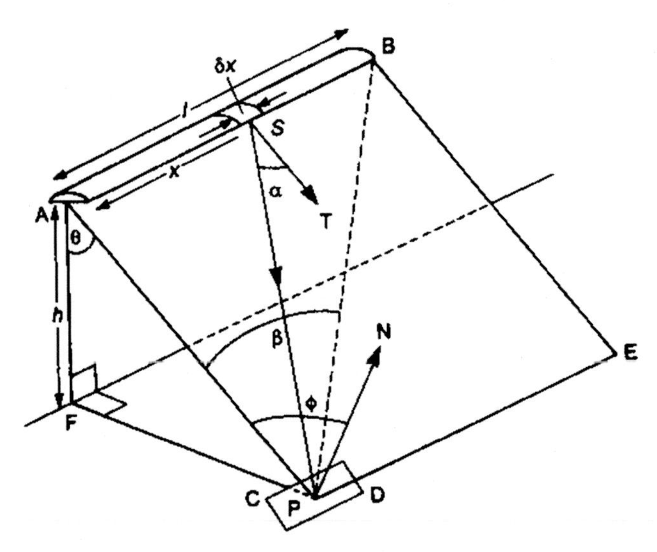

FIGURE 2.14.3

Illumination effect of a linear source of light.

Consider an element δx at S, distant x from A. ST lies in the axial plane and is normal to the axis of the source. Let $I(\alpha,\theta)$ be the intensity of the source AB in the direction parallel to SP. The illuminance δE at P on the plane CD due to element δx of the source is then given by:

$$\delta E = \frac{\delta x I(\alpha,\theta)}{l} \cdot \frac{\cos\alpha\cos\varphi}{(x/\sin\alpha)^2}. \tag{2.14.2}$$

$$\text{Since AP} = \frac{h}{\cos\theta} = \frac{x}{\tan\alpha}$$

$$\text{and } x = h\tan\alpha\sec\theta$$

$$\delta x = h\sec^2\alpha\sec\theta\,\delta\alpha$$

Substitution of these for x and δx in Eq. (2.14.2) gives

$$\delta E = \frac{I(\alpha,\theta)}{lh}\cos\theta\cos\phi\cos\alpha\,\delta\alpha$$

For the whole length of the source,

$$E = \frac{I(0,\theta)}{1h}\cos\theta\cos\phi\int_0^\beta \frac{I(\alpha,\theta)}{I(0,\theta)}\cos\alpha\,d\alpha \tag{2.14.3}$$

The integral known as the *parallel plane aspect factor* is a function of the shape of the intensity distribution in the axial plane inclined at θ to the horizontal and the angle β (the aspect angle) subtended by the luminaire at P, which must be opposite one end of the luminaire, as in Fig. 2.14.3.

The shape of the intensity distribution is similar in any axial plane for most fluorescent luminaires, and for a given luminaire, the aspect factor is independent of θ, varying only with β, so that it may be denoted by AF(β).

Most diffuser cross-section shapes can be expressed mathematically, so that AF(β) can be found by integration. For a uniform diffuser, for which $I(\alpha)$ equals I(0) cos α:

$$\begin{aligned} AF(\beta) &= \int_0^\beta \cos 2\alpha \, d\alpha \\ &= (\beta + \sin\beta \cos\beta) \end{aligned} \tag{2.14.4}$$

A numerical method of integration may be used, such as dividing the axial curve into a number of equiangular zones.

To calculate the illuminance produced at a point opposite the end of the luminaire on a plane parallel to the luminaire (CD in Fig. 2.14.3) the aspect angle β is determined and the value of AF(β) is entered in Eq. (2.14.3).

$$E = \frac{I(0,\theta)}{lh} \cos\theta \cos\phi \, AF(\beta) \tag{2.14.5}$$

Should the point concerned not be opposite the end of the luminaire, the principle of superposition can be applied. If it is opposite a point on the luminaire, then the illuminance due to the left and right hand parts of the luminaire are added. If it is beyond the luminaire then the illuminance due to a luminaire of extended length is reduced by the illuminance due to the imaginary extension.

Fig. 2.14.4 shows a typical calculation sheet for linear source lighting.

The same method may be used for calculating the illuminance on a plane perpendicular to the axis of the luminaire (AFP in Fig. 2.14.3). The derivation followed above applies, with sin α replacing cos α cos ϕ in Eq. (2.14.2), resulting in a final expression for the illuminance at P of:

$$E = \frac{I(0,\theta)\cos\theta}{lh} \int_0^\beta \frac{I(\alpha,\theta)}{I(0,\theta)} \sin\alpha \, d\alpha \tag{2.14.6}$$

The integral expression is now the perpendicular plane aspect factor AF(β).

Again this is substantially independent of θ in most practical situations, resulting in:

$$E = \frac{I(0,\theta)}{lh} \cos\theta \, AF(\beta) \tag{2.14.7}$$

which determines the illuminance on a plane perpendicular to a linear luminaire.

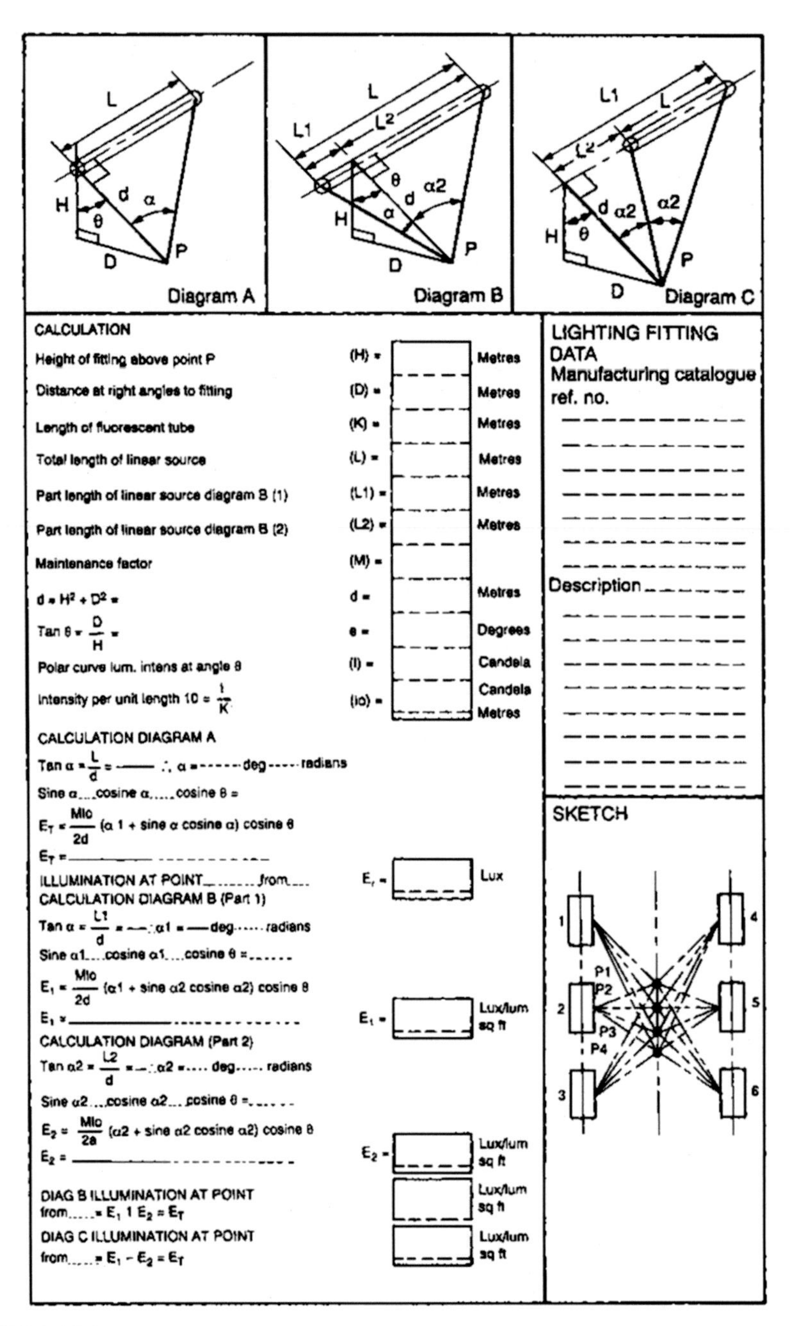

Diagram A

Diagram B

Diagram C

CALCULATION

Height of fitting above point P	(H) =		Metres
Distance at right angles to fitting	(D) =		Metres
Length of fluorescent tube	(K) =		Metres
Total length of linear source	(L) =		Metres
Part length of linear source diagram B (1)	(L1) =		Metres
Part length of linear source diagram B (2)	(L2) =		Metres
Maintenance factor	(M) =		
$d = H^2 + D^2 =$	d =		Metres
$\text{Tan } \theta = \frac{D}{H} =$	θ =		Degrees
Polar curve lum. intens at angle θ	(I) =		Candela
$\text{Intensity per unit length } 10 = \frac{I}{K}$	(io) =		Candela Metres

CALCULATION DIAGRAM A

$\text{Tan } \alpha = \frac{L}{d} =$ ——— ∴ α =------deg-----radians

Sine α....cosine α.....cosine θ =

$E_T = \frac{MIo}{2d} (\alpha\ 1 + \text{sine } \alpha \text{ cosine } \alpha) \text{ cosine } \theta$

$E_T =$ ——————————

ILLUMINATION AT POINT........from.... $E_r =$ [ ] Lux

CALCULATION DIAGRAM B (Part 1)

$\text{Tan } \alpha = \frac{L1}{d} =$ ——∴α1 = ——deg......radians

Sine α1....cosine α1....cosine θ =......

$E_1 = \frac{MIo}{2d} (\alpha 1 + \text{sine } \alpha 2 \text{ cosine } \alpha 2) \text{ cosine } \theta$

$E_1 =$ —————————— $E_1 =$ [ ] Lux/lum sq ft

CALCULATION DIAGRAM (Part 2)

$\text{Tan } \alpha 2 = \frac{L2}{d} =$ ——∴α2 =..... deg..... radians

Sine α2....cosine α2....cosine θ =......

$E_2 = \frac{MIo}{2a} (\alpha 2 + \text{sine } \alpha 2 \text{ cosine } \alpha 2) \text{ cosine } \theta$

$E_2 =$ —————————— $E_2 =$ [ ] Lux/lum sq ft

DIAG B ILLUMINATION AT POINT from.....= E_1 1 $E_2 = E_T$ [ ] Lux/lum sq ft

DIAG C ILLUMINATION AT POINT from.....= $E_1 - E_2 = E_T$ [ ] Lux/lum sq ft

LIGHTING FITTING DATA

Manufacturing catalogue ref. no.

Description

SKETCH

FIGURE 2.14.4

Typical calculation sheet for linear sources.

UTILISATION FACTORS FOR INTERIORS AND AVERAGE ILLUMINANCE

The utilisation factor UF(S) for a surface S is the ratio of the total flux received by S to the total lamp flux of the installation. Utilisation factors are used to calculate the number of luminaires needed to provide a given illuminance on a surface. UFs vary according to the light distribution of the luminaire, the geometry of the room, the layout of the luminaires and the reflectance of the reflecting surfaces.

The average horizontal illuminance $E_h(S)$ produced by a lighting installation, or the number of luminaires required to achieve a specific average illuminance, can be calculated by means of utilisation factors using the 'lumen method' formula as follows:-

$$E(S) = \frac{F \times n \times N \times MF \times UF(S)}{A_h(S)}$$

where F is the initial bare lamp flux, n is the number of lamps per luminaire, N is the number of luminaires, MF is the maintenance factor associated with the deterioration due to dirt, dust and lamp ageing, $A_h(S)$ is the area of the horizontal reference surface S and $UF(S)$ is the utilisation factor for the reference surface S.

Although utilisation factors may be calculated by the lighting designer (see Bibliography), most manufacturers publish utilisation factors for standard conditions of use and for three main room surfaces. The first of these surfaces, the C surface is an imaginary horizontal plane at the level of the luminaires, having a reflectance equal to that of the ceiling cavity. The second surface, the F surface, is a horizontal plane at normal working height which is usually assumed to be 0.85 m above the floor. The third surface, the W surface, consists of all the walls between the C and F planes. A typical table of utilisation factors is shown in Fig. 2.14.5.

The *room index* is a measure of the angular size of a room, and is the ratio of the sum of the areas of the F and C surfaces to the area of the W surface, each rectangular area of room being treated separately. For rectangular rooms, the room index is given by:-

$$RI = \frac{LW}{(L+W)H}$$

where L is the room length, W is the room width, and H is the height of luminaire plane above the horizontal reference plane. The *effective reflectances* (ratio of reflected flux to incident flux) are needed for the wall surface, ceiling cavity and the floor cavity.

The wall surface will consist of a series of areas A_1 to A_n of different reflectances R_1 to R_n respectively. The effective reflectance of a composite surface is the area-weighted average R_a, given by:

$$R_a = \frac{\sum_{k=1}^{n} RkAk}{\sum_{k=1}^{n} Ak}$$

TEST 1

AREA 1

Date: 28 January 1992

Scheme: G MPD0998

Ref: 09980006/ 1

AIMING DATA

AIM No	TOWER X Pos	TOWER Y Pos	TOWER Z Pos	XAIM	YAIM	ELEV deg	HORZ deg	FLUX klm	No Flds	CAT No
1	15.0	25.0	8.0	15.0	10.0	61.9	90.0	28.0	1	OT 250

Illuminance: On Horizontal Plane

Grid: Horizontal at Z = 0

	.0	2.5	5.0	7.5	10.0	12.5	15.0	17.5	20.0	22.5	25.0	27.5
30.00	0	0	0	0	0	0	0	0	0	0	0	0
27.50	0	0	0	1	1	2	3	2	1	1	0	0
25.00	0	1	4	10	23	43	56	43	23	10	4	1
22.50	1	3	7	14	30	50	59	50	30	14	7	3
20.00	2	4	8	16	28	39	44	39	28	16	8	4
17.50	3	5	11	21	35	48	54	48	35	21	11	5
15.00	3	8	16	29	44	59	66	59	44	29	16	8
12.50	5	10	19	31	43	54	58	54	43	31	19	10
10.00	6	11	19	26	34	40	43	40	34	26	19	11
7.50	6	10	14	18	22	25	26	25	22	18	14	10
5.00	5	8	10	12	14	16	16	16	14	12	10	8
2.50	4	6	7	8	9	10	10	10	9	8	7	6
.00	3	4	5	5	6	6	6	6	6	5	5	4

x direction ------> Dimensions in m

Minimum illuminance: 0 lux

Mean illuminance: 16 lux

Maximum illuminance: 66 lux

Utilisation factor: 0.48

Uniformity (Min/Max): 0.00

Uniformity (Min/Avg): 0.00

Note:- Aiming direction and location of floodlight shown by ▽

FIGURE 2.14.5

Illuminance chart produced by computer for the floodlight in Fig. 2.14.6.

Courtesy Thorn Lighting Ltd., Borehamwood.

The CIBSE technical memorandum TM5 provides a table of effective reflectances R_e, but an approximate value can be obtained from:

$$R_e = \frac{CI \times R_a}{CI \times 2\,(1 - R_a)},$$

where R_a is the area-weighted average reflectance of the surfaces of the cavity and CI is the cavity index, defined as

$$CI = \frac{LW}{(L + W)\,d}$$

where d is the depth of the cavity.

The ratio of the spacing between luminaires and their height above the horizontal reference plane, the spacing-to-height ratio (SHR), affects the uniformity of illuminance on that plane. The nominal and maximum values of SHR should be quoted in the UF tables (Table 2.14.1).

CALCULATION PROCEDURE

AVERAGE ILLUMINANCE

The following sequence of calculations is recommended in order to calculate the number of luminaires required to achieve a given average reflectance:

1. Calculate the room index *RI*, the floor cavity index *CI*(F), and the ceiling cavity index *CI*(C).
2. Calculate the effective reflectances of the ceiling cavity, walls and floor cavity. This must include the effect of furniture and/or equipment.
3. Determine the *UF*(F) value from the manufacturer's tables, using the room index from the effective reflectances calculated.
4. Calculate the maintenance factor (*MF*) by multiplying together the various factors obtained from manufacturer's curves which reduce effective luminaire output.
5. Use the lumen method formula to obtain the number of luminaires required.
6. Arrange the lamps in a suitable configuration for the room. If the furniture or equipment in the room requires variations in spacing, refer to the method given in the CIBSE Technical Memorandum No.5 and recalculate the number of luminaires using the modified *UF*(F) obtained.
7. Check that the proposed layout does not exceed the maximum SHR.
8. Finally, calculate the illuminance that will be achieved by the final layout. The values of illuminance required should be above the following minimum values for offshore locations:-(Table 2.14.2).

Table 2.14.1 Utilisation Factors and Initial Glare Indices

Utilisation Factors UF[F]			Nominal SHR = 2.00								
Room Reflectances			**Room Index**								
C	**W**	**F**	**0.75**	**1.00**	**1.25**	**1.50**	**2.00**	**2.50**	**3.00**	**4.00**	**5.00**
0.70	0.50	0.20	0.44	0.55	0.59	0.63	0.68	0.72	0.74	0.77	0.80
	0.30		0.38	0.49	0.54	0.58	0.64	0.68	0.70	0.74	0.77
	0.10		0.34	0.45	0.50	0.54	0.60	0.64	0.67	0.71	0.74
0.50	0.50	0.20	0.40	0.50	0.54	0.57	0.61	0.64	0.66	0.69	0.71
	0.30		0.35	0.46	0.49	0.53	0.58	0.61	0.63	0.66	0.69
	0.10		0.31	0.42	0.46	0.49	0.55	0.58	0.61	0.64	0.67
0.30	0.50	0.20	0.36	0.45	0.48	0.51	0.55	0.57	0.59	0.61	0.63
	0.30		0.32	0.42	0.45	0.48	0.52	0.55	0.57	0.59	0.61
	0.10		0.29	0.39	0.42	0.45	0.50	0.53	0.55	0.58	0.60
0.00	0.00	0.00	0.25	0.34	0.36	0.39	0.42	0.44	0.46	0.48	0.49
BZ class			4	3	3	3	4	4	4	4	4
Glare Indices											
Ceiling reflectance		0.70	0.70	0.50	0.50	0.30	0.70	0.70	0.50	0.50	0.30
Wall reflectance		0.50	0.30	0.50	0.30	0.30	0.50	0.30	0.50	0.30	0.30
Floor reflectance		0.14	0.14	0.14	0.14	0.14	0.14	0.14	0.14	0.14	0.14

Room Dimension											
X	Y	Viewed Crosswise					Viewed Endwise				
2*H*	2*H*	7.0	8.4	8.0	9.5	10.8	6.8	8.2	7.8	9.2	10.5
	3*H*	8.9	10.2	10.0	11.3	12.6	8.6	9.8	9.6	10.9	12.2
	4*H*	9.9	11.1	10.9	12.2	13.5	9.4	10.6	10.4	11.7	13.0
	6*H*	11.0	12.1	12.0	13.2	14.5	10.3	11.4	11.3	12.5	13.8
	8*H*	11.6	12.6	12.6	13.7	15.1	10.7	11.8	11.7	12.9	14.2
	12*H*	12.2	13.2	13.2	14.3	15.7	11.1	12.1	12.1	13.2	14.6
4*H*	2*H*	7.7	8.9	8.7	10.0	11.3	7.5	8.7	8.5	9.8	11.1
	3*H*	10.0	11.0	11.0	12.1	13.5	9.6	10.6	10.7	11.7	13.1
	4*H*	11.2	12.1	12.2	13.2	14.6	10.6	11.6	11.7	12.7	14.1
	6*H*	12.5	13.4	13.6	14.5	15.9	11.8	12.6	12.8	13.7	15.1
	8*H*	13.3	14.0	14.4	15.2	16.6	12.3	13.1	13.4	14.2	15.7
	12*H*	14.0	14.8	15.1	15.9	17.3	12.8	13.5	13.9	14.7	16.1
8*H*	4*H*	11.8	12.6	12.9	13.7	15.2	11.4	12.2	12.5	13.3	14.7
	6*H*	13.5	14.2	14.6	15.3	16.8	12.8	13.5	13.9	14.6	16.1
	8*H*	14.4	15.0	15.5	16.1	17.6	13.5	14.1	14.6	15.2	16.7
	12*H*	15.4	16.0	16.6	17.1	18.6	14.2	14.8	15.4	15.9	17.4
12*H*	4*H*	12.0	12.7	13.1	13.8	15.3	11.6	12.3	12.7	13.5	14.9
	6*H*	13.7	14.3	14.9	15.5	17.0	13.1	13.7	14.3	14.9	16.4
	8*H*	14.8	15.3	16.0	16.5	18.0	14.0	14.5	15.1	15.7	17.2
	12*H*	15.7	16.2	16.8	17.3	18.8	14.6	15.0	15.7	16.2	17.7

Conversion Terms		
Luminaire length (mm)	1500	1800
Wattage (W)	1 × 65	1 × 75
Conversion factor *UF*	1.00	1.00
Glare indices conversion	0.63	0.00

Table 2.14.2 Minimum Illuminance Values Offshore

Illuminance (Lux)	Location
50	External areas (floodlit)
75	External walkways
100	Normally unmanned modules
100	Escape routes (normal lighting)
0.2 (Minimum)	Escape routes (battery-backed escape lighting only)
10–30	Illumination of safety and firefighting controls (battery-backed escape lighting only)
30	Areas earmarked for casualty treatment and triage (battery-backed escape lighting only)
30	Emergency muster stations including lifeboat boarding (battery-backed escape lighting only)
100	Unmanned switchrooms
100	Manned areas and accommodation except:-
500	Galleys [Note: use lamps with good colour rendering]
300	Laundries and offices
50	Cabins
150	Bedheads
300	Process control rooms and navigation bridges [Note: use lamps with good colour rendering and avoid glare/reflection on computer screens]

ILLUMINANCE AT A POINT

The majority of situations offshore demand other methods of illuminance calculation, since congested or odd shaped areas and specific lighting of structures will render average illuminance calculations inadequate or meaningless.

In such circumstances it is necessary to calculate the illuminance at particular points using one of the following:

Basic photometric data Basic photometric data can be used in conjunction with Eqs (2.14.1–2.14.7) to calculate the illuminance at particular points by hand. If the number of point calculations are few, this method will usually provide sufficient information for the selection and positioning of several luminaires but will become laborious and time-consuming if a significant number of points and/or luminaires are involved.

Pre-calculated manufacturer's design aids, such as isolux diagrams Isolux diagrams, where contours of equal illuminance on a specified plane are plotted, offer a faster method of performing these calculations. A typical isolux diagram is shown in Fig. 2.14.6. The calculation process can be accelerated using a spreadsheet type computer program for the repetitive calculations.

Specific Computer programs Although some limited design programs do exist, most programs simulate the illuminance pattern produced by a chosen layout of

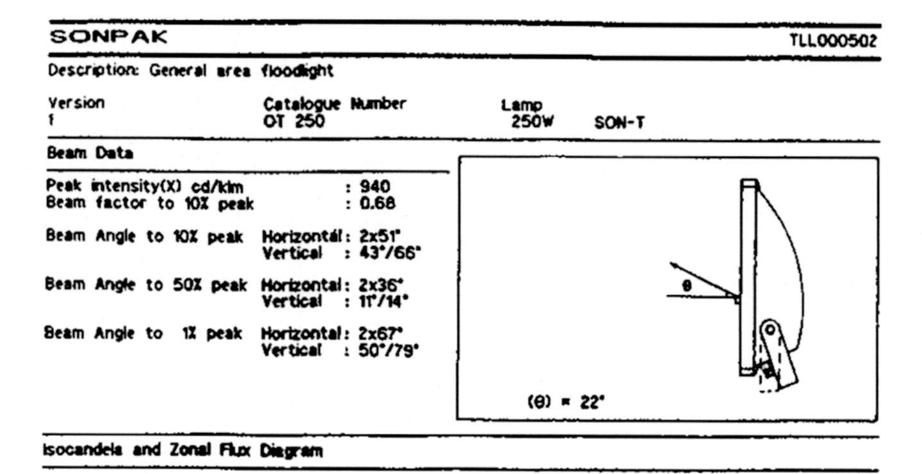
SONPAK TLL000502

Description: General area floodlight

Version	Catalogue Number	Lamp	
1	OT 250	250W	SON-T

Beam Data

Peak intensity(X) cd/klm : 940
Beam factor to 10% peak : 0.68

Beam Angle to 10% peak Horizontal: 2x51°
Vertical : 43°/66°

Beam Angle to 50% peak Horizontal: 2x36°
Vertical : 11°/14°

Beam Angle to 1% peak Horizontal: 2x67°
Vertical : 50°/79°

Isocandela and Zonal Flux Diagram

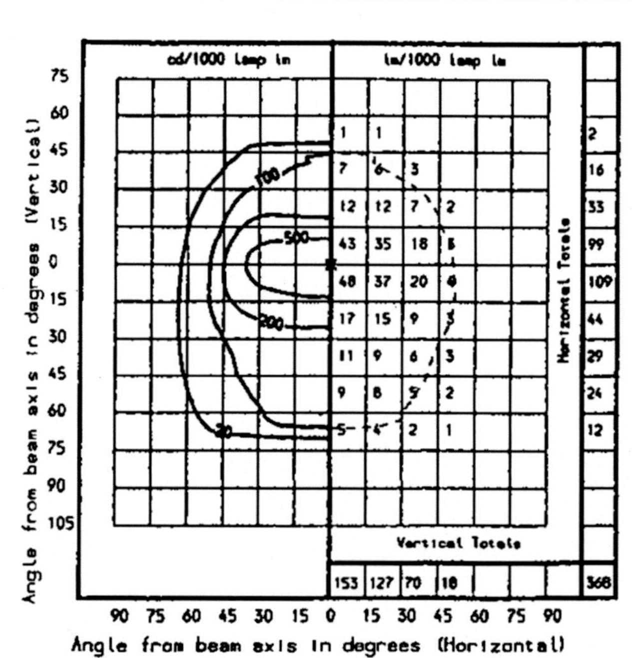

X-indicates the position of peak intensity. The pecked line shows 10% peak contour.
The direction of the peak intensity in the vertical plane is 22 degrees above the normal to the front face of the floodlight.

FIGURE 2.14.6

Isocandela and zonal flux diagram for hazardous area floodlight.

Courtesy Thorn Lighting Ltd., Borehamwood.

luminaires. A simple point-by-point lighting programme is shown in Appendix 3. Point-by-point calculations are required particularly for floodlighting schemes in order to examine the uniformity of illuminance. Eq. (2.14.8) may be applied for the illuminance on a surface due to a source of intensity I cd at distance r metres away when the light falls at an angle 0 to the normal to the surface, as follows:

$$E = \frac{I\cos\theta}{r^2} \qquad (2.14.8)$$

A programme is available for such calculations from Chalmit Ltd, called ChalmLite.

FLOODLIGHTING

General information on floodlighting can be obtained from the CIBSE Technical Report TR13, Industrial Area Floodlighting.

In floodlighting, the effective source intensity is the photometric value multiplied by the maintenance factor and by the atmospheric light loss factor. The most useful form of luminaire intensity data is one using the vertical/horizontal (V–H) coordinate system, such as the isocandela contours shown in Fig. 2.14.6. This accompanies a zonal flux diagram deliberately, for ease of point-by-point calculation after a lumen calculation.

Eq. (2.14.1) can be rewritten as:

$$E_H = \frac{(I\cos 3\theta)}{H^2}$$

where E_H is the horizontal illuminance due to a source mounted H metres above the horizontal.

When an external area such as a drillpipe laydown area is to be lit, the contribution of each floodlight installed can be summated at the points of minimum intensity (usually the corners) and the minimum value of illuminance found. The change in illuminance over the area to be lit is a measure of uniformity and is quoted in the CIBSE Memorandum 5 in terms of the following:

Criteria	Value for Working Areas
1. Maximum to minimum	20:1
2. Average to minimum	10:1
3. Minimum distance over which a 20% change in illuminance occurs	2 m

ACCOMMODATION LIGHTING

NORMAL LIGHTING

It is not normally possible to provide lighting of the quality and flexibility of a good hotel, but it is important that the normal lighting is restful and blends with a hopefully pleasant decor. Supplementary lighting will be required for reading and writing areas. Where televisions are viewed, care must be taken to prevent distracting reflections appearing on the screen. Galleys require a much higher level of lighting with good colour rendering for safety and hygiene. It is usual in the sleeping cabins, for each bed to be fitted with a combined lamp and entertainment bedhead unit located conveniently for operation by the bed occupant, such that he or she may read without shadow and without head injury when getting up.

ACCOMMODATION EMERGENCY LIGHTING

Suitably hazardous area certified emergency lighting must be provided for the adequate lighting of all rooms and escape routes. Each luminaire should be complete with a battery, charger and inverter. Schemes where all the emergency lighting is operated from a single battery are not recommended, for reliability reasons. Each sleeping cabin must be provided with a hazardous area certified emergency luminaire located in such a position as to illuminate the exit route sufficiently to overcome sleep disorientation. Suitable non-industrial designs of certified luminaire, with an acceptable appearance for use in accommodation areas are now available. With the exception of the sleeping cabin luminaires, emergency lighting is left operating continuously. The sleeping cabin units are provided with a mains failure sensing device and are switched on automatically during mains failure. The same arrangement may be used in communal TV rooms if required. (If such rooms are earmarked for triage in emergencies, the luminaires must also be manually switchable.)

PROCESS AREA LIGHTING

To operate satisfactorily and safely in hazardous process areas, luminaires or required to be mechanically strong in order to resist accidental blows from passing equipment, tools or scaffolding poles. Plastic cases or lenses must not crack due to vibration fatigue and the casing should be as free from crevices as possible to prevent the trapping of dirt, oil, etc., and for ease of cleaning. The luminaire must be certified for use in hazardous areas, and is usually of the Ex 'e' increased safety type. Luminaires should be hazardous area certified to meet ATEX requirements by an ATEX Notified Body. Some hazardous area certification does not meet the relevant ATEX harmonised standards and may be rejected by the operating oil company. Preferably the luminaire should incorporate a terminal box such that the luminaire enclosure remains sealed when terminals are accessed, otherwise a separate 'looping-in' terminal box will be required, adjacent to each luminaire. Problems have been experienced with Ex 'e' luminaires, due to water ingress possibly leading to arcing. This is particularly a problem in areas subject to deluge firefighting systems. One of the many possible scenarios for this maybe as follows:

1. Gas is detected in a hazardous module.
2. The ESD System initiates a shutdown of the affected plant.
3. As a consequence of this shutdown, either directly due to the shutdown logic or indirectly due to fuel gas starvation and failure to switch to diesel fuel, for example, the main generation shuts down.
4. With no power available, the air compressors stop, leaving air operated devices dependent on their reservoirs.
5. One by one as each air reservoir becomes depleted, the deluge sets release sea-water in the process areas they cover.
6. Water enters the luminaires in the module where gas is leaking. Luminaires with integral batteries allocated to escape lighting will still be live. Essential lighting may also be live in this situation, fed from the emergency generator, but this may be too small to run an air compressor.
7. Arcing may now take place within the affected luminaires, providing an ignition source for the released gas.

Similar problems have been experienced with other Ex 'e' equipment such as junction boxes, but unlike those with luminaires (unless clear Denso tape can be obtained!), the problem can largely be overcome using a layer of 'Denso tape over joints. Newer designs of luminaire with better water ingress protection are now available and these, combined with good safety-critical focused maintenance should reduce the ignition risk. Another possible improvement is to provide fire and gas system logic which shuts down all nonessential electrical equipment within the affected module should high-level gas be detected and deluge release occurs. Escape lighting would still need to remain energised, however.

In congested areas, floodlighting may be required in order to provide effective lighting, to avoid contamination from chemicals and oil and consequently the need for constant cleaning, and also to avoid obstructing the operator's line of sight with the luminaire itself.

DRILLING AREAS

It is likely that the most difficult lighting conditions on the offshore installation occur in the vicinity of the drilling derrick. The actual process of drilling involves large quantities of water and drilling mud, which may be oil based and difficult to clean from the luminaires. Mud returning from the well may be very hot which, apart from reducing visibility by producing a mist, may cause deformity in plastic luminaire mouldings. The drill floor and the monkey board need to be well-lit, but due to the movement of the travelling block, swivel, kelly etc., and the drillpipe itself, it is necessary to carefully select suitable locations for luminaires on the derrick so as to minimise the risk of mechanical damage. Apart from fluorescent luminaires for illuminating access ladders, landings and equipment on the derrick itself, the derrick may be used for positioning drill floor and monkey board floodlights. Designers should remember the discomfort and danger involved in maintaining derrick lighting when selecting the type and number required, as once installed, the luminaires must be regularly cleaned and re-lamped. Permanent safety-hooped access ladders with caged landings at appropriate intervals must be provided.

If there is a suitably high structure nearby which overlooks the derrick, on which floodlights may be mounted, the above problem may be alleviated to some extent. Drilling and wellhead areas present a flammable gas hazard, and luminaires within this area must be certified for use in the hazardous zone in which they are located.

LAYDOWN AREAS

Because of the continual movement of containers and equipment around these areas, and the crane operations involved, laydown areas should be floodlit rather than locally lit. While illuminance calculations are in progress, it is important to study crane operating loci in order to establish suitable locations for floodlights, to ensure that as well as providing good lighting of the area, they are unlikely to come into contact with parts of any crane or their moving loads.

HELIDECKS

As a result of the G-REDU accident in February 2009, a new CAP437 lighting scheme has been developed, which includes the following elements:

- **Green** helideck perimeter lights:
- The white painted 'H' should be outlined with steady **green** lighting, forming a lit 'H'.
- The Touchdown/Positioning (TD/PM) circle should be fully outlined with steady **yellow** lighting
- Omni-directional low intensity steady **red** obstacle lights should be fitted to the highest points of the installation, crane booms, lattice towers, etc.
- Omni-directional low intensity steady **red** obstacle lights should be fitted to objects which are higher than the landing area and close to it or the LOS boundary (150 degrees Limited Obstacle Surface cone above helideck level).
- Status lights (for night and day operations, indicating the status of the helicopter landing area) consisting of flashing **red** lights indicating 'landing area unsafe: do not land'.
- Installation/vessel emergency power supply design should include the entire landing area lighting system (see CAP437 Appendix C) fed from a suitable battery-backed secure power (UPS) supply system.

The detailed lighting requirements are given in Chapter 4 of CAP437 and the specification in Appendix C. These requirements must be read in conjunction with the comprehensive facilities for the safe landing and reception of helicopters given in full CAP437 document.

JACKET AND LEG LIGHTING

It is important that the lower parts of oil platforms and the immediate area of sea around them are lit. This lighting is required for the following reasons:

1. In case of a 'man overboard' situation.
2. To aid in the manoeuvring of supply and service vessels close to the platform.
3. To aid in the detection of leaks or inadvertent spillages into the sea.

This lighting is usually provided by floodlights, located on a lower deck, preferably in a nonhazardous area, positioned so that they may be directed at the jacket and adjacent area of sea. The floodlights should be contained in corrosion resistant metal (stainless steel or gunmetal) or impact resistant plastic enclosures. If located in a hazardous area, both the luminaire and its igniter/ballast unit will need to be certified for such use.

NAVIGATIONAL LIGHTING

It is a statutory requirement that every platform is equipped with navigation lights which may operate independently for 96 h following loss of platform electrical power. The details below are based on the Standard Marking Schedule for Offshore Installations issued by IALA.

The light system normally consists of two main, two secondary and various subsidiary flashing lights, an electronic controller, a battery charger and battery of required capacity.

The main lights are operated from the normal platform supply; operate in unison exhibiting the Morse letter 'U'. The composition of this Morse character must conform to the following:

1. The duration of each dot must be equivalent to the duration of darkness between the dots and of that between the dot and the dash.
2. The duration of the dash must be three times the duration of one dot.
3. The darkness interval between successive Morse characters must not be less than 8 s or more than 12 s.

The lights must be arranged to give an unobstructed white light normally visible for 15 nautical miles in any direction. The lights must be mounted at a height of normally less than 30 m (but exceptionally 35 m) above Mean High Water Spring (MHWS) and the beam axis directed at the horizon.

The apparent intensity of each light assembly after all losses should not be less than 14,000 cd. The total beam width in the vertical plane should not be less than 2.5°degrees at the point on the curve of intensity distribution where the intensity is 10% of maximum.

The lamp assembly is normally equipped with a two-filament lamp in a rotating lamp changer unit (charged with up to 6 lamps) which automatically ensures that a functioning filament is in circuit and in the correct focal position. The failure of the light assembly for any reason should cause an alarm to be initiated and the secondary battery operated lights be switched on.

The secondary and main lamps are normally positioned one above the other, with the secondary positioned below. The secondary lamp assemblies are of similar design to the main lamp assemblies, but the lamps are less powerful, having a range of 10 nautical miles.

Subsidiary, red flashing lights are required at the corners not marked by the main and secondary lights. These have the same characteristics as the secondary lights and are also battery-backed, but have a nominal range of 3 nautical miles.

WALKWAYS, CATWALKS AND STAIRWAYS

Installation personnel need to be able to make their way to most parts of the platform, at any time of the day or night, and in any weather. Therefore good lighting of all accessways and particularly stairways is vital for their safety. Most if not all installation lighting outside the accommodation areas will remain on continuously day and night.

EMERGENCY ESCAPE LIGHTING

Throughout the installation at least one in every four luminaires should be of the integral battery emergency type. These luminaires remain lit at all times, and will provide illumination for at least 1 h following a power failure. These lamps should be placed at strategic locations on escape routes such as inside process modules at ladders and exit doors to ensure adequate escape lighting is available in emergencies. They should also illuminate the controls to fixed firefighting equipment. This may require the one-in-every-four ratio to be increased, and in congested areas all luminaires may require to be of the integral battery type.

Certain requirements are laid down in PFEER Guidance Notes and are therefore mandatory on offshore installations. Notably, these requirements include the following:

1. The minimum duration of escape lighting after failure of normal lighting is now based on the Installation Safety Case, but is normally at least 1 h.
2. Adequate lighting of Firefighting Equipment Controls during emergencies.

ROUTINE MAINTENANCE

The importance of regular maintenance and testing of all offshore lighting equipment cannot be overstressed. Maintenance scheduling software is essential for ensuring that equipment is not an ignition hazard, or otherwise unsafe. Such software can also provide evidence to external surveyors and inspectors that adequate maintenance is being carried out, and can be linked to DCR Performance standards (see Part 9 Chapters 1 and 2).

IEC 60079: Part 17 provides guidance on inspection and this should be adhered to. Inspection intervals should be monitored and adjusted to suit the level of weather exposure and degree of deterioration found after each interval. Escape lighting should be regularly tested, and intervals between testing reduced if more than one or two luminaires are found to have failed in any one area. Escape lighting will receive particular attention during ICP and HSE surveys.

CHAPTER

Process Control and Monitoring Systems

15

INTRODUCTION

In the times before Piper Alpha, process control and safety systems were designed on the basis of a tick box exercise as described in API RP 14C. This changed following the disaster and resulting public inquiry. The Cullen Report highlighted a number of failings in the existing safety regime for the offshore oil industry. 'Much of today's offshore regulatory regime is a direct legacy of Piper Alpha. The HSE assumed responsibility for regulating the industry, and four new sets of regulations were introduced, including the Offshore Installations (Safety Representatives and Safety Committees) Regulations and the Offshore Installations (Safety Case) Regulations' (Allen, B., Lessons to learn from the Piper Alpha disaster (Online)). 'The Offshore Installations (Safety Case) Regulations came into force in 1992. By November 1993 a safety case for every installation had been submitted to the HSE and by November 1995 all had had their safety case accepted by the HSE' (Oil & Gas UK, 2008). The Safety Case Regulations (SCR) have been revised in 2005 and again in 2015. The objective of the revisions was to improve the effectiveness of the regulations in light of experience, reflecting the new philosophy of the safety case being a living document, subject to continual revue and revision.

Now all fixed and mobile offshore installations in the UK sector of the North Sea must have a Safety Case produced in accordance with SI 2015 398, The Offshore Installations (Offshore Safety Directive) (Safety Case etc.), Regulations 2015. This is the UK implementation of directive 2013/30/EU of the European Parliament and of the Council of 12 June 2013 on safety of offshore oil and gas operations and amending Directive 2004/35/EC. These legislative documents cover the basic requirements for the design, operation, maintenance and removal of offshore oil and gas installations to minimise risk to the personnel involved in their operation and the surrounding environment. They advocate a holistic approach to whole life design, installation, commissioning, operation, maintenance, modification, decommissioning and removal of all types of offshore installation based on continuous hazard analysis to ensure that the risk to personnel and the environment is kept as low as is reasonably practicable.

In order to implement this safety philosophy the installation must be considered as a whole, not as a collection of separate stand-alone systems or pieces of equipment as changes to any item on the installation will affect the performance of other parts of the installation. One way to appreciate this is to use the paper envelope analogy; the basic support structure for the installation is protected by design and in the case

Offshore Electrical Engineering Manual. https://doi.org/10.1016/B978-0-12-385499-5.00017-0

of a platform the corrosion monitoring system, which is the first paper envelope. Adding the process equipment requires additions to the jacket to protect it from the additional risks introduced by the process equipment, even though the process vessels and pipework are adequately designed, which is the second paper envelope. The process equipment also needs protection to control the process variables within the operational envelope of the installed equipment; the third paper envelope is the control and monitoring equipment. The fourth paper envelope is the process shutdown system to contain any extreme process disturbances. The fifth paper envelope is the gas detection equipment to detect and alarm escapes of gas that put either the personnel or the installation fabric at risk. The sixth paper envelope is the fire detection equipment, automatic extinguishant release and alarm system.

The process being constrained and contained on the installation can be viewed as an amorphous substance that grows spiky extensions and sharp edges, which can easily perforate a paper envelope, requiring additional envelopes to maintain containment. This example also demonstrates that a change to one part of the installation affects the other parts, control and safety systems and emphasises the need to consider the installation as a composite whole and not as isolated stand-alone systems.

PROCESS CONTROL SYSTEMS

The process control system should be viewed as the first level of the safety protection for the installation, as its function is to control the process within the design limits of the process plant and make the operator aware of any process excursion beyond these limits. It also performs the main data acquisition and display function, collecting information from all areas of the installation and displaying it in a central location in an easily comprehensible manner.

Modern process control systems have the capability of adjusting the individual control loops automatically to optimise loop performance leading to a safer and more efficient operation. 'Predict & Control' (P&C) provides multivariable model predictive control to ensure that process performance remains optimal and repeatable in all operating conditions and across all product changes. It uses state-of-the-art state space modelling to reduce process variations, increase throughput and reduce production costs by operating safely at the closest possible process constraint limits. P&C has an unrivalled success rate in improving productivity in a wide variety of industrial processes (ABB Advanced Process Control). They can also analyse the data coming back from the plant instruments to automatically monitor plant operation and schedule preventative maintenance activities thus enhancing operational efficiency and therefore safety by reducing the probability of breakdowns.

The information generated by the control system is presented to the operations personnel at a central location on the installation in a readily understandable way, using multiple graphical screens. As always the information presented by intelligent systems is only as good as the data available for processing, and to maximise the benefit of these control systems, intelligent instruments must be used. 'FOUNDATION fieldbus (Ff) devices deliver predictive alerts, millisecond data capture, validated data, field-based control, diagnostics, and asset information bi-directionally with the DeltaV system. DeltaV Ff I/O communicates digitally with field devices, increases your input/output capacity, and provides access to more information about your process than conventional I/O subsystems. DeltaV Ff I/O enhances device diagnostics that affect the control strategy and alert operators to device malfunctions' (Emerson Process Management DeltaV system overview). These instruments have the ability to self-diagnose faults and report back to the control system the nature of the fault and the time of occurrence. The facility to time-stamp in real time at the instant an alarm level being reached is invaluable in tracing the cause of any process upset and is a basic requirement of an effective alarm management system. Alarm management systems are necessary to prevent operator information overload and assist in the detection of the cause of the alarm event to ensure safe and speedy rectification of the problem. The ability to store real-time process information and display in detail the chronological sequence of events leading up to an alarm after the event is also helpful in analysing the root cause of the problem.

Information on plant status can also be made available at a remote location such as the shore support base, engineering centre or company management offices. Details such as real-time process operations, historical data, and operational efficiency can be made available to the relevant departments to assist with problem solving, maintenance planning and resource availability without having to visit the installation. This reduces the number of personnel required on the installation, thus increasing safety by reducing the risk to personnel and reduces the time taken to identify and rectify problems, further reducing risk and maximising the availability of the plant. 'Honeywell's Experion ® Process Knowledge System (PKS) with unique distributed system architecture and digital video management enables real-time remote monitoring, maintenance and operation of offshore facilities from a centralized onshore control centre. With this solution, you can leverage expertise across remote sites, make faster and more effective decisions, and achieve greater productivity. Additionally, you can reduce helicopter flights, ship movements and supply of material to your offshore platform, and above all, improve safety in hazardous environments by removing personnel from remote locations' (Honeywell Offshore Oil and Gas Solutions).

These sophisticated process control systems are available from several sources; those that are most commonly found in North Sea oil and gas installations are ABB's Advanced Process Control System, Emerson Process Management's DeltaV Digital Automation System and Honeywell's Experion Process Knowledge System. All systems offer similar performance and expansion capabilities; all systems can also offer an integrated process shutdown capability with fire and gas detection as part of the package.

PROCESS SHUTDOWN SYSTEM

This by definition is a safety-related system, as it is there to shut down the process when predetermined conditions occur that risk a loss of containment. As a safety-related system, it should be designed, operated and maintained in accordance with IEC 61508, 'functional safety of electrical/electronic/programmable electronic safety-related systems', which is the generic standard, and IEC 61511, 'functional safety - safety instrumented systems for the process industry sector', which is the derivative standard specifically written for process plants. These standards follow the same philosophy as the Offshore SCR mentioned earlier in that they are goal setting, requiring implementation during the whole life cycle of the plant and are based on continuous risk assessment to ensure that an adequate safety integrity level of protection is provided.

Some requirements of the standard relate to developmental activities where the implementation technology may not yet have been fully decided. This includes development of the overall safety requirements (concept, scope definition, hazard analysis and risk assessment). If there is a possibility that electrical and/or electronic and/or programmable electronic (E/E/PE) technologies might be used, the standard should be applied so that the functional safety requirements for any E/E/PE safety-related systems are determined in a methodical, risk-based manner.

Other requirements of the standard are not solely specific to E/E/PE technology, including documentation, management of functional safety, functional safety assessment and competence. In order to comply with the full intent of the standard, the hazard analysis and risk assessment must be continually reviewed and revalidated during the full life cycle of the system.

FIRE AND GAS CONSIDERATIONS

The Health and Safety Executive (HSE) published the third edition of prevention of fire and explosion, and emergency response (PFEER) L65 entitled 'Offshore Installations (Prevention of Fire and Explosion, and Emergency Response) Regulations 1995 - Approved Code of Practice and Guidance' in 2016. In which page 2 has the following definitions:

> ***Approved Code of Practice***
>
> *This Code has been approved by the Health and Safety Executive, with the consent of the Secretary of State. It gives practical advice on how to comply with the law. If you follow the advice you will be doing enough to comply with the law in respect of those specific matters on which the Code gives advice. You may use alternative methods to those set out in the Code in order to comply with the law.*
>
> *However, the Code has a special legal status. If you are prosecuted for breach of health and safety law, and it is proved that you did not follow the relevant provisions of the Code, you will need to show that you have complied with the law in some other way or a Court will find you at fault.*
>
> ***Guidance***
>
> *This guidance is issued by the Health and Safety Executive. Following the guidance is not compulsory, unless specifically stated, and you are free to take other*

action. But if you do follow the guidance you will normally be doing enough to comply with the law. Health and safety inspectors seek to secure compliance with the law and may refer to this guidance.

Page 7 has specific reference to the SCR and how following the Code of Practice and Guidance can help in gaining acceptance by the HSE or other competent authority of the Safety Case for the installation.

Safety Case Regulations

14 *SCR 2015 came into force on 19 July 2015. These Regulations apply to oil and gas operations in external waters (the territorial sea adjacent to Great Britain and designated areas within the continental shelf (UKCS)), and replace SCR 2005 in these waters, subject to certain transitional arrangements. Activities in internal waters (e.g. estuaries) will continue to be covered by SCR 2005.*

15 *SCR 2005 and SCR 2015 require a safety case to be submitted for acceptance by HSE or the competent authority, as appropriate, for each installation.*

16 *Regulation 12(1)(a) of SCR 2005 and regulation 16(1)(a) of SCR 2015 require a demonstration in the safety case of the adequacy of the management system for controlling risks to people on the installation or engaged in connected activities, and of arrangements for independent audit of the management system, as appropriate. The organisation and arrangements provided to meet the requirements of PFEER form part of the management system for the purposes of SCR 2005 and SCR 2015 (whichever is applicable).*

17 *Regulation 12(1)(c) and (d) of SCR 2005 and regulation 16(1)(c) and (d) of SCR 2015 require a demonstration in the safety case that major accident hazards have been identified, their risks evaluated and that suitable measures have been, or will be, taken to control those risks to ensure the relevant statutory provisions are complied with. Regulation 2(1) of SCR 2005 and SCR 2015 defines the relevant statutory provisions for the relevant regulations, and for SCR 2015 those are relevant statutory provisions that relate to offshore oil and gas operations. Regulation 5 of PFEER specifies requirements for a fire and explosion, and evacuation, escape and rescue assessment. The results of this assessment will contribute to the demonstration required by SCR 2005 and SCR 2015 (whichever is applicable).*

18 *In addition, PFEER specifies goals for preventive and protective measures for managing fire and explosion hazards, and emergency response. Complying with PFEER, and taking account of the practical guidance contained in this book, will facilitate HSE's or the competent authority's acceptance of the safety case. The Regulations are in general expressed as broad goals rather than specific requirements, allowing dutyholders the flexibility to develop detailed arrangements in the light of hazards, plant configuration and other circumstances specific to the installation.*

PFEER follows the same philosophy as the SCR and Functional Safety Standards in that it encourages a continual review of risk and the methods of detection and management of a major hazard to ensure that the risk to personnel and the environment is as low as is reasonably practicable. This philosophy should be followed from the initial design through manufacture, installation, commissioning, operation, maintenance, decommissioning and removal of the installation, continually reviewing risk, hazard and escape management to comply with the latest revisions of both PFEER and SCR.

All equipment used in the process control system should be suitable for the area it is in, so as to minimise the possibility of a loss of process containment or an ignition hazard. Detectors should be fitted throughout the plant to detect flammable and toxic gas escapes and flammable liquid spillages. These detectors should be placed and set to detect the presence of the hazardous substance in a timely fashion in order to allow automatic action to mitigate the hazard to be taken before escalation. Detection systems should be reliable with high availability. The equipment monitoring the detectors and initiating the automatic responses should be designed in accordance with the appropriate safety standards to perform the required actions with alacrity.

EMERGENCY SHUTDOWN SYSTEM

The emergency shutdown system is responsible for initiating the automatic actions needed following an emergency; these include taking action to mitigate the effects of the emergency and ensuring as far as possible that the equipment installed to carry out this duty remains operational for the duration of the emergency. This should include alarm equipment, communications equipment, temporary refuge and passage ways to the means of personnel evacuation. These provisions are included in both the SCR and PFEER and in common with the overall safety philosophy; their provision should be regularly reviewed and assessed for compliance.

All these regulations and standards are constructed to encourage engagement in the safety regime by all involved with the design, construction, operation and maintenance of the installation. From the outset, representatives from all these life phases should be intimately involved in all hazard and operability studies, risk assessments and therefore the risk reduction performance required of the safety system, referred to as the safety integrity level in IEC 61508 and associated standards, as they are the ones most at risk. This approach to safety consisting of a group of people gathered together to come to a common reasoned conclusion on the risks involved in the operation of an offshore oil and gas production facility and deciding on the level of risk reduction to be provided by the safety system is not perfect. It relies on the expertise and experience of those involved in the process being capable of identifying every possible risk and their competence and confidence in their ability to put forward the case for each in order to arrive at a reasoned consensus. But because it makes those involved think about the possibilities and consequences of their decisions, it is infinitely better than having a predefined safety device for a given plant item.

There have also been questions asked about the definitions and understanding of what is meant by safety integrity level. 'The concept of safety integrity levels (SILs) is now prevalent in the field of safety-critical systems, and a number of standards advocate its use in the design and development of such systems. However, not only do the various standards derive SILs differently, but none provides a clear and detailed explanation of how they are derived and applied. The result is that SILs are not well understood' (Redmill, F., 2000). There are also doubts about the veracity and accuracy of the calculation of the probability of failure on demand of the safety system, as this figure is derived by operating on historical data by statistical means in order to try and predict the future. Often a probability of failure on demand is taken as a mathematically accurate figure, but it should only be used as a guide in the comparison of different protective circuits, always bearing in mind that the demand could result in the failure to act.

The standard referred to at the beginning of the chapter has recently been upgraded; the eighth edition of API RP 14C Analysis, Design, Installation, and Testing of Safety Systems for Offshore Production Facilities was published in February 2017, 'On the face of it, it seems to be a well overdue update with an element of catch up. But it still follows the prescriptive line i.e. no allowance for risk based approaches' (Wild, G. 2017).

CHAPTER

Transformers

16

INTRODUCTION

On oil and gas installations, offshore distribution transformers are usually of either sealed silicon-oil-filled or encapsulated resin type. Standard mineral-oil-filled types are too great a fire hazard, and askarel-based insulants are prohibited because they are a health hazard owing to the presence of toxic/carcinogenic polychlorinated biphenyls.

Air-cored types are not recommended unless located in pressurised areas because of the ignition hazard they can pose and also the salt-laden environment would tend to lead to insulation problems. The sealed silicon-filled type has the advantages that Buchholz pressure sensing for winding faults can be fitted.

Being heavy devices, transformers need to be checked against switchroom maximum floor loadings.

If a transformer is used to interconnect the main system with the drilling system, the heating effect of silicon controlled rectifier harmonics will need to be considered.

On Offshore Renewable Energy Substations (e.g., for wind farms), the large step-up transformers required for transmission are usually silicon-oil-filled, but a 'dump tank' is provided that is capable of taking the full volume of oil in the transformer. This is located below the transformer and can be used to store the oil well away from the transformer in case of an internal transformer fault. Alternatively, SF6 insulated step-up transformers are available now. Although the transmission transformers tend to be the biggest fire hazard on such substations, storage of fuel for emergency and standby generator prime movers has to be catered for in the same way as with oil and gas installations with fire detection, firefighting and shutdown measures.

TRANSFORMER CONSTRUCTION

A three-phase transformer consists of a ferromagnetic core (shaped like the number '8' on its side – see Fig. 2.16.1) made up of a series of low-hysteresis steel laminations. On the three limbs of the core are wound the six (or more) windings. Although the special ferromagnetic steel laminations are low hysteresis, there are still losses, and these generate eddy currents in the laminations. This I^2R loss is the result of the need to make and collapse the magnetic field in the core for every cycle. Some of the energy is also lost in producing the 50 Hz hum that all transformers produce.

Offshore Electrical Engineering Manual. https://doi.org/10.1016/B978-0-12-385499-5.00018-2

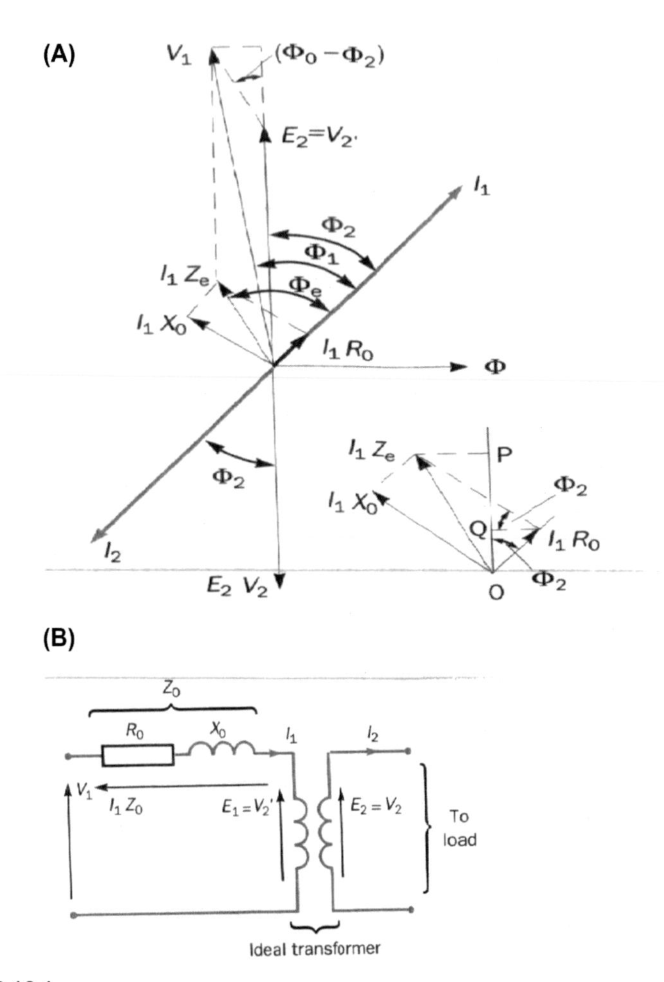

FIGURE 2.16.1

(A) Transformer phasor diagram. (B) Transformer equivalent circuit.

Dry-type transformers are used offshore, but they must be well protected inside air-conditioned and pressurised modules, to avoid both the external (salt-laden) atmosphere and the risk of flammable gases (see Fig. 2.16.2).

Oil-filled transformers are also used, but as mentioned earlier, they do create a fire hazard, especially if an internal winding fault occurs. Please refer to PART 4 Chapter 2 for further information (see Fig. 2.16.4).

Cast resin transformers are considered the most suitable for offshore use, although the very large transmission transformers utilised on wind farm substation are too large for resin casting, and oil dump tanks are used, as mentioned earlier (see Fig. 2.16.3).

FIGURE 2.16.2

Dry-type distribution transformer with cabinet removed.

FIGURE 2.16.3

Cast resin transformer.

Courtesy OTDS.

FIGURE 2.16.4

Oil-filled transformer.

Courtesy OTDS.

TRANSFORMER REGULATION

It is important to ensure that the secondary voltage of the transformer does not drop below the specified range, and therefore the transformer regulation needs to be checked, based on the reactance and expected maximum loading, as follows:

$$V_{pu} = \frac{I_1\,(Re\,.\,\cos\Phi_2\ + Xe\,.\,\sin\Phi_2)}{V_1}$$

where E_1, primary electromotive force (EMF).; E_2, secondary EMF; I_1, primary current; I_2, secondary current; R_o, no-load resistance; V_1, primary voltage; V_2, secondary voltage; X_e, equivalent reactance of the primary and secondary windings referred to the primary circuit; X_o, no-load reactance; Z_e, equivalent impedance of the primary and secondary windings referred to the primary circuit; Φ_1, phase angle between I_1 and V_1 and Φ_2, phase angle between I_1 and V_2.

CHAPTER

Telecommunications – Internal and External 17

INTERNAL COMMUNICATIONS

The communications system installed on an offshore installation provides the necessary facilities to operate the installation safely in normal conditions and also to provide communication facilities to control an emergency by:

1. Enabling messages and alarms to be broadcast throughout the installation. Note that there will be the need for secure communications between the control rooms, lifeboat and helideck muster stations and any secondary (fall-back) locations where communications have been provided for emergency situations when the main control point (process control room, bridge, etc.) has been compromised.
2. Communicating with the neighbouring installations (if any).
3. Communicating with vessels carrying out installation-related activities, and the standby vessel.
4. Portable communications for platform personnel.
5. Emergency communication for key personnel (e.g., OIM) to shore incident rooms.

A typical package would consist of the following:

An emergency intercom system designed for use in hazardous areas. The system should be installed in a ring, to provide communication between control points and muster areas even if any one of the cables interconnecting the stations is severed, or the platform ac power supply is not available. The system would normally provide:

- Two channel communications (one channel each for control points and muster points) with the option of 'All Call' simultaneous transmission on both channels.
- For reliability, full battery backup should be provided for each outstation, with outstation health and audio line continuity monitoring.

A UHF (FM) radio repeater system installed to enhance full duplex radio communications between personnel using hand portables. In the event of failure, the hand portables are normally equipped with simplex channels to allow communications, hand portable to hand portable, directly between personnel on the facility.

A radio paging system with facility for sending verbal messages that may be accessed from the installation telephone system(s), enabling key personnel to be paged.

Offshore Electrical Engineering Manual. https://doi.org/10.1016/B978-0-12-385499-5.00019-4

A duplicated, Public Address/General Alarm (PA/GA) system, fully certified for use in hazardous areas is provided to advise personnel of an emergency and the actions to be taken. Each system can be accessed from the control room, radio room, helideck and lifeboat stations. The PA/GA systems on bridge-connected platforms should be fully integrated to ensure all personnel are aware of the current situation and can be contacted individually. The PA/GA system should be able to carry routine and emergency speech and alarm on both halves of the system simultaneously. It is a statutory requirement that in an emergency, the PA system can broadcast a tone indicating the nature of the emergency (muster and prepare to evacuate) and to issue instructions to all areas where personnel may be present, including lifeboat, cabins and muster stations (Please refer to Reference 1).

REDUNDANCY, DIVERSITY, SURVIVABILITY AND SAFETY ISSUES

1. The internal communications must all be provided with secure battery-backed supplies to ensure that they will work if the main power systems have been disabled. The two (main and standby) UPS supply batteries and their chargers should be rooms located in geographically separate rooms. Chargers should be supplied from separate switchboards.
2. Cables for alternative telecom systems and their electrical supplies should be routed separately in order to minimise the possibility of common mode-failures.
3. All internal communication equipment requires to be suitable for operation in at least zone 2 hazardous areas, so that the risk of igniting uncontrolled gas releases is minimised.
4. Public address and telephone systems on bridge-connected platforms will require fall-back modes of operation to cater for scenarios such as loss of the bridge. Damage due to fire or explosion on one platform should not take out the whole system (For further guidance, please refer to the Bibliography).

EXTERNAL TELECOMMUNICATIONS

The normal and emergency roles of the external communications system installed on an offshore installation are to provide the necessary facilities to safely operate the installation in normal conditions, and also to supervise and inform personnel for control of emergencies by:

1. Communicating with shore-based incident rooms
2. Communicating with external emergency support personnel (coast guard, helicopters, rescue vessels, etc.)
3. Communicating with the neighbouring installations

4. All lifeboats (TEMPSCs) are also to be equipped with the statutory telecommunications fit

A typical package would consist of the following:

1. A dual multichannel VHF (FM) transceiver operating in the International Maritime Mobile Band, is provided for communication with supply vessels, standby vessels and shuttle tanker. Private channels are also used for communications with the shuttle tanker, supply vessels, crane cabs and hand portables. Each crane is equipped for VHF communication.
2. Lifeboats are equipped with a VHF radio telephone, working on channel 16 and at least one other channel, complete with suitably certified battery chargers. They also carry an EPIRB (Emergency Position-Indicating Radio-Beacon station) and SART (Search and Rescue Transponder) to satisfy SOLAS requirements.
3. A duplicated (main/standby) VHF (AM) aeronautical transceiver with control electronics and a number of aeronautical hand portable radios are available to communicate with aircraft.
4. A radio beacon also provides for navigation of helicopters on approach to the platform.
5. All normal communications to shore are usually through dual (K-band) satellite communication systems.
6. The INMARSAT[1] unit provides emergency communication direct to shore. The system provides two-way communications as a backup to the installation's multichannel communication links, fax and SOLAS facilities.

REDUNDANCY, DIVERSITY AND SAFETY ISSUES

1. The external communications must all be provided with secure battery-backed supplies to ensure that they will work if the main power systems have been disabled. The two UPS supply batteries should be located in geographically separate rooms.
2. Cables for alternative telecom systems and their electrical supplies should be routed separately in order to minimise the possibility of common mode-failures.
3. Unless it is isolated on a confirm gas alarm, all external communication equipment requires to be suitable for operation in at least zone 2 hazardous areas, so that the risk of igniting uncontrolled gas releases is minimised.
4. Care must be taken to avoid inadvertent ignition of flammable gases due to microwave radiation from radar antennae or radio navigation beacons (Reference 6).

[1] International Marine/Maritime Satellite (organization).

NOTES

1. Telecoms-based shutdown systems – where an unmanned installation is controlled from a local manned installation – two radio-telemetry channels may be utilised, provided the loss or corruption of the signal leads to the automatic operation of the unmanned platform's shutdown system after a short, timed period, unless the (maintenance) signal is restored during that time.
2. On 'not-usually-manned installations (NUIs) and renewable substation platforms, the telecom package requirements will obviously be simpler, but facilities must still meet national statutory and SOLAS requirements.

See also: PART 5 Chapter 1 – Installation Practice.
PART 5 Chapter 4 – Hazardous Area Installation.

PART

Design Project Organization

3

CHAPTER

Notes on Detailed Design Project Organisation and Documentation

1

INTRODUCTION

When a new offshore design and construction project gets under way, there is obviously a great deal of organisation involved. A good, disciplined organisation structure with clear, practical office procedures assists the engineers involved in producing the design, selecting the materials and equipment and supervising the construction. A detailed examination of design and construction office technical documentation is beyond the scope of this book, but a brief discussion is provided below for those not yet familiar with the subject.

TECHNICAL ORGANISATION OF THE DESIGN PROGRAMME

PROJECT PROCEDURES

The end product of any design office is to produce a set of documents which perform the following functions:

1. A design document which fully describes the design of the new system. Each stage of this description must be adequately supported by logical reasoning, calculations and diagrams, sketches, etc.
2. A material list which identifies every material and equipment component (down to the level of cable clips, nuts, bolts and washers, if need be).
3. An installation work scope document which fully describes in text, drawings, schedules and diagrams how and in what sequence the equipment and material are to be installed and commissioned.

This end product is not arrived at in one long session but is broken up in a series of submissions or packages. The level of detail will also increase as the project progresses.

Typically, this would occur as follows:

1. Conceptual or 'front-end' engineering study. This document should put forward technical and economic arguments for and against the feasibility of various alternative design schemes with recommendations as to which approach should be adopted.

Offshore Electrical Engineering Manual. https://doi.org/10.1016/B978-0-12-385499-5.00020-0

2. A detailed cost estimate, aimed for an accuracy of about +20% to −10% of the actual project cost. At this stage, a planning network for the whole project and an installation procedure will need to be worked up to a significant degree of detail to achieve the estimated level of accuracy.
3. Having produced a skeleton scheme for the accomplishment of the whole project, detailed work on the project proper can then commence. The final documents produced may be loosely divided into two categories: those required for construction, which contain material lists and commissioning and installation procedures, and those required for technical approval, which contain detailed descriptions and design calculations for each item of equipment, including information on its intended location on the offshore installation. These may be sent to a certifying authority such as Lloyd's Register, ABS or DNV GL for approval/verification.
4. Whilst the preparation of design packages is under way, procurement specifications for all large items of equipment must be prepared. It would be of great assistance to the engineers preparing the specifications if this particular task could be delayed until the associated design packages have reached an advanced state of preparation. Unfortunately, this is invariably not the case because manufacturers' delivery times for generators, large motors and switchgear are usually in excess of 6 months, making it imperative that orders are placed early to adhere to the project programme. It is therefore unavoidable that changes to switchgear to cater for changes in motor and distribution equipment ratings will be required, and these may reach cost figures of the same order of magnitude as the original switchgear order. It is hoped that project accountants bear this in mind. Note that this work is carried out in the normal commercial environment, with tendering and award of contracts for the various stages by the oil company concerned to various design contractors.

OFFICE ENVIRONMENT AND PROCEDURES

The project working conditions should be such that the team of people involved is assisted in their efforts to ensure that the end product design functions correctly, is as safe as practicably possible, has the least environmental impact and is cost effective.

In my experience, both the design team's physical environment and the administrative and quality assurance (QA) procedures adopted will significantly affect the degree of excellence of the end product.

Office working environment: Many books, both serious and humorous, have been written on this subject, but the principle points are

1. provision of good heating, ventilation and lighting;
2. sufficient work and storage space for each person, and the maintenance of uncluttered accessways throughout the working areas – the premises chosen to house the project should be large enough to cater for the project team at the peak manning level, otherwise the team will be broken up into portacabins, etc., at the most critical period of the project;

3. good sound insulation from outside noise, and the segregation or soundproofing of noisy office machines such as photocopiers and printers;
4. adequate provision of telephones, computers, software, drawing boards, catalogues, standards, codes of practice, etc.

OFFICE PROCEDURES

This subject should be discussed in detail by the QA manual of the company concerned. The manual should be based on the guidance given in the ISO 9001:2015 and information security management standard BS EN ISO/IEC 27001:2017. The general principles are outlined in the following.

1. The project organisation should appoint a QA-experienced management representative whose sole responsibility is to implement and maintain a satisfactory QA system. The representative must be given the necessary authority to carry out his or her duties, otherwise the 'quality procedures manual' will not be taken seriously by the design team, especially at the beginning of the project when the need for strict procedures is not obvious.
2. Every design document must be signed as checked and approved by those officially authorised to do so.
3. The circulation of all project documentation must be such that those whose responsibilities are affected by a particular document receive such documents in good time to take any necessary actions. This can be difficult to accomplish if the document has to be commented on by a series of people within a very limited period.
4. A quality control system is necessary in order that all items of equipment are inspected at the manufacturer's works and appropriate tests witnessed to ensure that the equipment is fit for its purpose before it is delivered and installed.

DRAWING REPRESENTATION

The following types of electrical drawing are required for most offshore design applications.

1. Single-line diagram
   This diagram is usually the prerequisite for any electrical system design. It will be developed through most of the design period, starting as a simple sketch and finally showing details of type and rating for circuit breakers, transformers, contactors, busbars, cables, protection and control relays, metering, interlocking and other safety devices. Indication and control circuitry will also be shown in an abbreviated form, although identical circuits will normally only be shown once, with appropriate references. A schedule may also be shown so that numerical information of a repetitive nature can be listed separately. Part of a typical fully developed single-line diagram is shown in Fig. 3.1.1.

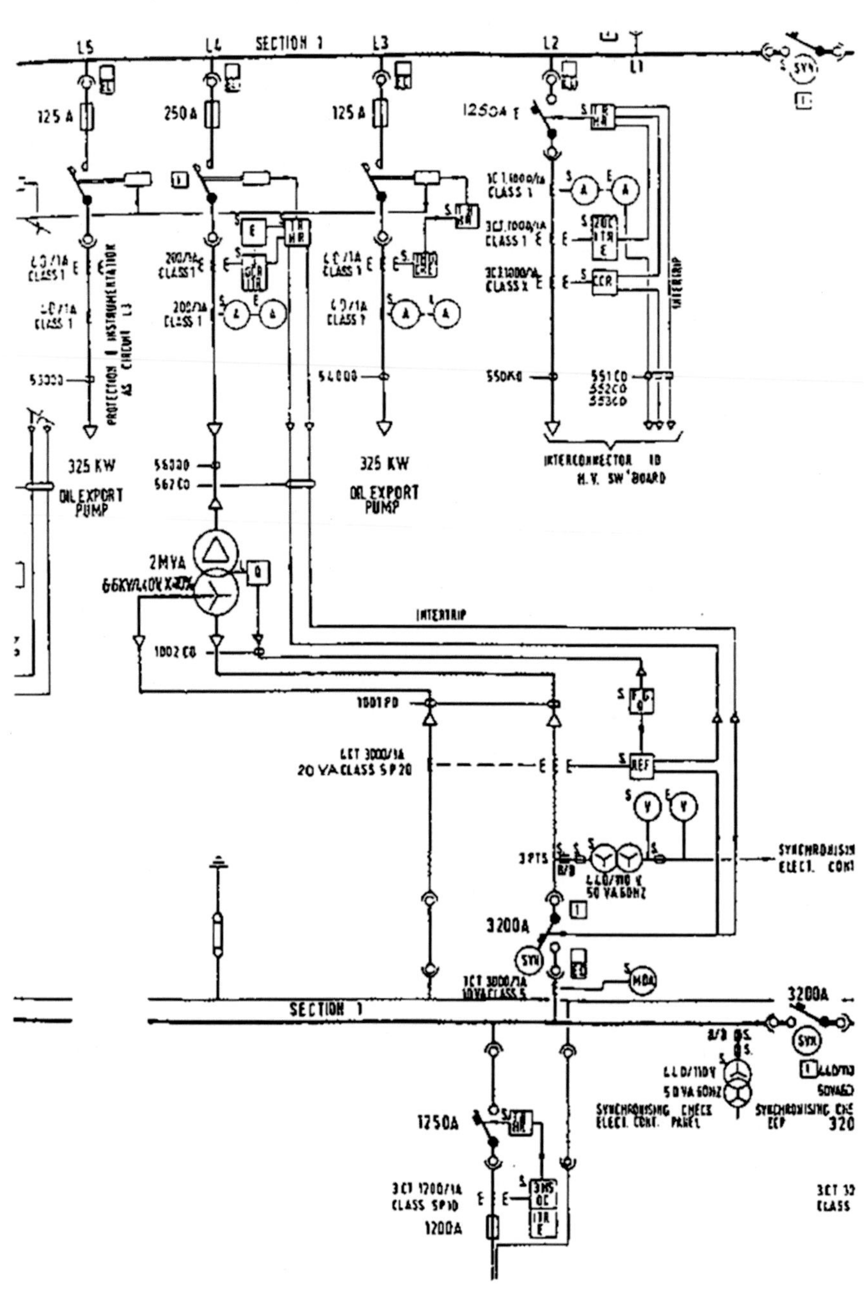

FIGURE 3.1.1

Part of a fully developed single-line diagram.

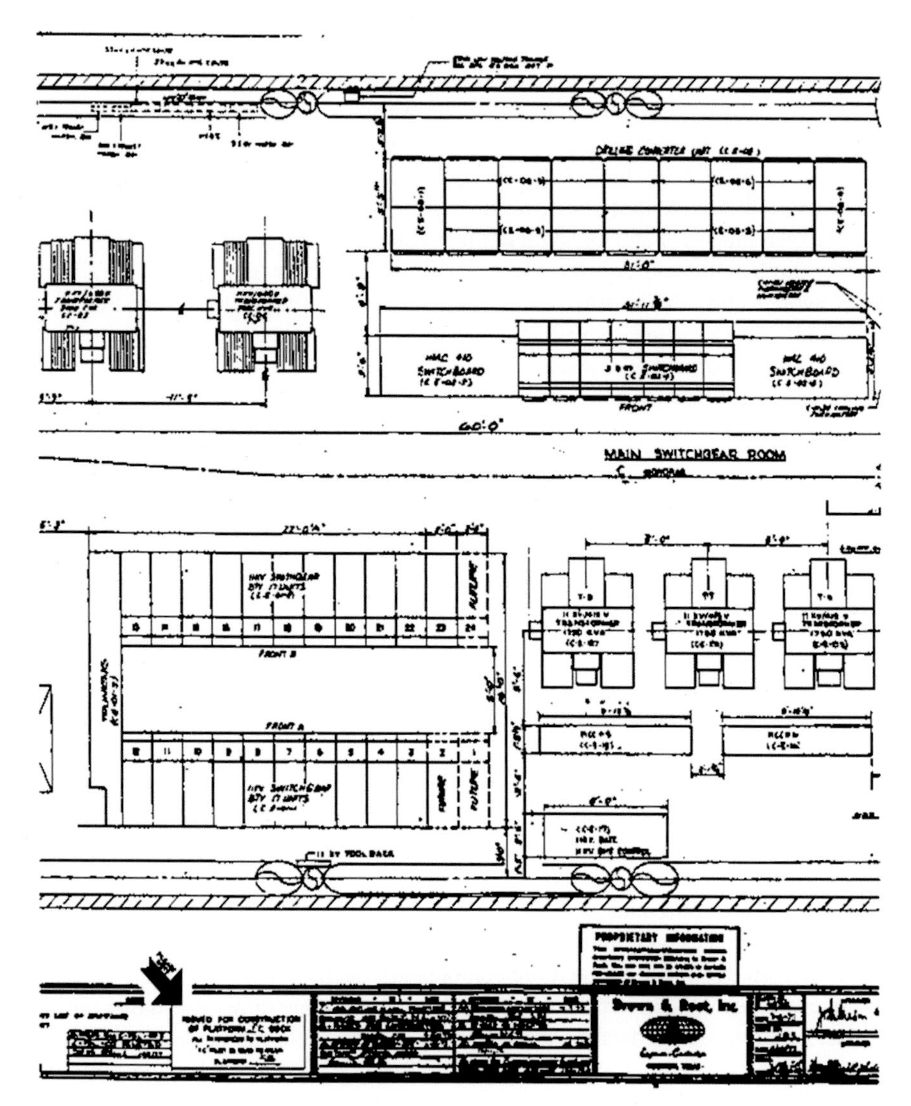

FIGURE 3.1.2

A typical equipment location diagram.

Courtesy of BP Exploration Ltd.

2. Equipment location diagram
   In the congested areas of an offshore installation, it is helpful to provide a separate drawing showing equipment locations only. Part of a typical equipment location drawing is shown in Fig. 3.1.2.
3. Cable rack routing diagram
   Particularly in the more complex process and the power generation modules, cable racking becomes a three-dimensional puzzle. This puzzle needs to be

solved in conjunction with the routing of pipework and ventilation ducting in areas already congested by the process or power equipment itself, the structural steelwork of the installation and all the other ancillary equipment such as lighting, communications and instrumentation.

It is no surprise that computer-aided design (CAD) systems are used extensively for the draughting representation of such areas, so that clashes between pipework and cable rack routes can hopefully be avoided.

Typical illustrations of cable rack diagrams are shown in Fig. 3.1.3. These drawings must also identify the width and number of tiers of rack and tray required between any two adjacent nodes in the cable network. Calculations must be performed to size this racking based on the following:

a. The degree of segregation required (see PART 2 Chapter 8 on cable installation).

b. The number and size of cables passing through this section.

c. The allowance to be made for growth in numbers of cables during the operational life of the tray/rack. This will depend on the location of the particular section of rack/tray, the age of the installation and the stage reached in the project, but allowances in the region of 300% have been known.

d. The dimensions of the space available for the racking. If multitier racking is necessary, it is important to ensure that there is sufficient space between each tier for reasonable access. Unless it can be guaranteed that all the cables are of small diameter, a minimum of 300 mm is recommended.

e. The maximum rack loading in kilogram per metre length of rack as quoted in the manufacturer's catalogue. The calculation for each section is best done on a prepared calculation sheet similar to that shown in Fig. 3.1.4.

4. Cable routing diagram

For the smaller less complex systems, racking and cable diagrams may be combined. However, in the more complex arrangements, it is more informative to present the information on two or more separate drawings so that the service, route, identification number, etc., of each cable can be easily identified. Separate diagrams may be produced for each service, such as power, instrumentation, communications and fire and gas. The disadvantage with such separation is that there will be less likelihood of spotting a clash if one exists, unless an integrated CAD system is being utilised which can identify such clashes. A typical example of this drawing is shown in Fig. 3.1.5.

5. Equipment schematic and block diagrams

These diagrams would normally be produced by the equipment manufacturer, but in some cases the manufacturer's drawings only describe the skid mounted, factory produced unit. These would need to be supplemented by a block diagram showing an overview of the whole platform-wide system and also, if necessary, a comprehensive schematic. Fig. 3.1.6 shows a typical example of an equipment schematic.

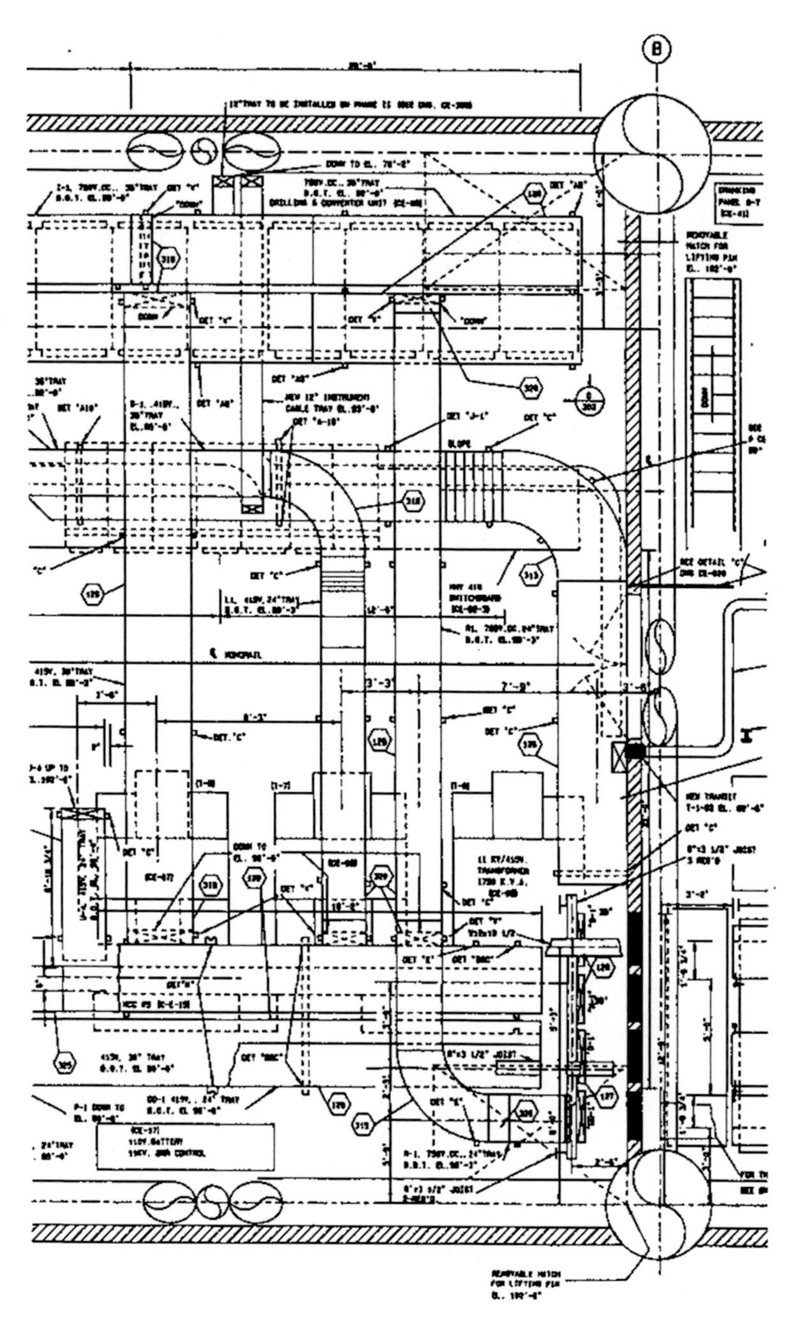

FIGURE 3.1.3

Part of a typical cable racking diagram.

Courtesy of BP Exploration Ltd.

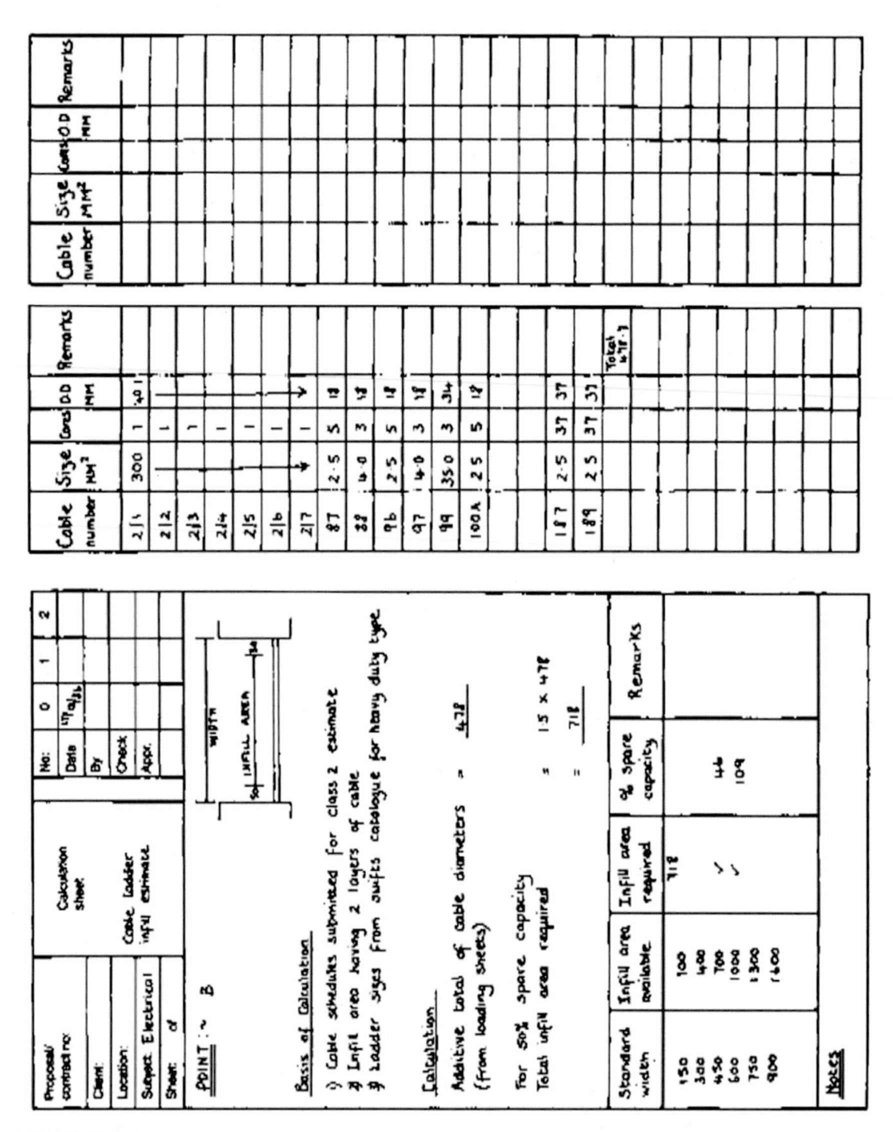

Proposal/ contract no:	Calculation sheet	No:	0	1	2
Client:		Date	17/9/86		
Location:	Cable ladder infill estimate	By			
Subject: Electrical		Check			
Sheet: of		Appr.			

POINT :~ B

WIDTH
50 INFILL AREA 50

Basis of Calculation

1) Cable schedules submitted for class 2 estimate
2) Infill area having 2 layers of cable
3) Ladder sizes from swifts catalogue for heavy duty type.

Calculation

Additive total of cable diameters = 478
(from loading sheets)

For 50% spare capacity
Total infill area required = 1.5 x 478
= 718

Standard width	Infill area available	Infill area required	% spare capacity	Remarks
		718		
150	100			
300	400			
450	700	✓	46	
600	1000	✓	109	
750	1300			
900	1600			

Notes

Cable number	Size MM²	Cores	O.D MM	Remarks
211	300	1	40.1	
212		1		
213		1		
214		1		
215		1		
216		1		
217	↓	1	↓	
87	2.5	5	18	
88	4.0	3	18	
96	2.5	5	18	
97	4.0	3	18	
99	35.0	3	34	
100A	2.5	5	18	
187	2.5	37	37	
189	2.5	37	37	
				Total 478.7

Cable number	Size MM²	Cores	O.D MM	Remarks

FIGURE 3.1.4

A typical cable rack loading calculation sheet.

6. Panel general arrangement
   Before panel wiring and schematic diagrams are produced, either by the manufacturer or by the design contractor, it is necessary to produce a drawing giving arrangement details for the panel. The layout of any control panel is important from both the ergonomic and safety point of view, and producing this drawing also assists the designer in the selection of switches, pushbuttons, lamps, annunciators, etc.

(A)

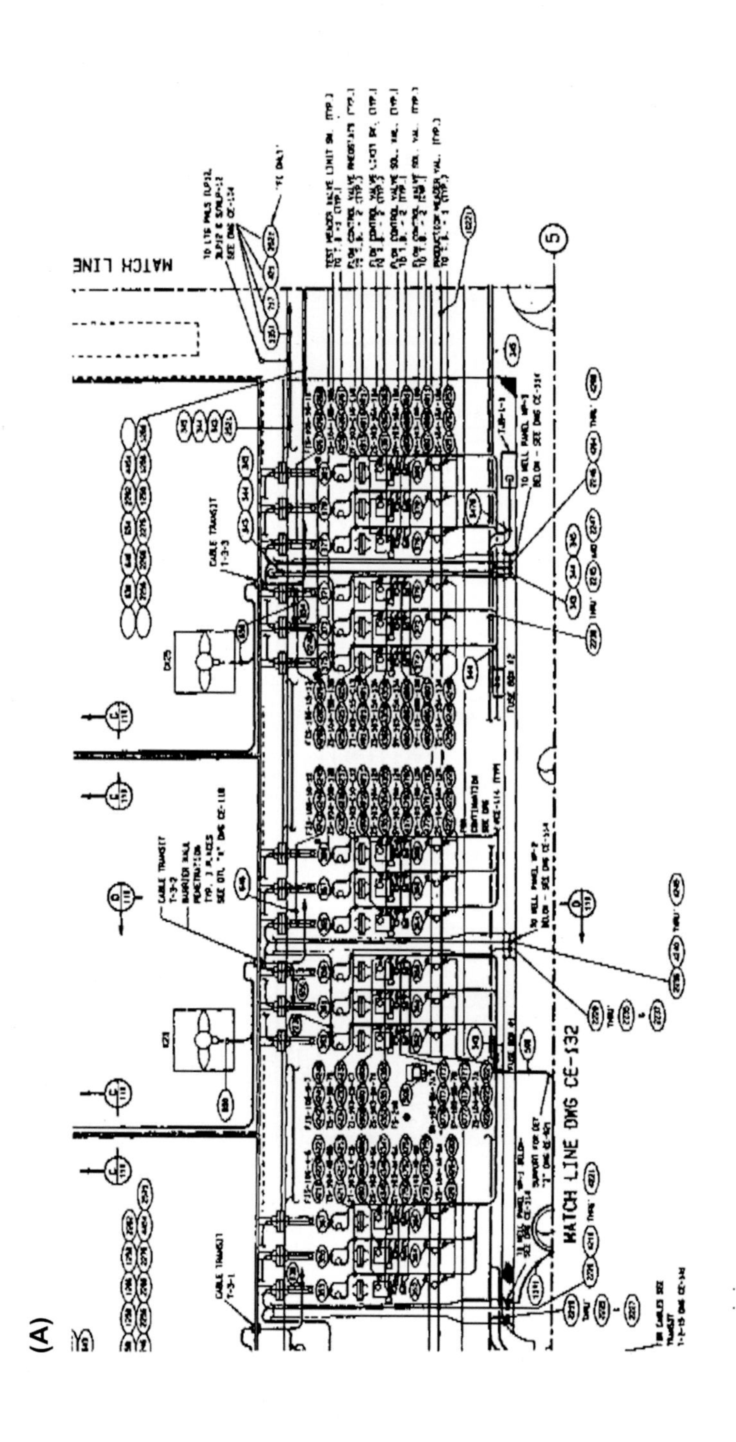

FIGURE 3.1.5

(A) A typical cable routing diagram. (B) Part of a typical cable routing diagram showing the section and transit detail associated with (A).

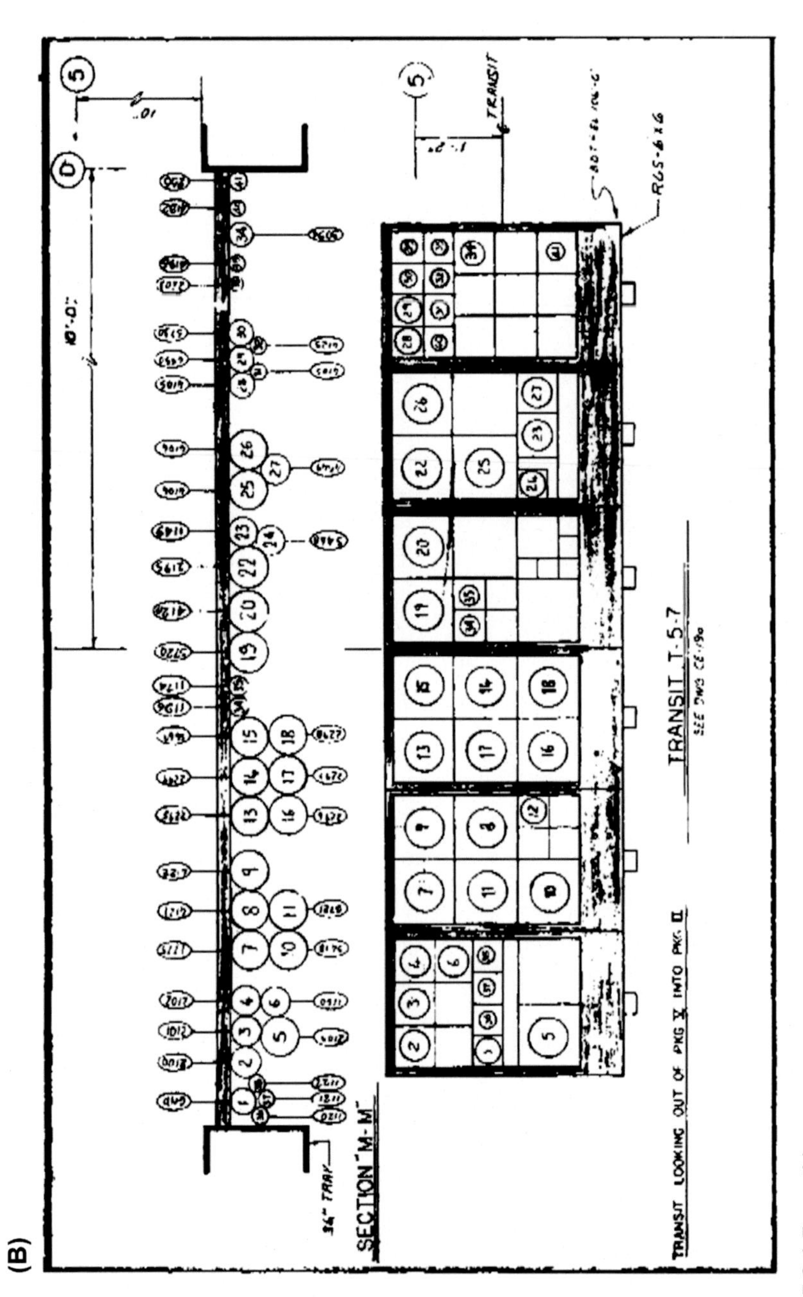

FIGURE 3.1.5, cont'd

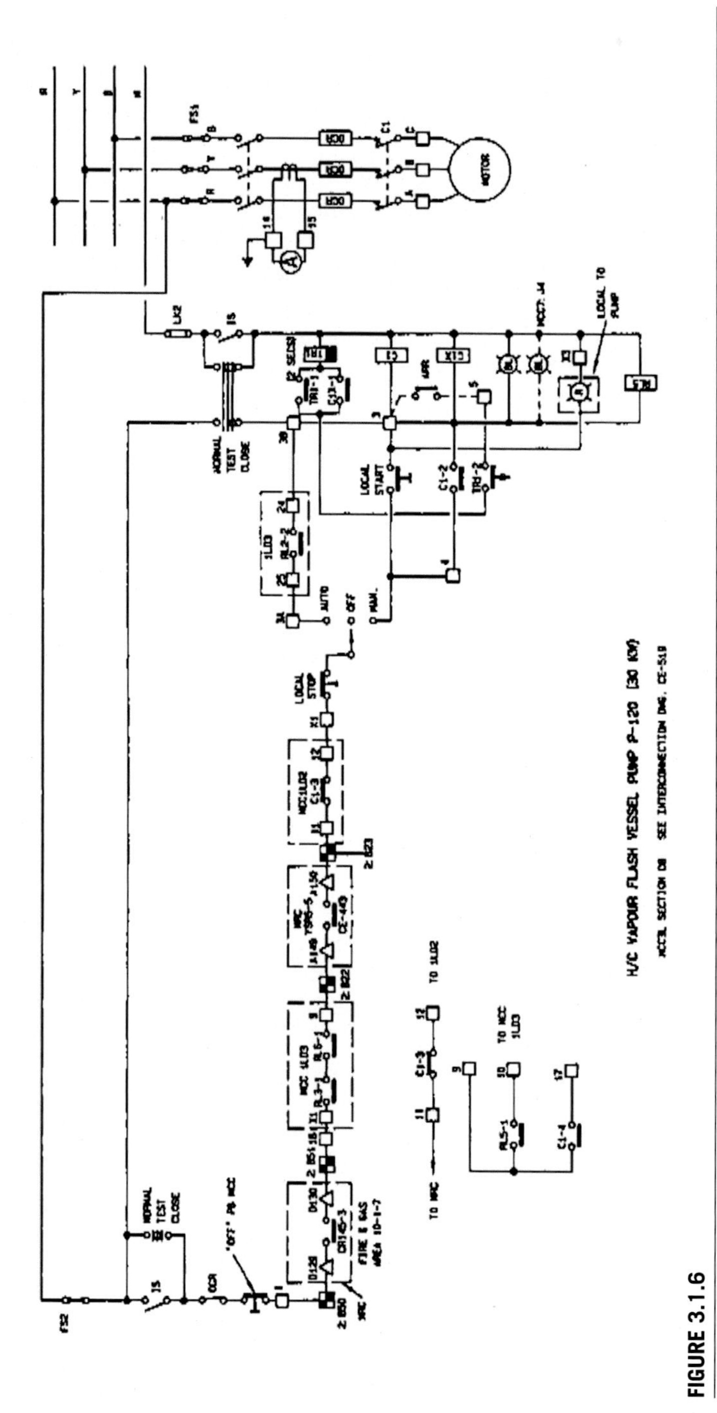

Courtesy of BP Exploration Ltd.

FIGURE 3.1.6
Part of a typical equipment schematic.

7. Panel wiring diagram
Again, these are normally produced by the equipment manufacturer, but may require supplementing in certain instances. Modifications may be required long after the manufacturer's original involvement and it may not be possible to call upon his assistance again.
8. Termination and interconnection diagrams
Having produced schematic and wiring diagrams, termination diagrams need to be produced. These diagrams are the key to platform cable routing because they not only highlight all equipment terminations but also identify all cables and their cores. Once the majority of these have been identified, cable and rack drawings can be produced in detail. Examples are given in Fig. 3.1.7.
9. Circuit breaker and starter schematic/wiring Diagrams
These are similar to other schematic and wiring diagrams, although the design contractor will normally have a greater input to these because they will interface with design information from a number of equipment manufacturers. One manufacturer may be responsible for the main switchboard and generators, in which case there will be less for the design contractor to do. However, this still leaves the design of interfaces between the main switchboard and the controls for all the large process drives, such as main oil line pumps and gas compressors. These diagrams must include the interfaces not only for the drive controls and instrumentation but also for platform monitoring systems such as emergency shutdown and fire and gas systems. Examples of these drawings are shown in Fig. 3.1.8.
10. Hook-up diagram
With offshore labour cost approximately five times higher than that of onshore, any means of presenting design information which will assist the offshore electrical personnel in their task is worth considering. The hook-up diagram is the electrician's equivalent of a motorist's route planning map. It follows an individual circuit for a pressure switch, for example, through all the cabling, back to the control panel and the source of electrical power, identifying every terminal and cable core on the way, just as the motorist's map would concentrate on a description of the route between the departure and destination points. These drawings tend to be used mainly for instrumentation, and each instrument connection would be drawn on a separate sheet. An example is shown in Fig. 3.1.9.
11. Junction box and distribution board interconnection diagram
Every offshore installation has a multitude of junction boxes, terminal boxes and distribution boards, and the interconnection of these must be detailed adequately for both installation use and maintenance records. These drawings will be similar to the interconnection diagram in Fig. 3.1.7.

DATABASES AND SCHEDULES

1. Cable schedules: A large offshore installation may have hundreds of kilometres of cables installed and it is important that during the design stage as much information on the identity, route size and type of each cable is retained. This

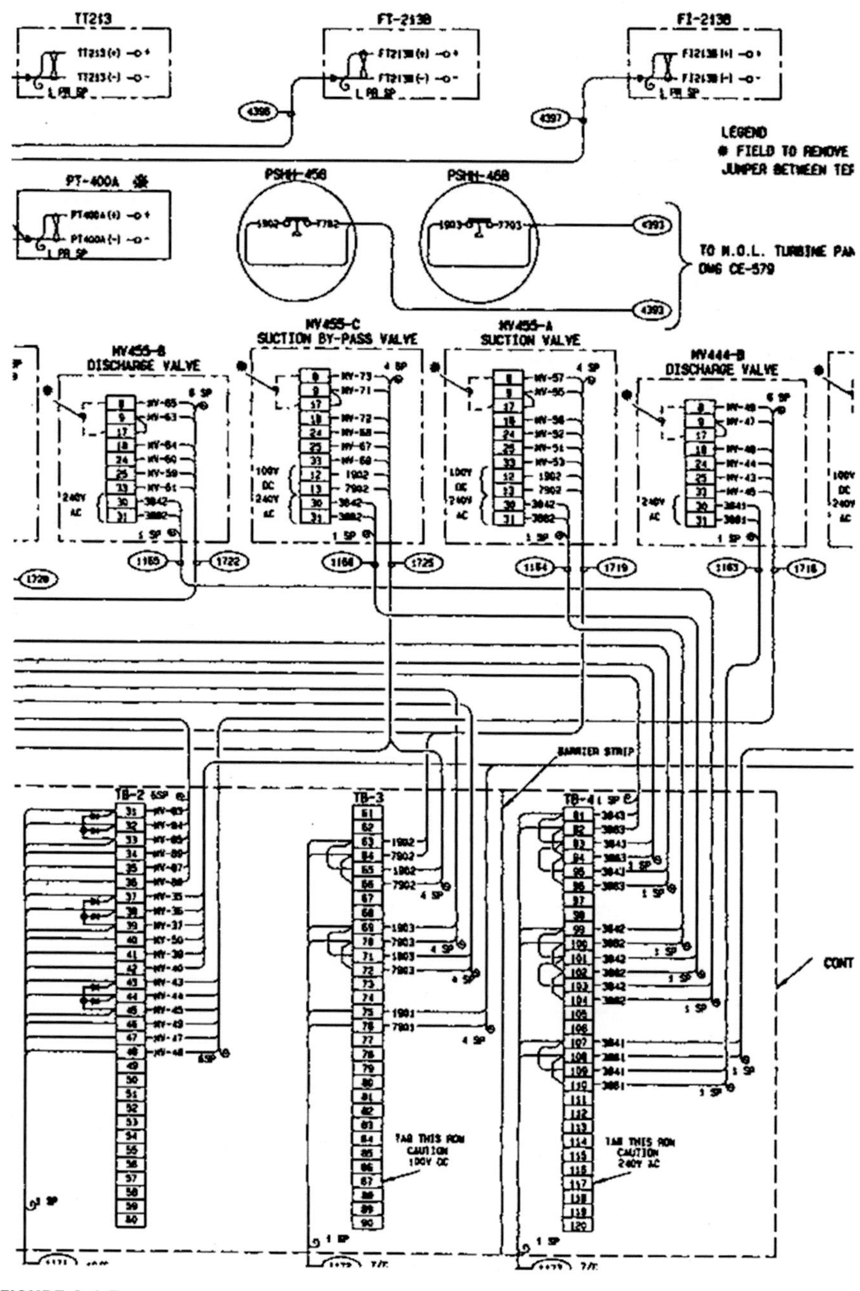

FIGURE 3.1.7

Part of a termination and interconnection diagram.

Courtesy of BP Exploration Ltd.

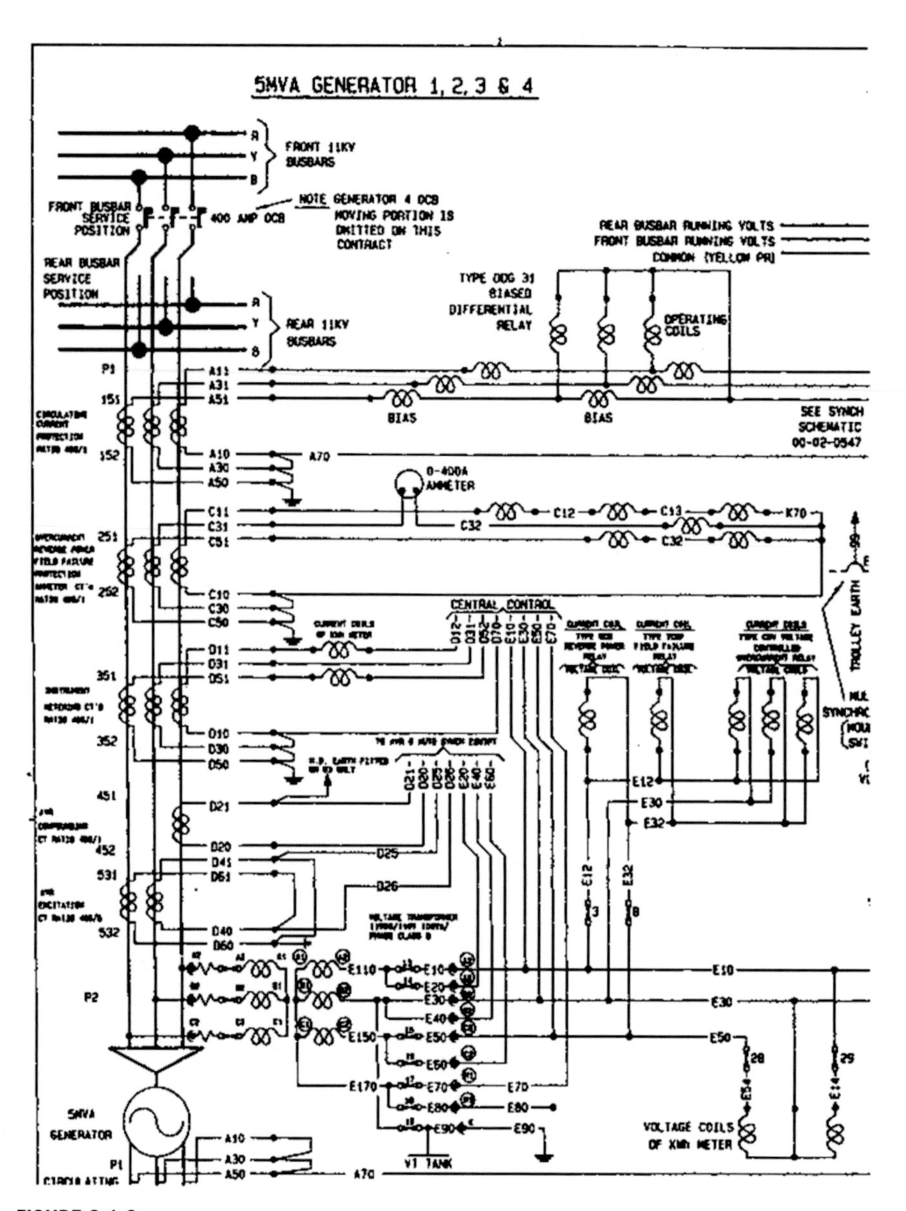

FIGURE 3.1.8

A typical circuit breaker and starter schematic/writing diagram.

information will obviously be needed for installation, but it is also vital to trace, reroute or replace a cable some time later because, for example, of fire or mechanical damage. If a particular motor is uprated, it will be necessary to look at the schedule to check whether the existing cable has the capacity for the

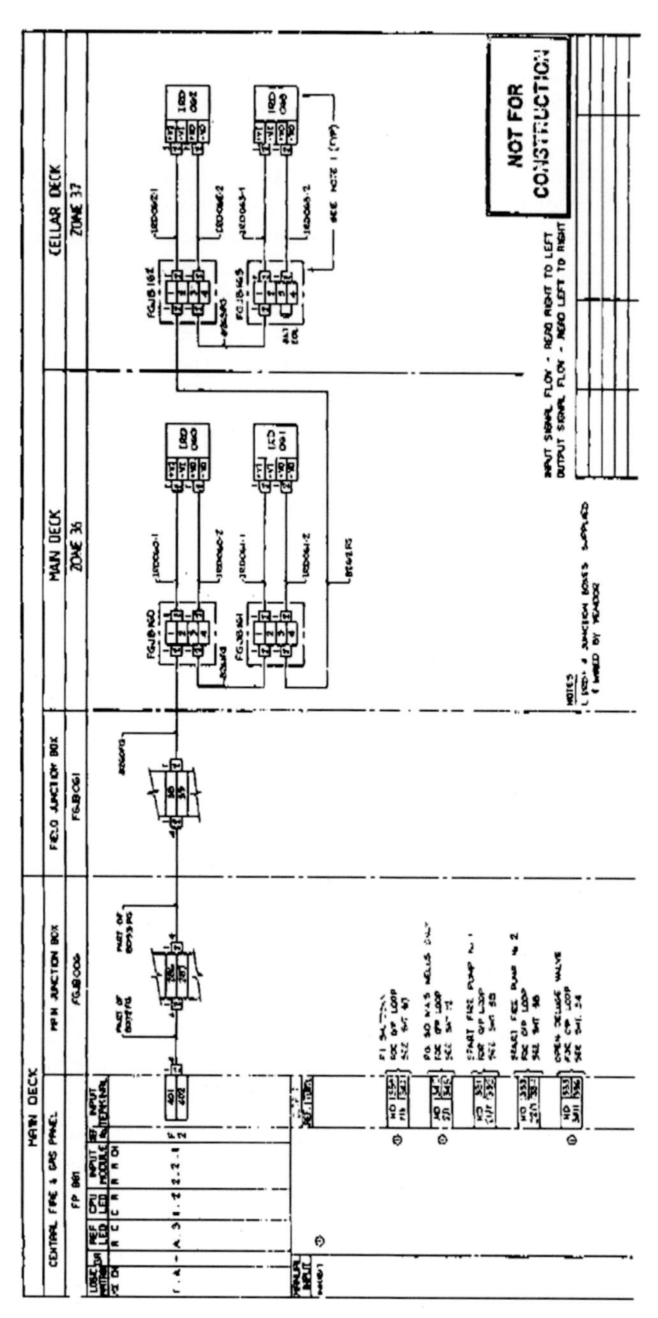

FIGURE 3.1.9

A typical hook-up diagram.

increased load. Small-project cable schedules may be produced manually using draughting blanks. However, a more practical medium is the computer database, especially during a large project when large numbers of cables are continually being added. For ease of identification, cables should be grouped into service and area. Insertion of extra cables into a schedule would be extremely laborious on a manual schedule but is relatively easy on a database. Other advantages are that duplicate cable numbers cannot exist and that most database programs have powerful sorting and search facilities. An (manually produced) example is given in Fig. 3.1.10A and B. The cable schedule should also contain a list of cable types giving a full description of each type, a type code that can be referred to in the main schedule and the total length for each type. Drum/cutting schedules should also be produced to minimise cable wastage during installation. The cable schedule must state how the individual cable lengths were derived. For example, do they already contain a margin for drawing inaccuracy, cutting wastage, installation errors, etc.?

2. Electrical equipment schedule: As an aid to draughting and ensuring that every node in the electrical distribution is catered for in the design, an electrical equipment schedule should be produced. This may be combined with a master equipment list later on in the project. Every item should be allocated a client tag number and the information presented should include the service, location, manufacturer, type and environmental details such as ingress protection number and suitability for hazardous areas. Motor schedules will give details of the controls and instrumentation required at the motor, at the starter and any other point of control. It should also indicate the motor full-load current and the fuse size and indicate any special requirements, such as earth fault relays or thermistors. The schedule may be split into types of equipment such as motors, junction boxes and luminaires, and in some cases, it will only be beneficial to produce schedules for particular types of equipment, depending on their populations and whether the same information has been produced on another document. An example of this type of schedule is shown in Fig. 3.1.11. Again, a computer database is recommended for the production of these schedules.
3. Plug and socket schedule: These schedules are often required for diving and subsea equipment, as they tend to be a prolific user of special underwater connectors. Each connector may have over a hundred connections. Pin current ratings may not all be identical on the same plug, and it may also be necessary to parallel several pins in one circuit to obtain the required rating. It may be necessary to monitor circuits for continuity as well as for earth leakage. To further complicate the matter, every socket of identical size will require a different orientation to prevent plugs being mated with the wrong sockets. Some form of schedule is essential to keep track of circuit routing through the various connectors. Each connector should be provided with a separate page (or pages) in the schedule and this will need to identify the service, the plug and socket manufacturer and the catalogue number and the orientation between the plug and socket. Each circuit should then be listed, with details of cable cores terminated in the plug and socket, circuit rating, purpose, etc.

(A)

CABLE No.	ROUTE FROM EQUIPMENT	ROUTE TO EQUIPMENT	CABLE	TYPE	No. OF CORES	SIZE mm²	WORKING VOLTS
1	TURBINE GEN. G1 5 MVA	11KV SWITCHGEAR CB-9		A.1.0	3 SINGLE	95	11 KV
2	G2 5 MVA	CB-10					
3	G3 5 MVA	CB-15					
4	11 KV SWITCHGEAR CB-8	11KV/600V TRANSF. T1 2 MVA		A.1.1	1 3/C		
5	CB-18	T2 2 MVA					
6	CB-6	11KV/3.3 KV T3 3.5 MVA					
7	CB-7	T4 3.5 MVA					
8	CB-19	T5 3.5 MVA					
9	CB-5	11KV/415V T6 1750 KVA					
10	CB-20	T7 1750 KVA					
11	CB-21	T8 1750 KVA					
12	11 KV/600V TRANSF. T1	DRILL CONV. INCOMING CUB INC.#1		A.3.4	12 SINGLE/C	300	600
13	AUX. TRANS.	#1		A.3.3	1-4/C	25	415
14	11 KV/600V TRANSF. T2	#2		A.3.4	12 SINGLE/C	300	600
15	AUX. TRANS.	#2		A.3.3	1-4/C	25	415
16	11 KV/3.3V TRANSF. T3	3.3 KV SWITCHGEAR CB-32		A.2.1	2-3/C	185	3300
17	T4	CB-34					
18	T5	CB-36					
19	11KV/415V TRANSF. T6	415V CB-41		A.3.4	17 SINGLE/C	300	415
20	T7	CB-45					
21	T8	CB-47					
22	DEISEL GEN. G6	CB-43			11 SINGLE/C	150	
23	G7	CB-49					
24	415 VSWGR CB-48 CUB. E	M.C.C. No. 4 IS 53 TIER 'E' UPPER			13 SINGLE/C		
25	CB-44 CUB. C	M.C.C. No. 4 IS 51 TIER 'E' LOWER					
26	M.C.C. - 1LK1 CB-56	M.C.C. No. 5 COMPT. E1					
27	M.C.C. - 2LA1 CB-57	M.C.C. No. 6 COMPT. A1					
28	TURBINE GENERATOR G1	NEUT EARTHING RESISTOR BOX NER 1		A.1.0	3 SINGLE	95	11 KV
29	G2	2					
30	G2	3					
31	NEUTRAL EARTHING RESISTOR #1	EARTHING SYSTEM		A.3.4	3-1/C		
32	NEUTRAL EARTHING RESISTOR #2	EARTHING SYSTEM			3-1/C		
33	NEUTRAL EARTHING RESISTOR #3	EARTHING SYSTEM			3-1/C		
34	11KV/600V TRANSF. 'T1'	NER 'T1'		A.3.4	1-1/C	300	
35	11KV/600V TRANSF. 'T2'	NER 'T2'		A.3.4	1-1/C	300	
36	NER 'T1'	EARTHING SYSTEM		D.I.F.	2-1/C	95	
37	NER 'T2'	EARTHING SYSTEM		D.I.F	2-1/C	95	
38							
39							

FIGURE 3.1.10

(A) Part 1 and (B) part 2 of a typical sheet from a cable schedule.

Courtesy BP Exploration Ltd.

(B)

GRADE	ROUTE FEET	DUTY	EQUIP. No.	[illegible] GLAND CABLE	CABLE ROUTED VIA TRAY	REMARKS
11 KV	375	POWER	C-J-01	D	B7 B5 B2 B3 & B1	RUN CABLES IN 1 TREFOIL GROUP
	350		C-J-02	D	B7 B5 B2 B3 & B1	
	280		C-J-03	D	A7 A5 A6 A3 & A1	
	70		T1	F	LOCAL PKG I	
	50		T2			
	85		T3			
	65		T4			
	65		T5			
	75		T6			
	55		T7			
	40		T8			
1000	50			C2		RUN CABLES IN 4 TREFOIL GROUPS & 3" BETWEEN GROUPS
	50			C		
	65			C2		RUN CABLES IN 4 TREFOIL GROUPS & 3" BETWEEN GROUPS
	65			C		
3.3 KV	70			F		3" BETWEEN GROUPS
	100			F		
	95			F		
1000	65			C2		5 TREFOIL GROUPS & 3" BETWEEN GROUPS (2 CABLES NEUT CONN)
	60			C2		
	60			C2		
	40		C-J-06	C		3 TREFOIL GROUPS & 3" BETWEEN GROUPS (2 CABLES NEUT CONN)
	40		C-J-07			
	300				A7 A5 A6 A3 & A1	2 TREFOIL GROUPS & 3" BETWEEN GROUPS (2 CABLES NEUT CONN)
	325				B7 B5 B2 B3 & B1	
	60				LOCAL PKG I	
	65					
11 KV	18	EARTHING		D	B7-1 PKG VII	NEUTRAL CONNECTION
	18			D	B7-2	
	18			D	A7-1	
1000	16			C	LOCAL	EARTHING
	11			C		
	16			C		
	40			C2	LOCAL PKG I	
	40			C2		
	50			C		
	50			C		

FIGURE 3.1.10, cont'd

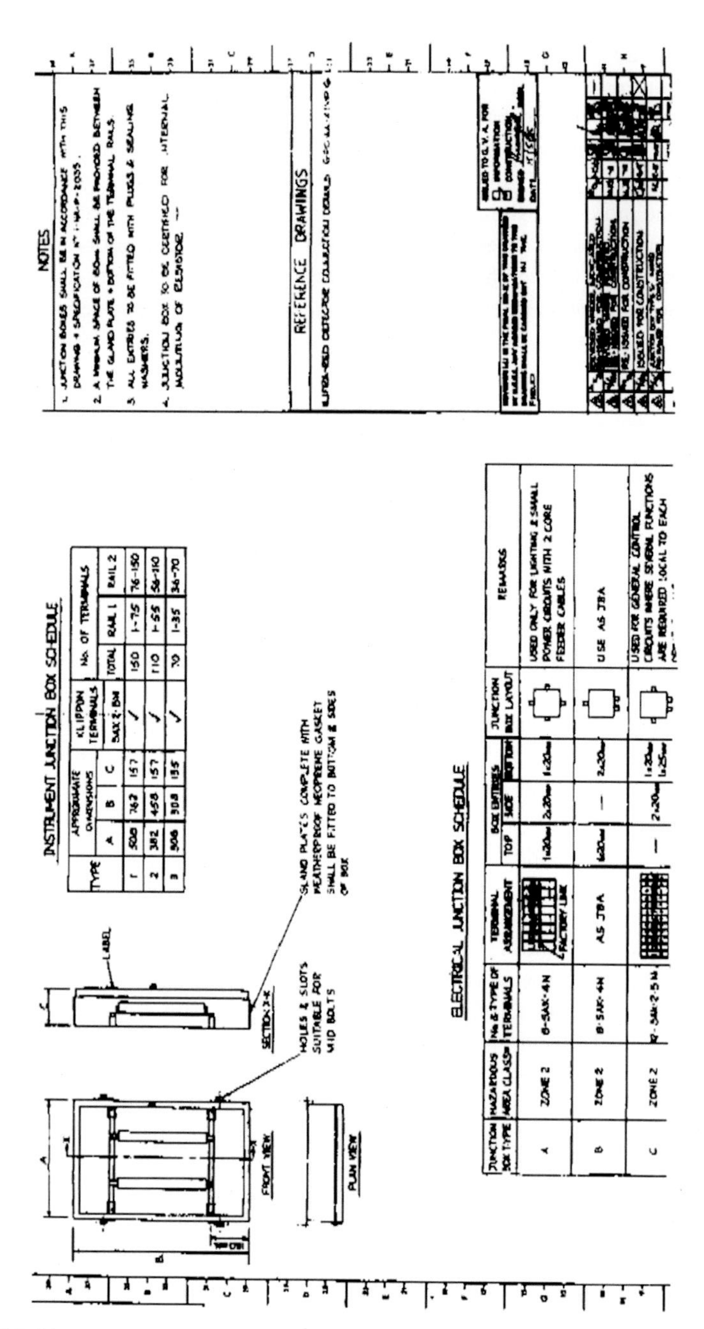

FIGURE 3.1.11

A typical sheet from an electrical equipment schedule for junction boxes.

Courtesy BP Exploration Ltd.

4. Electrical load list: As with any project where electrical power will be required, a continuous review of the projected loading is necessary. This is particularly important where the electrical system is isolated and dependent on its own power sources to support the load. The preparation of an accurate load list often becomes a priority early in the project, as the less accurate the system load figures are the more the risk involved in purchasing suitable generators. In any case a good margin should be kept over estimated loadings if generator outputs are not already restricted by weight and space limitations on the installation. The schedule should indicate connected and diversified loads for all the system and generation operational states envisaged. A typical load list is shown in Fig. 3.1.12.

MATERIAL AND EQUIPMENT HANDLING AND STORAGE

Adequate, secure, warm and dry storage needs to be provided for equipment before being called offshore. The storage facility should be local to the offshore supply base if construction delays are to be avoided.

A goods inspection procedure is important, as it should detect damaged or incorrect items which can then be repaired or replaced before they are sent offshore. An item or equipment whose damage or unsuitability remains undetected until it is being installed can cause problems in the construction programme.

The contractors carrying out the installation work will need to store materials offshore, so suitable containers will need to be located on the installation. The areas where such containers are stored are known as laydown areas. Laydown areas need to be adjacent to platform cranes so that containers and equipment can be off- and onloaded easily from supply vessels. Crane operations need to be studied to ensure that the risk of dropping or impacting equipment or containers on modules, particularly those containing hazardous production equipment, is minimised.

Every item sent offshore should be clearly identified and marked with its gross weight. This reduces the incidence of loads being dropped by overloaded cranes, or mysterious packing cases arriving on some other oil company's offshore installation when the equipment inside is vital to the next stage of your client's construction programme.

Once equipment has reached the offshore installation, particularly if heavy structural work is going on, it must be protected from physical damage, shock, vibration, ingress of dust, moisture, welding sparks and any other foreign matter until it is permanently protected.

The manufacturer's storage and preservation procedures must be adhered to if warranties are to remain valid.

ERECTION PROCEDURE

An installation specification should have been produced in the design office and copies of this should be made available to those carrying out the installation and to the

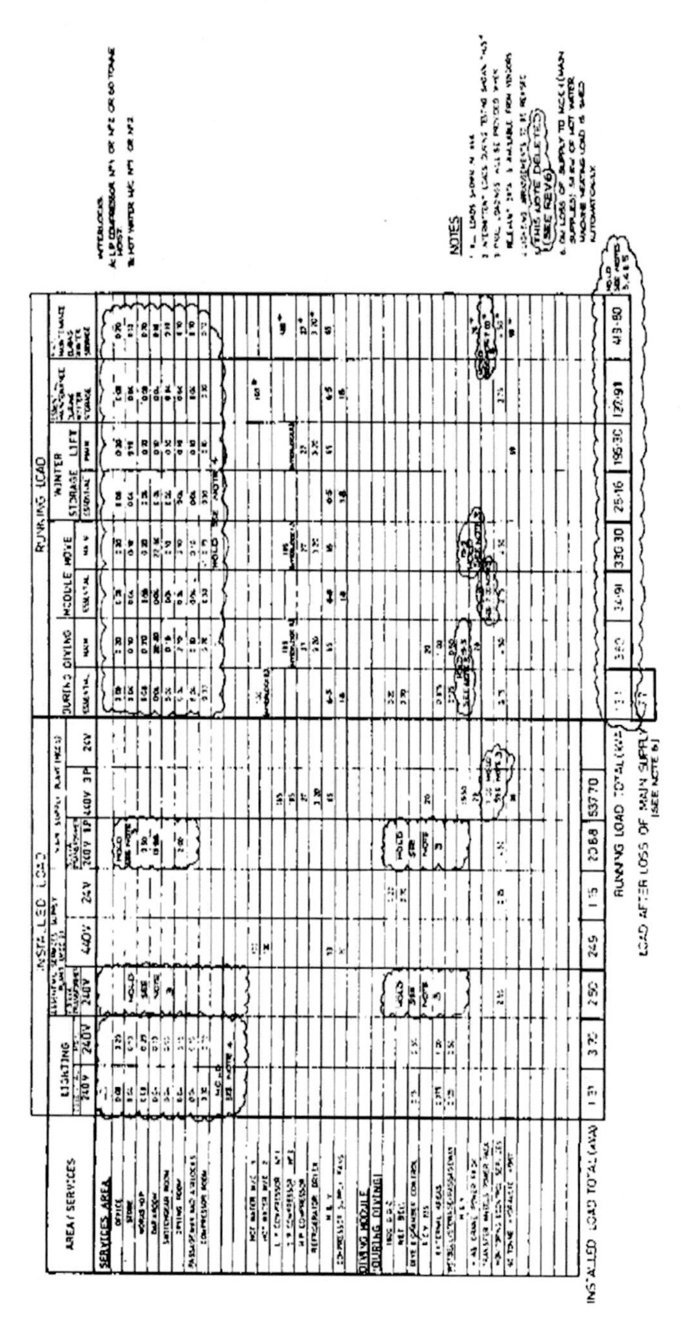

FIGURE 3.1.12

Part of a typical sheet from an electrical load list.

installation inspectors. In addition to an installation general code of practice, this document should also contain a series of blank test sheets for recording all electrical equipment precommissioning tests.

The arrangements for preservation of the equipment will need to continue while erection is completed. It is important to produce or obtain from the manufacturer a written erection procedure well before the contractor starts working. This will allow the procedure to be checked for compatibility with the relevant construction package and for any operation that might be dangerous to personnel or risk damaging the equipment.

The procedure should identify every item in the plant the installation contractor will need to install the equipment, so that space may be allocated for it and, if it is a source of ignition, electromagnetic interference, etc., the relevant precautions can be taken in advance. When cables are being installed or electrical equipment commissioned, the installation contractor should be required to mark up a set of drawing prints showing any changes to cable routes, terminations, etc. which prove necessary. These 'as-built' drawings are then returned to the design office for review and if accepted, the changes should be incorporated in the drawing masters. Providing the work is done to the inspector's satisfaction, the contractor will receive a certificate of acceptance, which he or she will need for invoicing purposes.

Commissioning and precommissioning tests are covered in PART 7 Chapters 1–5.

Practical notes on the installation of particular equipment are given in PART 7 Chapter 1.

Please also refer to PART 2 Chapter 8 for electrical cable selection, sizing, etc.

PART

Electrical Faults and Protection Devices

4

CHAPTER

Alternator Faults and Protection Devices

1

INTRODUCTION

As discussed in earlier chapters, the isolated location, onerous operating conditions and harsh environment make it particularly important that provisions are made in the design in order to protect electrical equipment from faults on the system and to remove faulty equipment from the system quickly and safely. Alternators must be protected from prime mover and fuel system faults. Motors must be protected from faults in driven equipment. In this chapter, common protection relay applications are discussed Worked examples are provided in PART 4 Chapter 7 for power system relay configurations likely to be met in offshore installations.

NEUTRAL EARTHING AND EARTH FAULTS

Medium-voltage generators are normally resistance earthed to limit fault currents. Stator windings should be designed to minimise third harmonic circulating currents, so that generators may be paralleled without extra weight and space being taken up with earthing circuit breakers. The short time ratings of any earthing resistors must be borne in mind when calculating earth fault relay settings. A current sensing relay of either inverse or definite time characteristic may be used for unrestricted earth fault protection. Alternatively, an attracted armature type relay may be used in conjunction with a time delay relay. In either case, the time delay must be within the time rating of the earthing resistor and the relay must coordinate with those downstream. The term 'back-up' earth fault protection is sometimes used to describe this scheme, as it supplements the generator differential protection. But the use of this term is not recommended, as unrestricted earth fault protection will not detect phase-to-phase faults and therefore the term 'back-up' or 'standby' leads to some confusion. The neutral earthing conductor arrangement must be rated for the prospective earth fault current flows available from the generator. If the generator is a low-voltage machine with bolted earth connection, this current is likely to be high enough to affect cable conductor sizes. With directly earthed low-voltage machines, the high currents are used to operate fuses and miniature circuit breakers usually found in low-voltage distribution systems. With larger high-voltage machines the damage caused to laminations is limited by the insertion of the neutral resistance, and more sensitive forms of protection may be used in the medium-voltage distribution system.

Offshore Electrical Engineering Manual. https://doi.org/10.1016/B978-0-12-385499-5.00021-2

OVERLOAD PROTECTION

The larger offshore alternators will not have overload protection as such, but resistance temperature devices (RTDs) are buried in the stator windings and will give alarm and trip signals at temperatures where overloading or other abnormal conditions such as excitation faults have occurred. The temperature protection will avoid unnecessary shortening of insulation life. With small emergency generators, the associated circuit breaker should be provided with a thermal element.

OVERCURRENT PROTECTION

To protect the generator from downstream faults, overcurrent relays of either the inverse definite minimum time (IDMT) or definite time type are required. The protection relay must provide sufficient time for downstream protection relays to operate to clear the fault, but this time must be within the time rating of the alternator. Larger generators are usually provided with permanent magnet pilot exciters to maintain a fault current of around three times rated current continuously after the transient short-circuit current has decayed away. This can be maintained usually for an absolute maximum of 10 s before the generator overheats. However, it should not be necessary to approach even 5 s in order to operate downstream protection, and the damage due to arcing faults if such a large amount of fault energy is let through can be extensive. It is therefore advisable to keep fault times to a minimum by good relay coordination. With small self-excited generators, it is important to specify a 'fault current maintenance kit' consisting of a voltage sensitive relay and compounding current transformers (CTs) designed to maintain excitation current when the output voltage collapses because of a fault. If reasonable coordination cannot be obtained using normal IDMT or definite time relays, voltage controlled or voltage restraint type relay can be used to take advantage of close-up fault voltage collapse. An example of this is given at the end of this chapter.

PHASE AND INTERTURN FAULTS

Stator and interturn faults are relatively uncommon but are more likely to occur at the ends of the windings or in the terminal box. Insulation failures between phases may be sensed by differential protection. Both types of fault will cause heating and possibly a fire, which should be detected by suitable smoke detection in the acoustic hood, if not by the RTDs or the differential protection.

WINDING PROTECTION

Restricted earth fault (if the individual phases are not accessible at the neutral end) or phase and earth fault circulating current type protection is usually provided on machines of rating 500 kW and larger, although this threshold will vary with the degree of criticality of the supply. A stabilising resistor must be provided in series with the relay to prevent CT saturation effects causing the relay to operate in faulty conditions. The method of calculating the resistance value is given in an example at the end of the chapter.

OVER-/UNDERVOLTAGE PROTECTION

This facility is often provided as part of the generator's automatic voltage regulator (AVR), but if this is not the case, it must be included in the generator control panel or switchboard. If fed from a separate voltage transformer (VT), it may duplicate the AVR facility, so that failure of one set of VT fuses cannot lead to loss of voltage control and/or monitoring.

OVER-/UNDERFREQUENCY PROTECTION

This facility is often provided by the prime mover governor, but as it is more critical on a generator isolated from a grid system, it is often duplicated electrically. This ensures that the circuit breaker trips before damage is done to motors or driven machinery, even if the prime mover overruns because of gas ingestion at the air intake (see PART 2 Chapter 4).

UNBALANCED LOADING AND NEGATIVE PHASE SEQUENCE PROTECTION

If the platform load is unbalanced, there will be a negative phase sequence component in the generator load current which, if excessive and over a long period, will cause overheating of the rotor. On the larger offshore generator packages, it is therefore necessary to fit negative phase sequence protection, which after a timed period corresponding to the thermal characteristic of the rotor, will trip the generator. As the source of negative sequence current is external to the generators and will lead to the tripping of all the generators if not removed, it is normal to provide an alarm at onset of the problem so that operators may have time to find the offending load before production shutdown occurs. A suitable setting value should be sought from the generator manufacturers in each case, but a value no greater than 30% is recommended.

ROTOR FAULTS

The exciter output current on brushless machines is monitored and should a rotating diode go short circuit, an alarm will be annunciated on the generator control panel. This facility is usually included in the AVR circuitry, and therefore, it is not likely to be included in the discrete protection relay suite.

FIELD FAULTS AND ASYNCHRONOUS OPERATION

If the generator field current fails and the generator is running as the sole supplier of power on the installation, the set will trip on undervoltage, causing a blackout until the emergency generator starts. However, if the generator is running in parallel with a second machine, it would continue to generate power as an induction generator, whilst demanding a heavy reactive power flow from the machine in parallel with it. Both machines will tend to heat up, and in some cases, it may appear as if the healthy machine is the offender. A field failure relay of the mho impedance type is normally used to protect generators from this condition. The generator reactance for a machine where the excitation has failed is not fixed but describes a circular locus, and so the relay characteristic should be set to enclose this locus as fully as possible.

PACKAGE CONTROL AND SUPERVISION, MASTER TRIP AND LOCKOUT RELAYS

Individual relay protection elements are usually provided without any form of lockout facility, although some means of electronically or mechanically (i.e., flag) recording relay operation is often included. Where a switchboard is fed by a number of generators, the operational logic of the system must be carefully studied to ensure every mode of operation possible is reliable and safe. It is usual to provide a master trip relay for each generator, which will operate from a signal derived from any of the fault-sensing relays and then lock the generator out of service. Without the master trip relay, indication of the fault on the control panel could well be lost, and there would be nothing to prevent an operator attempting to restart the machine before the fault is cleared. If failure of one generator leads to automatic starting of another, it is important that the control logic differentiates between generator failure and uncleared switchboard faults, in order to avoid the second generator circuit breaker closing on to the same (switchboard) fault. The generator circuit breaker tripping coil and its circuit should be monitored by a trip supervision relay, which provides an alarm on failure of coil continuity or loss of tripping supply. Because of the risk of generator, switchboard or transformer fires due to sustained fault conditions if a generator circuit breaker fails to trip, and their possibly serious consequences offshore, it is advisable to fit (to main generator circuit breakers at least) trip supervision relays which continuously monitor both tripping supply and trip circuit continuity.

For fault calculations see PART 4 Chapter 6.

For relay discrimination topics see PART 4 Chapter 7.

CHAPTER

Transformer Faults and Protection Devices

2

TRANSFORMER FAULTS

EARTH FAULTS

The magnitude of any fault current flowing in a transformer will be a function of the winding arrangement (and hence leakage reactance) and the type of earthing (solid or impedance). Most offshore distribution transformers, however, are of the delta star (usually Dy11) type, the only exceptions normally being those used for supplying the drilling silicon controlled rectifiers. With the usual solidly earthed star point arrangement, the fault current will be highest when the fault is near neutral. There is no zero-sequence path through such a transformer and therefore coordination is not required with any earth fault protection on the primary side. The magnitude of any earth fault on the delta primary winding will be governed by the type of earthing in that part of the system. If it is fed from the main switchboard, it will probably be limited by the generator earthing resistors.

PHASE-TO-PHASE FAULTS

Faults between phases on offshore transformers are relatively rare, especially with the newer sealed or encapsulated types of transformer. Such a fault is likely to produce a substantial current large enough to be detected quickly by upstream overcurrent protection.

CORE AND INTERTURN FAULTS

Although interturn faults are unlikely on offshore distribution transformers because of the relatively low voltages (LVs), the fire hazard is such that their probability cannot be ignored. A fault in a few turns will cause a high current to flow in the short-circuited loop and produce a dangerous local hot spot. A conducting path through core laminations will also cause severe local heating. It is difficult to detect such a fault in a resin-encapsulated transformer. Offshore distribution transformers are less likely to suffer line surges or steep fronted impulse voltages, as cabling is relatively short and lightning strikes are rare. However, moisture ingress is more likely in the offshore environment, and mechanical vibration levels are higher, making chafing and cracking of insulation more likely, so the possibility of interturn faults cannot be ignored. If the transformer is housed in a tank containing insulating oil, there is always a danger that the tank will corrode in the salt-laden atmosphere, or will be damaged by crane operations or the

Offshore Electrical Engineering Manual. https://doi.org/10.1016/B978-0-12-385499-5.00022-4

like. Offshore transformers are as likely to suffer from abnormal system operation as their onshore counterparts. The maintenance of system voltage and frequency may be dependent for long periods on the satisfactory operation of one generator package. The governor or automatic voltage regulator of this package may be set low or high, or may be subject to drift within the limits of the voltage and frequency protection. Therefore, transformers may be subject to variations in voltage and frequency. High voltage and low frequency may together cause shifting of flux in structural parts of the transformer, which will heat up and destroy insulation. The transformer must also be protected from overloads and overcurrents due to downstream faults.

MAGNETISING INRUSH

Magnetising inrush is a normal healthy but transient condition associated with the establishment of linking flux between the windings. However, if it is forgotten when setting upstream overcurrent protection relays, it may cause nuisance tripping. Values quoted by transformer manufacturers are in the region of 12 times full load for 15 ms.

OVERCURRENT PROTECTION

A common arrangement offshore is to protect the primary winding of the transformers between the main (medium voltage) and the production (LV) switchboards with inverse definite minimum time overcurrent relays graded with the main generator overcurrent protection. The relay may have an earth fault element that detects earth faults in the primary winding only.

RESTRICTED EARTH FAULT PROTECTION

With the restricted earth fault scheme, the residual current obtained from a current transformer (CT) in each line is balanced against the current from a CT in the neutral. The neutral is usually solidly earthed and therefore a healthy fault current will be produced even from a fault at the last turn of the winding (i.e., closest to the neutral). As faults are only detected between the line CTs and the neutral CT, only the star secondary is protected, using a sensitive instantaneous relay.

DIFFERENTIAL PROTECTION

Differential protection is rarely used on transformers offshore. However, it does have the advantage over restricted earth fault protection in that both primary and secondary windings are protected from both earth and phase-to-phase faults. The relay will only operate for faults appearing in the protection zone between the sets of sensing CTs at the primary and secondary terminals. Under normal load conditions the CT secondary currents are equal and no current flows in the relay operating coil. If these currents become unequal because of a fault current being sensed in the transformer windings, the resulting current energises the operating coil. The relay contacts close when the ratio of this differential current to the through current exceeds the slope of the relay operating characteristic. This characteristic may be altered by adjusting the turns ratio of the operating and restraint coils. The bias slope is chosen so that the relay is insensitive to unbalanced external lead burdens which normally give a lower ratio of differential current to through current than an internal fault. In addition, a fairly high bias slope is required to prevent maloperation by CT differential currents arising from

1. tap changing on transformers giving CT mismatch,
2. different CT ratios and hence saturation levels giving differential currents under through-fault conditions,
3. magnetising inrush giving secondary currents in one set of CTs only.

To prevent maloperation by magnetising inrush, the relay operation is delayed by a selected time delay which allows the initial current peaks to decay below the relay setting value determined by the percentage bias.

OIL AND GAS-OPERATED DEVICES

Pressure or flow switches of various kinds may be fitted to oil-filled transformers in order to detect faults in the windings which give rise to sudden pressure increases in the tank or sudden flow to the conservator (i.e., Buchholz devices). As conservators are rarely fitted offshore, because sealed silicon oil designs are now preferred to reduce fire risk, a rate-of-rise pressure sensing device is usually fitted on the tank. Protection of this kind which minimises the risk of transformer fires should be considered for all but the smallest of offshore distribution transformers because of the serious consequences such a fire would have.

PARALLEL TRANSFORMERS AND INTERTRIPPING

MOMENTARY PARALLELING SCHEMES

In order to allow for transformer and associated equipment maintenance and for unplanned outages, it is usual to provide two transformers in parallel as feeders to the production LV switchboard. In most systems, each transformer is rated for the full production switchboard load, but the production switchboard is only fault rated for operation with one transformer connected. In such a case, interlocking and paralleling facilities are provided which only allow momentary paralleling as the load is transferred from one transformer to the other. The risk that a fault occurs during the changeover period is usually very small and therefore acceptable.

Where the LV switchboard has a bus section switch, a third operating alternative is available where each section is fed from one transformer. This has the benefit that one transformer fault will only require the reinstatement of one LV bus section and will not disrupt supplies to the bus section fed by the healthy transformer.

INTERTRIPPING

A faulty transformer must be isolated by the circuit breakers on both the primary and secondary sides, wherever the fault is sensed, for two reasons.

First, a fault sensed by protection only on the secondary circuit breaker would leave the transformer energised in a dangerous state. Second, a fault sensed by protection only on the primary circuit breaker could allow the transformer to be back fed from the secondary circuit, again leaving the transformer energised in a dangerous state.

A scheme where the primary and secondary circuit breaker trip circuits are linked is known as intertripping. Momentary paralleling and intertripping schemes are often combined, and the scheme may have additional facilities, such as those required for the emergency back feeding of a main switchboard air compressor when main generation is not available.

CHAPTER

Motor Faults and Protection

3

The various types of motor used offshore are discussed in Chapter 4. In this chapter, the various types of motor protection devices and their application offshore are discussed. In general, there are seven main categories of drive system fault which need to be protected against. These are:

MOTOR WINDING ELECTRICAL FAULTS

Such faults are usually catered for by means of fuses or circuit breakers fitted with instantaneous overcurrent relays. Sensitive earth fault relays set at about 20% of full-load current are normally provided on motor starters rated at more than about 40 kW.

MOTOR MECHANICAL FAULTS

Bearing failures:

Without some form of vibration monitoring, it is not very practicable to detect the incipient failure of a ball or roller bearing and shut down the machine before the bearing disintegrates. However, monitoring devices are being developed which detect the effect of various forms of abnormal vibration on a machine's magnetic 'signature'. This information is then processed to provide various alarms and trips. The incipient failure of a sleeve bearing can usually be detected by a resistance temperature device or similar device which detects the rise in bearing temperature before seizure or disintegration.

ABNORMALITIES IN THE DRIVEN MACHINERY

The control logic of the driven machinery package should prevent the occurrence of severe changes in torque, negative torques or reverse running before the motor is energised. An example of such a fault would be if an extra water injection pump was being brought into service and seawater, pressurised by the running pumps, was allowed to flow backwards through the pump being started, thus driving the pump and the motor backwards. The extra time at elevated current required by the motor to decelerate before reaccelerating in the forward direction may be sufficient to overheat the machine. It may be necessary to fit a tachometric device to the machine such that reverse running will result in an 'inhibit start' signal. Although valve actuator interlocking logic should be designed to prevent the water valves being opened in the wrong sequence, it is worth checking the driven equipment control scheme to ensure that this or any other likely cause of motor failure is catered for by one or more protective measures.

Offshore Electrical Engineering Manual. https://doi.org/10.1016/B978-0-12-385499-5.00023-6

High winds blowing directly into ventilation fans, causing them to run backwards at high speed ('windmilling'), is a similar problem. In this case, nonreturn dampers or brakes which release when the motor is started can cure the problem.

ABNORMALITIES IN THE SUPPLY SYSTEM

The supply system depends on a limited number of platform generators which are, in turn, dependent for their fuel, cooling water, etc. on the correct operation of other equipment on the installation. Therefore, there is a significant probability that frequency and voltage fluctuations of sufficient magnitude will occur to affect the running of motors. With low-frequency problems, stalling protection will adequately deal with the situation. High-frequency problems should be catered for by overload protection devices. However, with large reciprocating compressors, it may be advisable to fit timed over- and/or underfrequency trips to avoid running the machine for long periods at a frequency close to tolerance limits. Generator under-/overfrequency trip settings should be a compromise between taking the most sensitive plant into account and introducing the likelihood of generator nuisance tripping. Running at abnormally low system voltage will cause motors to draw larger currents and increase heating effects in motor windings and supply cables. A lower system voltage will also increase the likelihood of stalling, as the motor torque is proportional to the square of the voltage at a given speed.

An abnormally high voltage will reduce heating effects in windings and supply cables and give faster starting times because of the increased torque. Because of this, it is the practice on some installations to run the system at a few percent above the nominal voltage. Unfortunately, prospective fault currents increase in proportion with the square of the voltage. Therefore, if the system is to be normally run at a voltage several percent above nominal, the prospective fault currents should be recalculated to reflect this and switchgear fault ratings checked accordingly.

OVERLOAD PROTECTION

The heating effects due to motor overloads can be simulated to some degree by a bimetal element or an electronic thermal replica (Fig. 4.3.2). Both the stator windings and the rotor will begin to heat up at the onset of an overload. At ratings below about 20 kW, depending on the design and type of machine, the stator windings will tend to suffer thermal damage in a shorter time than the rotor, in which case the motor is termed 'stator critical'. Stator critical motors are easier to protect than rotor critical types because the stator windings are more accessible than the rotor and can be fitted with temperature sensors if the thermal overload device is felt to be inadequate. If the motor is driven by a variable-speed drive (VSD), where it is likely to run for some time at low speed, the thermal effects of reduced ventilation must be taken into account in the choice of motor and the setting of the overload. The VSD unit is usually fitted with sophisticated overload protection as a standard feature. Electronic replica type protection relays are normally reserved for the protection of larger machines. However, a drive of only a few tens of kilowatts which has a particularly critical function, or is located at some part of the platform where no form of overheating could be tolerated, for example, an Ex 'e' motor in a hazardous area, may also require such a device to be fitted. In ATEX-certified machines, the overload is certified as part of the motor certification.

STALLING PROTECTION

When the elevated current drawn by a motor during starting is prolonged by, for example, a mechanical fault in the motor or driven machinery, then the motor will very rapidly overheat, unless quickly disconnected from the supply. If the probability of a motor stall is considered significant for a particular drive package, a stalling relay should be fitted. This device is a thermal element similar to the overload element but designed to respond to the higher short-term starting currents and to trip if the starting current continues for more than 5–10 s beyond the normal starting time. Normal starting times vary depending on the characteristic of the driven equipment, and as they increase, discrimination between starting and stalling conditions becomes more and more difficult, until the point where it becomes necessary to fit a tachometric device to the drive shaft in order to detect the absence of rotation or the failure to achieve normal running speed.

PHASE UNBALANCE PROTECTION

The loss of one phase of the supply or unbalanced supply voltages will have two effects. First, it will create a negative phase sequence current which will cause additional rotor heating. Second, it will cause excessive heat in the stator windings if the current in one or more phases exceeds the normal full-load value. The heating effect due to the negative sequence current can pose a greater threat than that due to the unbalanced phase currents, and therefore it is advisable, especially with larger machines, to provide a relay with negative phase sequence detection. The operation of the phase imbalance facility on the P&B Golds relay is described in the next section.

CONVENTIONAL RELAY TYPES

The most common of the conventional thermal overcurrent relay offshore is the P&B Golds relay. The relay consists of three tapped saturating core current transformers (CTs) which are fed from the main protection CTs. The secondary of each saturating core CT feeds a helical heating coil element which operates its own moving contact. The construction of the basic thermal overload element is shown in Fig. 4.3.1. The centre phase heater is used to provide '% Running Load' indication, overload trip when the moving contact touches the '% Load to Trip' contact and unbalanced load trip when contact with either of the outer phase contacts is made. Under healthy operating conditions, the out-of-balance contact cradle is horizontal so that no contacts are touching. However, if the motor load becomes unbalanced or single phasing occurs, the differential heating effect will cause the cradle to tilt over until the contacts touch and the motor contactor is tripped.

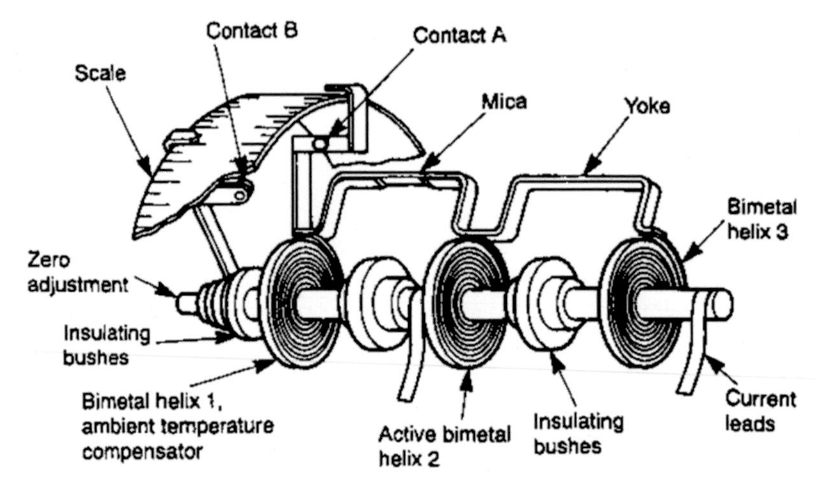

FIGURE 4.3.1

Basic thermal overload element of a P&B Golds relay.

Courtesy P&B Engineering Ltd.

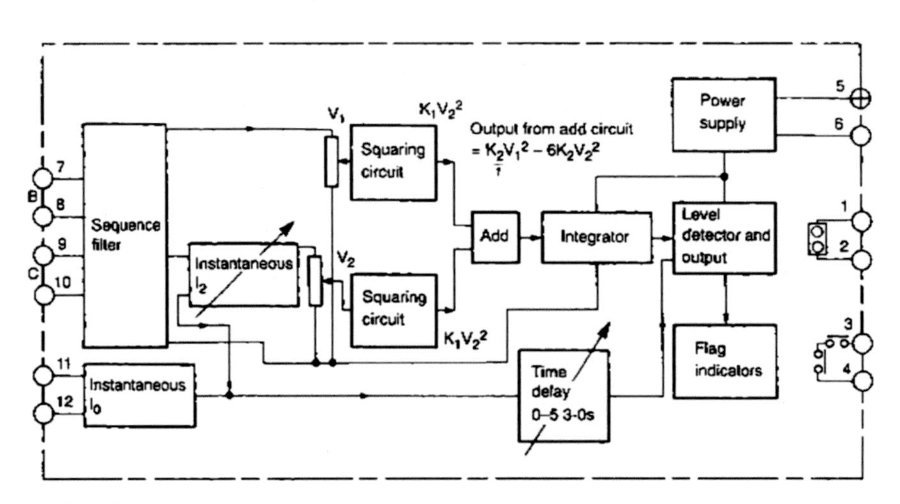

FIGURE 4.3.2

Schematic of a typical electronic thermal replica relay.

STATIC AND MICROPROCESSOR-BASED RELAY TYPES

Examples of static and microprocessor-based relay types currently available (in 2017) are shown in Table 4.3.1.

Table 4.3.1 Examples of Static and Microprocessor-Based Relay Types

Relay	Manufacturer
REM series, 49/50/51 series	ABB Limited, Daresbury Park, WA4 4BT, Warrington, Cheshire, UK (http://new.abb.com/medium-voltage/distribution-automation/numerical-relays).
A21M, ADR24 series, A21M	Ashida Electronics Pvt. Ltd., Ashida House, Plot No. A-308, Road No. – 21, Wagle Industrial Estate, Thane 400604, India (http://www.ashidaelectronics.com/index.php/portofolio/motor-protection-relay/).
BE1 series, ES series	Basler Electric (https://www.basler.com/SiteMap/Products/Protective-Relay-Systems/).
GMS7 series, IMM7 series, IMM8 series, NPM800, TMS series	CEE Relays, 87C Whitby Road, Slough, SL1 3DR, UK (http://www.ceerelays.co.uk/motorprotection.htm).
Multilin 469, M60,MM300	General Electric (http://store.gedigitalenergy.com/ProductCategory.aspx?CatId=3&ProdCatId=26).
MP8000 series, MPU-32 series, MPS series	Littelfuse, 8755 West Higgins Road, Suite 500, Chicago, Illinois 60631, USA (http://www.littelfuse.com/products/protection-relays-and-controls/protection-relays/motor-and-pump-protection.aspx).
MPC3000, MicroMotor, Motorvision, Primacon	P&B, Belle Vue Works, Manchester, M12 5NG, UK (http://www.pbsigroup.com/).
Micom series, Sepam series, Vamp series	Schneider Electric, Stafford Park 5, Telford, Shropshire, TF3 3BL, UK (http://www.schneider-electric.com/en/product-category/4700-Protection%20Relays%20by%20Range?filter=business-6-Medium%20Voltage%20Distribution%20and%20Grid%20Automation).
DMP	Orion Italia (http://www.orionitalia.com/protection-relays).
SEL7 series	SELINC, 2350 NE Hopkins Court, Pullman, WA 99163, USA (https://selinc.com/products/?Protective%20Relays.Motor%20Protection).
SIPROTEC series, Reyrolle 7SR Rho series	Siemens AG, Freyeslebenstrasse 1, 91058 Erlangen, Germany (http://w3.siemens.com/smartgrid/global/en/products-systems-solutions/Protection/motor-and-generator-protection/Pages/Overview.aspx).
NG10	Thytronic, Piazza Mistral 7, 20139 Milano, Italy (http://www.thytronic.it/index.asp?menu=3&submenu=7).

ADDITIONAL PROTECTION FOR SYNCHRONOUS MOTORS

The following additional protection devices may be required for synchronous motors:-

1. Pull-out protection
   If the compression torque for a reciprocating gas reinjection compressor, for example, exceeds the motor pull-out torque, the motor comes out of synchronism and stalls. To prevent the motor continuing to draw a heavy stator current, both the motor supply and the field supply are tripped.
2. Damper winding protection
   The damper windings are often designed to allow the machine to be run up as an induction motor. However, if an excessive number of starts are attempted in succession, the windings will overheat. A thermal relay should be provided, which will trip the motor supply before any damage is done.
3. Reverse power
   If there is a disturbance in the supply system or if the supply fails, the machine will generate using the inertia of the drive train for power. In this situation, the motor should be disconnected if the stability of the system is not regained after a short (timed) interval.

DETECTION OF MOTOR FAULTS ON LARGE MOTORS WITH THE ROGOWSKI COILS

The basis of motor current signal analysis is that the stator current contains current components directly linked to rotating flux components caused by electrical or mechanical faults. These harmonic current components caused by faults can be used for early failure detection. The Rogowski coil is a uniformly wound coil on a non-magnetic former of constant cross-sectional area formed in a closed loop. The coil surrounds the primary current conductor to produce a voltage that is proportional to the time derivative of the primary current and the mutual inductance of the coil. The Rogowski coils do not have ferromagnetic material in their cores, which means that they will never be saturated. An integration module may be used for fast Fourier transform analysis, where the large motor operates at a single speed. However, this is a disadvantage with a VSD. A band-pass filter is utilised to concentrate analysis on the harmonics which indicate the presence of faults, such as broken rotor bars. Refer to Reference 7.

CHAPTER

Busbar Protection

4

BUSBAR FAULTS

Busbar faults in medium- and low-voltage metalclad indoor switchgear are very rare in onshore installations. However, several incidents have occurred offshore, and in one instance where the risk of busbar faults was considered too low to warrant the installation of specific protection, tripping was initiated by fire detection rather than upstream electrical protection devices. Water from leaking pipes, or from driving rain or sea spray penetrating module walls, may enter switchgear enclosures, giving rise to explosive faults or serious fires. Arcing faults on the busbars of low-voltage switchgear will cause extensive damage to the switchboard and may lead to fire in an area of the installation likely to be close to accommodation modules or the process control room. However, the offshore installation of a full bus-zone protection scheme with check and supervision relays may only be advisable on the main switchboards of the largest platforms because of the weight of all the current transformers (CTs) required and the extra space taken up by such a scheme. Nevertheless, it is important that every part of the power system is adequately covered by the protection scheme, and the busbars of switchboards are no exception to this rule.

OVERCURRENT AND DIRECTIONAL OVERCURRENT PROTECTION

The overcurrent fault protection relays on the primary circuit of the transformer will provide some protection to the secondary circuit, but earth fault protection devices on the upstream side will not provide any secondary circuit protection because there is normally no zero-sequence path through the transformer. If the busbars are sectionalised, the bus-section switch may be fitted with overcurrent or directional overcurrent protection. Plain overcurrent relays operating on bus-section switches cannot, of course, remove the faulted section from the supply, unless it is being fed via another section. However, if the switchboard is fed by plain feeders or directly by generators with overcurrent protection, such an arrangement can provide reasonable overcurrent protection whilst preserving the supplies on healthy bus sections (see Fig. 4.4.1). If the switchboard has three sections, directional overcurrent relays, arranged to detect fault currents flowing outwards, may be fitted to the bus-section switches. Such an

Offshore Electrical Engineering Manual. https://doi.org/10.1016/B978-0-12-385499-5.00024-8

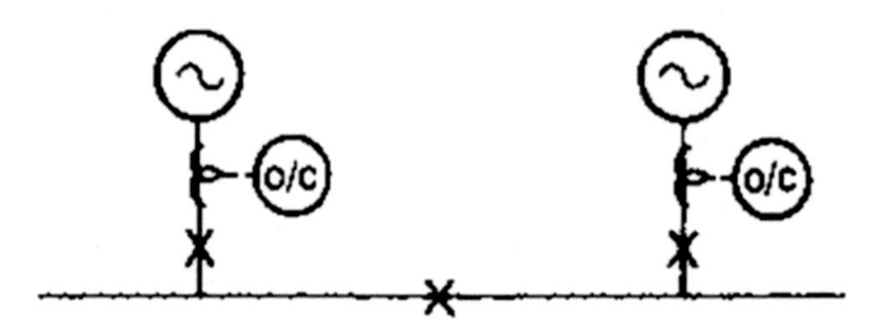

FIGURE 4.4.1

Simple plain overcurrent busbar protection scheme.

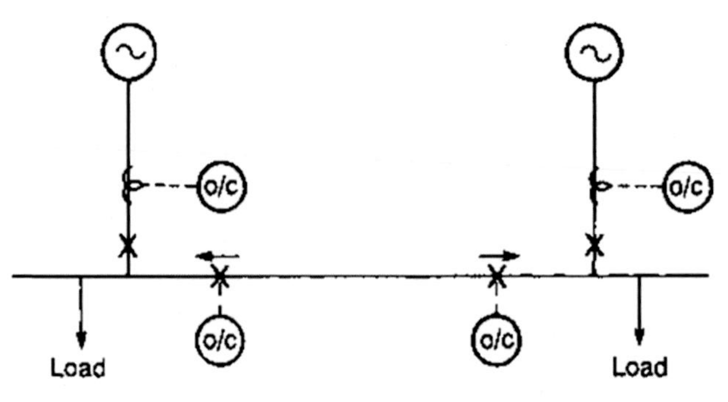

FIGURE 4.4.2

Simple directional overcurrent busbar protection scheme.

arrangement will preserve the two healthy sections, provided the fault is not on the central section. As central section faults rely on upstream protection for clearance, this section should be kept as short and uncomplicated as possible for good reliability. This arrangement could be used for the connection of a third generator, with all outgoing feeders (except third generator auxiliaries) fed from the end sections (see Fig. 4.4.2).

UNRESTRICTED EARTH FAULT PROTECTION

On main generator switchboards, either inverse definite minimum time or definite time earth fault characteristics may be used, but the use of definite time is often preferable, as fault current magnitudes are restricted by the generator neutral earthing resistors. Any relay used should be designed so as not to be sensitive to third harmonic currents, if the generator winding configuration is such that significant third harmonic currents may circulate. Low-voltage switchboards are more likely to suffer arcing faults than medium-voltage switchboards because with increasing voltage, the reduction in fault current due to arc resistance becomes less pronounced as the arc voltage ceases to be a significant proportion of the total fault circuit driving voltage (see Reference 1). Hence, the importance of earth fault protection for low-voltage switchboards is stressed.

FRAME EARTH PROTECTION

This type of system has been used successfully offshore and consists of a frame earth relay and neutral check relay arranged to monitor earth fault current flows into and out of the switchboard. A simple frame earth system is illustrated in Fig. 4.4.3. Care must be taken when installing the switchboard into the module such that it is and will remain insulated from the steel of the module construction. The generator earthing resistor and the switchboard frame should be earthed at the same point to avoid both series earth connections and increasing the risk of the fault current being limited below the operating setting of the relay by a poor connection due to corrosion of the decking steel. This is less likely to be a problem in an onshore substation using earth electrodes.

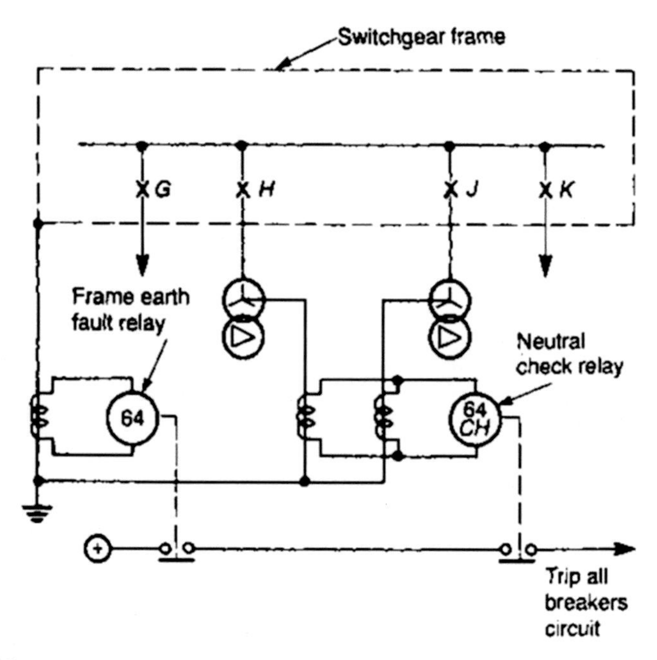

FIGURE 4.4.3

Busbar frame earth protection scheme.

Courtesy GEC Alsthom Measurements, now part of GE Grid Solutions.

However, earth connections should be maintained and checked for resistance value on a routine basis, particularly to prevent dangerous potentials appearing on the switchboard frame during fault conditions. An earth or bonding bar interconnects the framework of each cubicle, but is only earthed via the conductor monitored by the frame earth relay. Where the circuit breakers are in trucks, the truck must also have a heavy current earth connection to the bonding bar to allow earth fault current flow from the circuit breaker. To prevent currents induced in cables from causing spurious tripping, the gland plates at the switchboard end should be insulated from the switchboard frame but earthed to a second separate earth bar which is directly earthed to the main earth point.

DIFFERENTIAL PROTECTION

Kirchhoff's first law may be directly applied by monitoring incoming and outgoing currents at all the switchboard external circuit connections, summating them and detecting the error, if an internal switchboard fault exists. The switchboard may be divided into a series of zones, and each zone boundary should be treated in the same way, as shown in Fig. 4.4.4. The relay settings must be such that the transient flux produced by through-fault currents will not cause tripping because of the imbalance due to unequal burden or saturation of CTs. By the use of high-impedance relays with series stabilising resistors, the voltage across the relay during through-faults, known as the stability voltage, may be kept below the relay operating voltage. The minimum value of the stabilising resistor for stability may be calculated as follows:

$$V_S = \text{Relay setting current} \times (R_{SR} + R_R)$$

where V_S, stability voltage; R_{SR}, resistance of stabilising resistor and R_R, resistance of the relay coil.

An example calculation is given in PART 4 Chapter 7.

For fault calculations see PART 4 Chapter 6.

For relay discrimination topics see PART 4 Chapter 7.

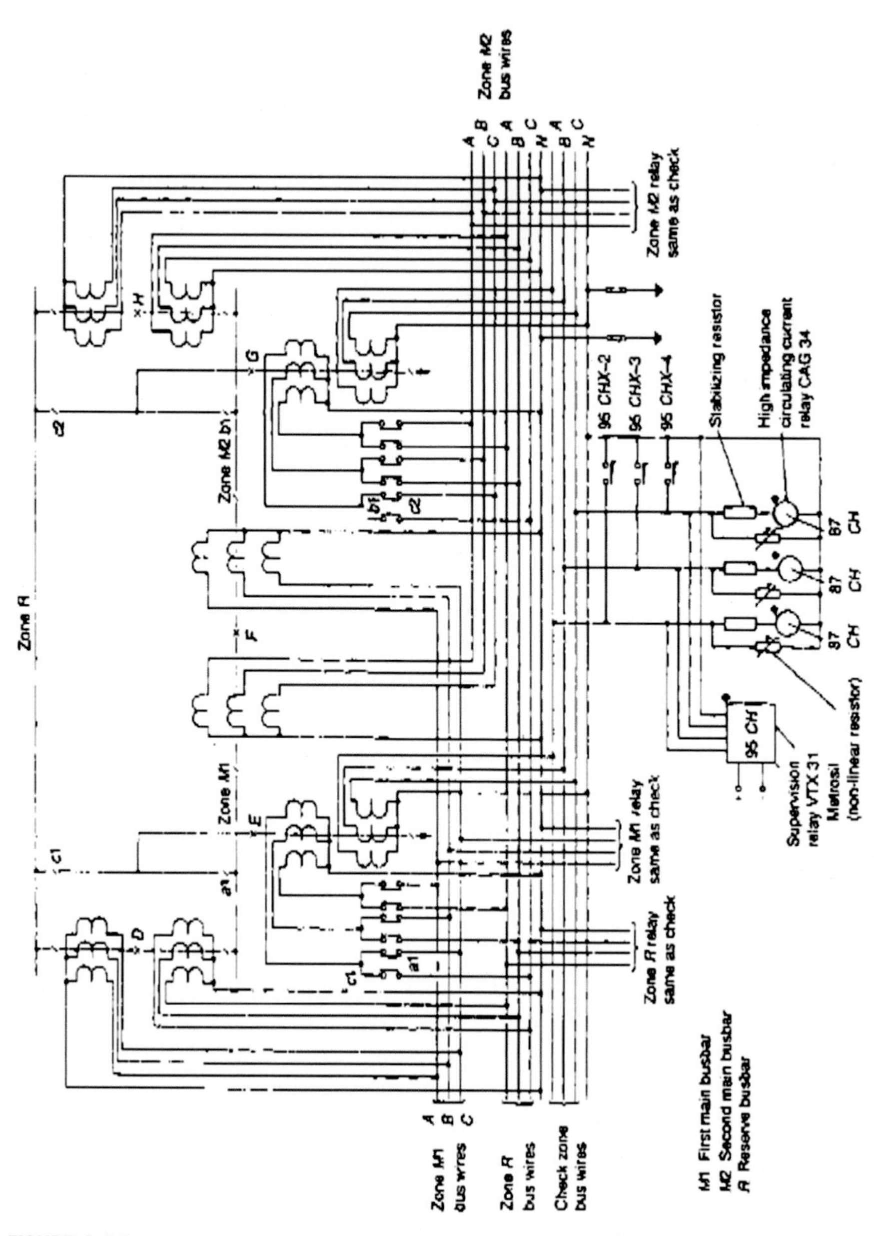

FIGURE 4.4.4

Typical busbar zone protection scheme.

Courtesy GEC Alsthom Measurements, now part of GE Grid Solutions.

CHAPTER

Feeder Protection, Conductor Sizing, Load Flow and Fault Calculation

5

FUSES

Low-voltage (LV) distribution systems in offshore installations are protected by fuses in the same way as their onshore counterparts. Although the Institution of Electrical Engineers (IEE) Wiring Regulations specifically excludes offshore installations from its scope (Part 1 paragraph II-3(vii)), the document that it refers to (*IEE Recommendations for the Electrical and Electronic Equipment of Mobile and Fixed Offshore Installations*) has been replaced by *BS IEC 61892, Mobile and fixed offshore units. Electrical installations*. The methods of calculation given in the *IEE Wiring Regulations* are normally adopted when designing lighting and small power distribution systems, particularly for accommodation modules.

The use and limitations of high rupturing capacity fuses are discussed in PART 2 Chapter 5.

MINIATURE CIRCUIT BREAKERS

In most situations, miniature circuit breakers (MCBs) may be used as an alternative to fuses where the fault rating of the MCB is sufficiently high. However, because of the different shape of the MCB tripping characteristic compared with that of a fuse (see Fig. 4.5.1), it is not advisable to mix fuses with MCBs in the same circuit if discrimination is required. The degree of discrimination between one MCB and another, and between MCBs and fuses, varies with the BS EN 60898-1 MCB type, but generally a better discrimination can be obtained using fuses.

MCBs with fault current ratings more than 16 kA are now available and are in common use offshore, where they save weight and reduce the quantity and type of spare fuses stocked. MCBs used offshore must provide positive indication of contact clearance and must be padlockable when in the open position.

OVERCURRENT AND EARTH FAULT PROTECTION

Most of the conventional forms of short cable feeder protection may be used offshore. The use of inverse definite minimum time (IDMT) overcurrent and residually connected earth fault relays may give discrimination problems if a large

Offshore Electrical Engineering Manual. https://doi.org/10.1016/B978-0-12-385499-5.00025-X

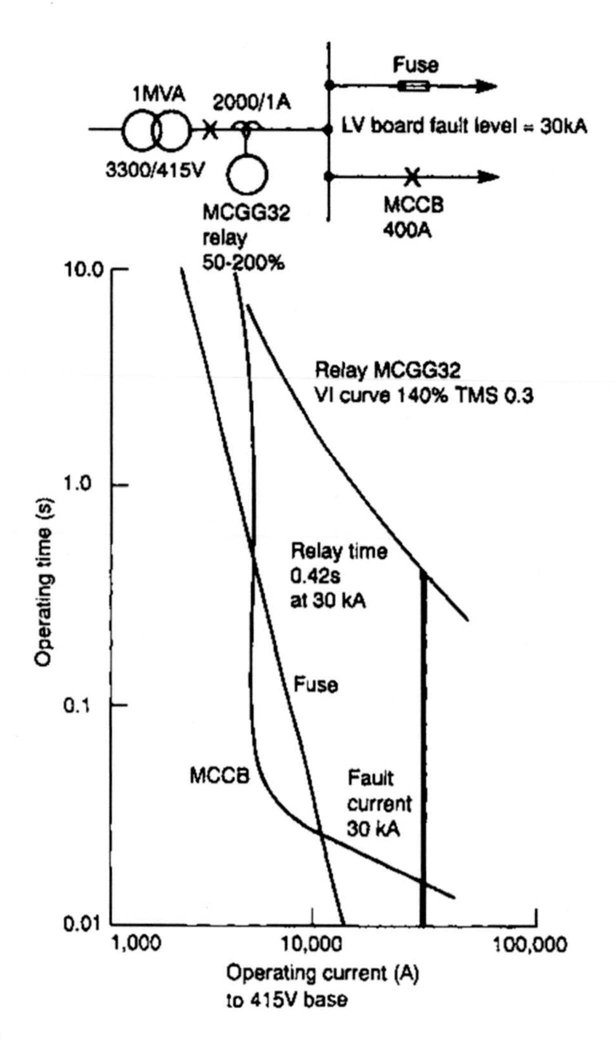

FIGURE 4.5.1

Miniature circuit breaker and fuse characteristic comparison. *LV*, low voltage; *MCCB*, moulded case circuit breaker.

Courtesy GEC Alstom Measurements, now GE Alstom Grid Solutions.

proportion of the platform load is on the switchboard being fed. This is because the steady-state fault current available from the platform generators will be of the order of three to four times the full-load current; therefore, if fault clearance times are to be kept short, generator and feeder relay current settings will need to be close. Voltage-controlled relays may be used on generator circuit breakers to overcome this problem, as discussed earlier for generator protection (see PART 4 Chapter 1).

An alternative is to use definite time relays. However, on LV systems, it is less feasible to use definite time relays because, if generator and main switchboard clearance times are to be kept reasonably short, there are likely to be discrimination problems with fuses and MCBs lower down in the system.

As cables are more exposed to mechanical damage than switchboard busbars, it is advisable to protect cables which interconnect switchboards by some form of simple unit protection, rather than IDMT relays with intertripping. This has the advantage of faster operation and may also relieve any discrimination problems associated with the unrestricted method of protection.

SIZING OF CONDUCTORS

LOAD FLOW

When the electrical distribution system has been configured for optimum convenience, safety and reliability, the various busbars and cables should be sized for the maximum continuous load in each system operating condition.

The first task is to ensure that the system 24-h load profile and the load schedule are as up to date as possible and that diversity factors and operating modes have been agreed by all parties and 'frozen'. If the system is simple, with few parallel paths, load flows may be manually calculated.

In either steady-state or transient conditions, the power system can be represented by a physical model, such as produced in a network analyser, or by a mathematical model using a digital computer. With the proliferation of desktop computers, the use of network analysers, even on small systems, is now rare.

On larger installations, with many parallel paths, computer load flow programs should be used in any case. Such programs are now available for use on desktop microcomputers at prices starting from a few hundred pounds. Load flow calculations by nodal analysis have become firmly established. Such methods involve

1. the solution of a set of linear simultaneous equations which describe the system configuration,
2. the application of restraints at each node to enable the required complex power and voltage conditions to be maintained.

The advantages in using nodal voltage analysis are that the number of equations is smaller than that with the alternative mesh current analysis method, and the system may be described in terms of its node numbers and the impedances of the interconnecting branches. In nodal analysis, the node voltages V are related to the nodal injected currents I by the system admittance matrix Y.

In the matrix form,

$$[I] = [\mathrm{Y}]\,[\mathrm{V}]$$

The voltage V refers to a value between node and earth, and the current *I* is the injected nodal current. The total nodal injected power S is obtained from the product of voltage and current conjugate, as follows:-

$$[S] = [V]\,[I]^*$$

By taking the current conjugate, the reactive power is given the same sign as active power for lagging current.

There are three basic types of nodal constraints:

1. fixed complex voltage
2. fixed complex power
3. fixed-voltage modulus with real power.

Type (1) constraint is given to the reference node, usually known as the 'slack' or 'swing' (in the United States) bus. The type (2) constraint represents a load bus; the type (3), a generation bus.

The formation of the nodal matrix and methods available for digital iterative sequences of solution are given in Bergen (1986).

BUSBAR SIZING

Switchboard main busbars must be rated to carry the maximum continuous load which can flow in any healthy power system operating condition. Transient conditions giving rise to higher currents, such as those due to large motors starting or downstream faults, may be tolerated momentarily, provided the protection devices are incorporated which will ensure that the outgoing equipment is removed from the system before the busbars get overheated.

The continuous current rating must be for the busbars as enclosed in their offshore environmental protection, with natural cooling only. This also applies to the switching and isolating devices in the switchboard.

CABLE SIZING

Cable sizing is generally carried out in accordance with current versions of the following:

IEC 61892: Mobile and fixed offshore units – Electrical installations:

Part 2: System design
Part 4: Cables
Part 6: Installation
IEC 60287: Electric cables – Calculation of the current rating.
IEC 92-201: Electrical installations in ships
Part 201: System design – General
Part 202: System design – Protection
Part 352: Choice and installation of electrical cables

For motor cables, the basis of the calculation is

$$L = \frac{(V_d \times 1000)}{(1.732 \times I \times (R\cos\Phi + X\sin\Phi)}$$

where L, cable length in metres; V_d, permitted steady-state volt drop in volts; I, motor full-load current; R, cable resistance in ohms per kilometre; X, cable reactance in ohms per kilometre and cosΦ, motor power factor at full load.

The permissible maximum steady-state volt drop is normally 2.5%, whilst the permissible volt drop during starting is 10%.

A typical computer spreadsheet generated motor cable sizing chart is shown in Fig. 2.8.1.

Derating factors for cables which pass through insulation and for bunching, application of protective devices, etc. need to be considered in accordance with the current edition of the IEE Wiring Regulations.

WORKED EXAMPLE: FAULT CALCULATION

The following calculations and information are not exhaustive but are intended to give the reader sufficient knowledge to enable switchgear of adequate load and fault current rating to be specified. The subject may be studied in more detail by reading the relevant documents listed in Appendix 1. The nomenclature used is generally as given in the GE Alstom Grid Solutions Network Protection and Automation Guide (NPAG) and the Electricity Council's Power System Protection (IET).

When a short-circuit occurs in a distribution switchboard, the resulting fault current can be large enough to damage both the switchboard and associated cables owing to thermal and electromagnetic effects. The thermal effects will be proportional to the duration of the fault current to a large extent, and this time will depend on the characteristics of the nearest upstream automatic protective device which should operate to clear the fault.

Arcing faults due to water or dirt ingress are most unlikely in the switchboards of land-based installations, but from experience, they need to be catered for offshore. For switchboards operating with generators of 10 MW or more, it is usually not difficult to avoid the problem of long clearance times for resistive faults. However, with smaller generators, clearance times of several seconds may be required because of the relatively low prospective fault currents available (see earlier section on busbar protection). With small emergency generators, pilot exciters are not normally provided and the supply for the main exciter is derived from the generator output. This arrangement is not recommended, as it allows the collapse of generator output current within milliseconds of the onset of a fault. With such small generators, even subtransient fault currents are small, and it is unlikely that downstream protection relays set to operate for 'normal' generation will have operated before the output collapse. It is usual to provide a fault current

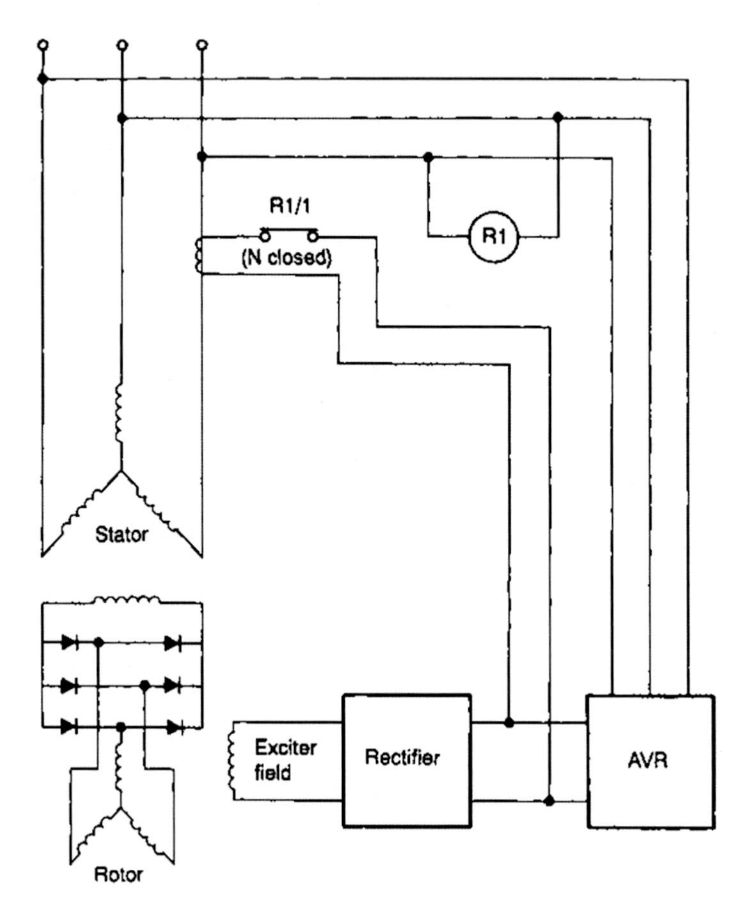

FIGURE 4.5.2

Schematic of small-generator fault current maintenance circuit. *AVR*, automatic voltage regulator.

maintenance unit as shown in Fig. 4.5.2. This device is basically a compounding circuit which feeds a current proportional to output current back to the exciter field. When the output current reaches a threshold value well above the normal load current, a relay operates, switching in the compounding circuit. Thus a high output current is maintained by this feedback arrangement until the operation of definite time overcurrent protection, which is set to prevent the generator thermal rating being exceeded.

MAIN GENERATOR FAULT CURRENTS

In this example, one generator is rated at 15.0 MW. The alternator has the following parameters: rating 17.7 MVA at a power factor of 0.85 and voltage of 6.6 kV;

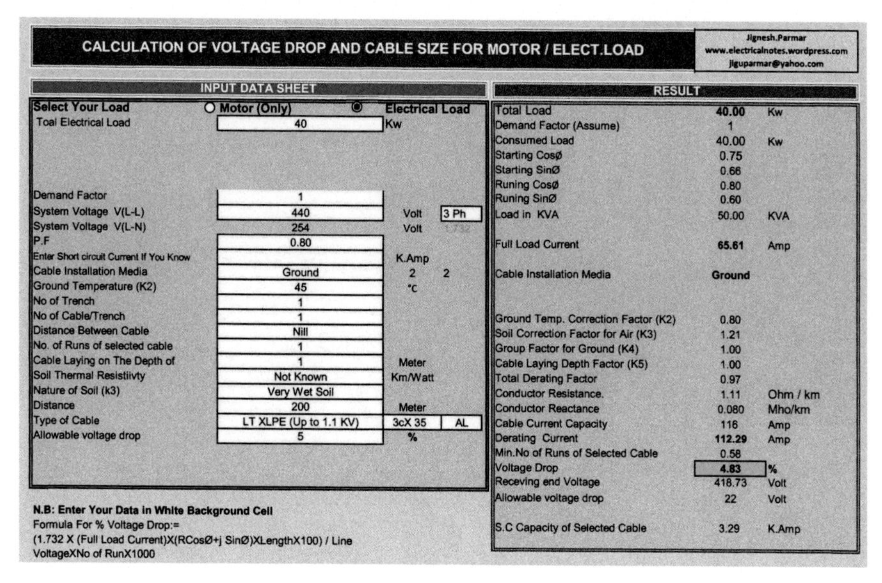

CALCULATION OF VOLTAGE DROP AND CABLE SIZE FOR MOTOR / ELECT.LOAD

Jignesh.Parmar
www.electricalnotes.wordpress.com
jiguparmar@yahoo.com

INPUT DATA SHEET

Select Your Load	Motor (Only) / Electrical Load	
Toal Electrical Load	40	Kw
Demand Factor	1	
System Voltage V(L-L)	440	Volt 3 Ph
System Voltage V(L-N)	254	Volt
P.F	0.80	
Enter Short circuit Current If You Know		K.Amp
Cable Installation Media	Ground	2 2
Ground Temperature (K2)	45	°C
No of Trench	1	
No of Cable/Trench	1	
Distance Between Cable	Nill	
No. of Runs of selected cable	1	
Cable Laying on The Depth of	1	Meter
Soil Thermal Resistiivty	Not Known	Km/Watt
Nature of Soil (k3)	Very Wet Soil	
Distance	200	Meter
Type of Cable	LT XLPE (Up to 1.1 KV)	3cX 35 AL
Allowable voltage drop	5	%

N.B: Enter Your Data in White Background Cell
Formula For % Voltage Drop:=
(1.732 X (Full Load Current)X(RCosØ+j SinØ)XLengthX100) / Line VoltageXNo of RunX1000

RESULT

Total Load	40.00	Kw
Demand Factor (Assume)	1	
Consumed Load	40.00	Kw
Starting CosØ	0.75	
Starting SinØ	0.66	
Runing CosØ	0.80	
Runing SinØ	0.60	
Load in KVA	50.00	KVA
Full Load Current	65.61	Amp
Cable Installation Media	Ground	
Ground Temp. Correction Factor (K2)	0.80	
Soil Correction Factor for Air (K3)	1.21	
Group Factor for Ground (K4)	1.00	
Cable Laying Depth Factor (K5)	1.00	
Total Derating Factor	0.97	
Conductor Resistance.	1.11	Ohm / km
Conductor Reactance	0.080	Mho/km
Cable Current Capacity	116	Amp
Derating Current	112.29	Amp
Min.No of Runs of Selected Cable	0.58	
Voltage Drop	4.83	%
Receving end Voltage	418.73	Volt
Allowable voltage drop	22	Volt
S.C Capacity of Selected Cable	3.29	K.Amp

FIGURE 4.5.3

Typical cable sizing spreadsheet (see Web Reference 2).

reactances $X_d'' = 0.195$ per unit (pu), $\times X_d' = 0.31$ pu, $X_d = 2.2$ pu and $X_q'' = 0.265$; full-load current, 1550A and neutral earthing resistor, 10Ω.

There are four more generators of a different type, each rated at 3.87MVA and having parameters as follows: reactances $X_d'' = 0.15$ pu, $X_d' = 0.18$ pu, $X_d = 2.05$ pu and $X_q'' = 0.15$; full-load current, 340A and neutral earthing resistor, 40Ω (Fig. 4.5.3).

The generator reactance values (pu) may be tabulated on their own base as follows:

MVA Unit	X_d''	X_q''	X_o''	R_n (Ω)
1×177	0.195	0.265	0.09	10
1×3.87	0.15	0.15	0.045	40

Referring to a common base of 17.7MVA, 6.6kV and considering the units in parallel:

MVA Unit	X_d''	X_q''	X_o''	R_n
1×3.87	0.686	0.686	0.2058	16.25
4×3.87	0.1715	0.1715	0.0515	4.06
1×17.7	0.195	0.265	0.09	4.06

Summing the reactances,

$$X_d'' = \frac{1}{1/0.195 + 1/0.1715} = 0.0912 \text{ pu}$$

$$X_q'' = \frac{1}{1/0.265 + 1/0.1715} = 0.104 \text{ pu}$$

$$X_o'' = \frac{1}{1/0.09 + 1/0.0515} = 0.0327 \text{ pu}$$

$$R_n = \frac{1}{1/4.06 + 1/4.06} = 2.63 \text{ pu}$$

Hence,

$$X_d'' + X_q'' = 0.0912 + 0.104 = 0.1952 \text{ pu}$$
$$X_d'' + X_q'' = 0.912 + 0.104 + 0.0327 = 0.2279 \text{ pu}$$

$$|Z_1| = (X_d'' + X_q'' + X_o'')^2 + (3R_n)^2$$

$$Z_1 = (0.2279)^2 + (3 \times 2.03)^2 = 6.09 \text{ pu}$$

Therefore the system fault currents (neglecting motor contribution) are

$$\text{Three-phase fault current} = \frac{17.7}{3 \times 6.6 \times 0.0912} = 16.97 \text{ kA}$$

$$\text{Phase-to-phase fault current} = \frac{17.7}{6.6 \times 0.1952} = 13.74 \text{ kA}$$

$$\text{Earth fault current} = \frac{3 \times 17.7}{6.6 \times 6.09} = 0.762 \text{ kA}$$

This method of calculation provides low values which, although not suitable for providing fault currents for the selection of switchgear, are useful for relay setting, as neglecting motor fault contributions provides a safety margin. The extra fault current provided by motors should significantly reduce the operating times of all overcurrent and earth fault protection devices, provided the current transformer saturation is not a problem. A check should be made, however, that generator cable reactances do not reduce prospective fault currents by more than a negligible amount. This reduction is likely to be significant for the smaller generators operating at LVs.

SWITCHBOARD FAULT CURRENTS

The prospective fault currents for medium-voltage switchboards can be obtained in a similar manner by summing generator and transformer reactances as follows. Take the base again as 17.7 MVA, and take as an example the generator switchboard connected to the LV switchboard by a 2 MVA transformer with 440 V secondary winding. Then the sequence reactances (pu) are as follows:

	X_1	X_2	X_0
At transformer rating	0.13	0.11	0.11
On the base MVA	1.1505	0.974	0.974

Summing generator and transformer reactances,

$$X_1 = 0.0912 + 1.1505 = 1.2417 \text{ pu}$$
$$X_2 = 0.104 + 0.974 = 1.078 \text{ pu}$$
$$X_0 = 0.974 = 0.974 \text{ pu}$$

Hence,

$$(X_1 + X_2) = 1.2417 + 1.078 = 2.3197 \text{ pu}$$
$$(X_1 + X_2 + X_0) = 1.2417 + 1.078 + 0.974 = 3.294 \text{ pu}$$

Therefore the LV system fault currents are

$$\text{Three - phase fault current} = \frac{17.7}{3 \times 0.44 \times 1.2417} = 18.7 \text{ kA}$$

$$\text{Phase - to - phase fault current} = \frac{17.7}{0.44 \times 2.3197} = 17.34 \text{ kA}$$

$$\text{Earth fault current} = \frac{3 \times 17.7}{0.44 \times 3.294} = 21.15 \text{ kA}$$

The workings required for protection settings of the same scenario are continued in PART 4 Chapter 7.

CHAPTER

Calculation of Load Flow, Prospective Fault Currents and Transient Disturbances

6

FAULT CALCULATION

The following calculations and information are not exhaustive but are intended to give the reader sufficient knowledge to enable switchgear of adequate load and fault current rating to be specified. The subject may be studied in more detail by reading the relevant documents listed in Appendix 1. The nomenclature used is generally as given in 'Power System Protection' (IET).

When a short circuit occurs in a distribution switchboard, the resulting fault current can be large enough to damage both the switchboard and associated cables due to thermal and electromagnetic effects. The thermal effects will be proportional to the duration of the fault current to a large extent and this time will depend on the characteristics of the nearest upstream automatic protective device which should operate to clear the fault.

Arcing faults due to water or dirt ingress are most unlikely in the switchboards of land-based installations, but from experience, they need to be catered for offshore. For switchboards operating with generators of 10 MW or more, it is usually not difficult to avoid the problem of long clearance times for resistive faults. However, with the smaller generators clearance times of several seconds may be required because of the relatively low prospective fault currents available. (See PART 4 Chapter 4) With small emergency generators, pilot exciters are not normally provided and the supply for the main exciter is derived from the generator output. This arrangement is not recommended, as it allows the collapse of generator output current within milliseconds of the onset of a fault. With such small generators even sub-transient fault currents are small, and it is unlikely that downstream protection relays set to operate for 'normal' generation will have operated before the output collapse. It is usual to provide a fault current maintenance unit as shown in Fig. 4.6.2. This device is basically a 'compounding' circuit which feeds a current proportional to output current back to the exciter field. When the output current reaches a threshold value well above normal load current, a relay operates, switching in the compounding circuit. Thus a high output current is maintained by this feedback arrangement until definite time overcurrent protection, set to prevent the generator thermal rating being exceeded, operates.

A worked example for use in setting Main Generator Protection is given in PART 4 Chapter 5.

Offshore Electrical Engineering Manual. https://doi.org/10.1016/B978-0-12-385499-5.00026-1

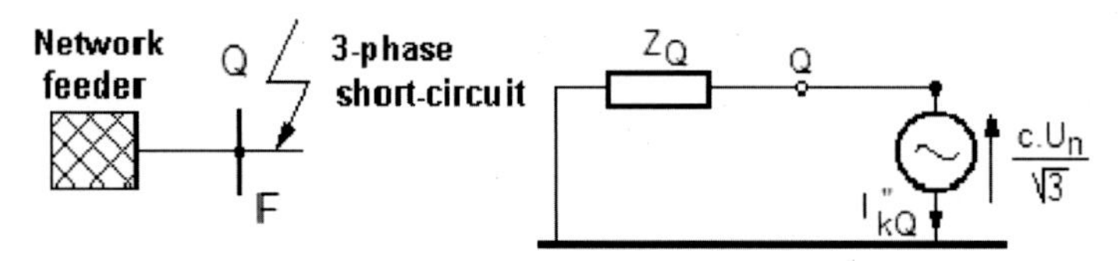

FIGURE 4.6.1

Equivalent voltage source method.

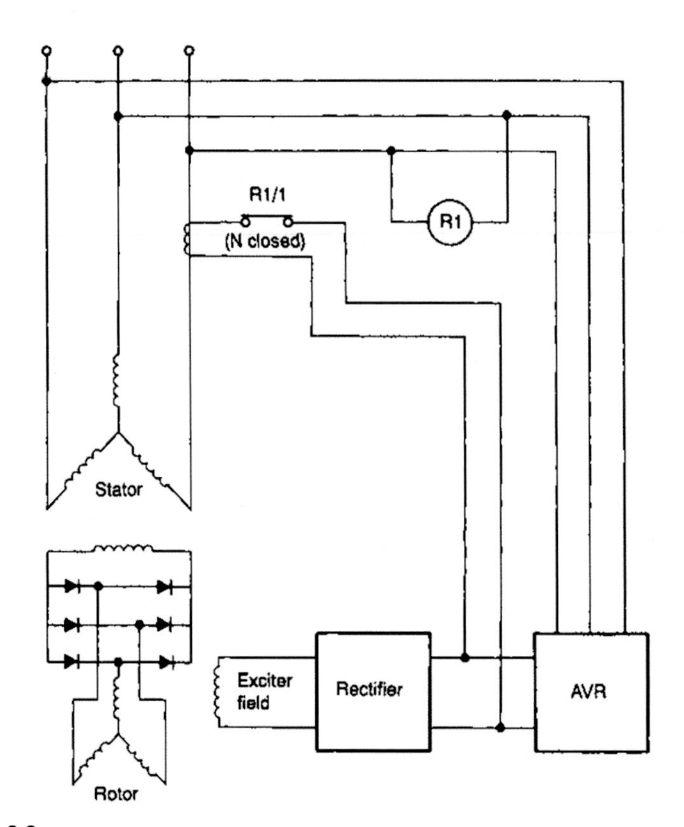

FIGURE 4.6.2

Schematic of small generator fault current maintenance circuit.

STANDARD METHODS OF CALCULATION

There are several ways to calculate the short circuit current for a marine electrical system, some very simple as the example in PART 4 Chapter 7, others more complex are listed in the table below. The complexity of the calculation is not always an indication as to the efficacy of the result.

Type of Circuit	Standard Specification	Method of Calculation	
		Breaking Current	Making Current
Systems of ships and mobile & fixed offshore units	IEC 61363	Ac decrement based on generator transient impedance	Dc and ac decrements calculated
Primarily for land-based systems	IEC 60909	Use of c and k factors. Refer to IEC TR 60909-1	
All systems	IEEE 141/ANSI C37	Use of impedance correction factors for rotating machines can significantly affect results.	

IEC 61363

Calculation methods that include generator and motor short circuit decrement will tend to produce the lowest acceptable values of short circuit current. As ship and offshore systems normally consist of 'Island' generation where machine decrement has significant effect, IEC Standard 61363-1, usually, produces the most accurate result, i.e., it produces study results that represent conditions that may affect typical marine or offshore installations more significantly than land-based systems, including more emphasis on generator and motor decay.

Where feeder cable impedance tends to reduce decrement effects, the study can take advantage of the Equivalent Generator approach outlined in IEC 61363-1 section 7. For more simple marine electrical systems, the equivalent generator method will involve extensive calculations and produce results little different from more simple methods. In this case, IEC 60909 may be a better choice for your study.

IEC 60909

The calculation method, used in the IEC Standard 60909 determines the short circuit currents at the location F using equivalent voltage source:

$$\frac{\mathbf{cUn}}{\sqrt{3}}$$

This source is defined as the voltage of an ideal source applied at the short circuit location in the positive sequence system, whereas all other sources are ignored. All network components are replaced by their internal impedances (see Fig. 4.6.1).

In the calculation of the maximum short circuit currents, a voltage factor c is assumed (from IEC60038) for $c_{max,}$ for Low (Table 1), Medium (Table 2) and High (Table 3) voltage levels. According to Fig. 4.6.1, in the case of balanced short circuits, the initial symmetrical short circuit current is calculated by:

$$I_k'' = \frac{cUn}{\sqrt{3}.ZQ}$$

IEEE141/ANSI37

The decrement factors used by the ANSI SCC give reasonable decays provided the X/R ratios are within the intervals specified in the IEEE *Application Guide for AC High Voltage Circuit Breakers Rated on a Symmetrical Current Basis*.

Your client may specify which study method is to be used, but if the installation is mobile, then it would be advisable to follow a global standard such as ABS or DNV GL for guidance on which study method to use.

DIGITAL METHODS OF FAULT CALCULATION

In digital fault calculation, an admittance matrix is formed which is extended to include the source admittances, and the matrix is then reduced to a single impedance connected between the neutral (zero node) and the point of fault.

DIGITAL SIMULATION OF SYSTEM DISTURBANCES

When designing large offshore power systems, it is vital that the response of the system to the starting of large motors, and the application of phase-to-phase or earth faults, have been analysed to a reasonable degree of certainty, so that the risk of dangerous operating conditions and system instability can be avoided as much as possible.

The digital computer programs used in such analysis have the following elements:

1. The program will obtain its basic system data, including models of all generators, governors, AVRs etc. and network configuration, from data stored in various input files. Models for such devices as AVRs, prime movers etc. are in the form of differential equations.
2. Having obtained the necessary input data and initialized the system, the program uses a step-by-step implicit trapezoidal method of simultaneously solving the large number of differential equations involved. Iterative methods are used at each step to obtain the desired accuracy (usually 99.95%).
3. After a defined simulated period, the program stops and creates an output file containing the values of all variables calculated at each step.
4. For faster and more user-friendly interpretation of this output data, graphics subroutines are used to plot the variables against time.

Suitable program suites are available commercially from the organizations listed in the Bibliography section.

TRANSIENT SIMULATIONS AND HARMONIC ANALYSIS

Most of the commercial programs available have other modules, including simulation of generator outages and transient system upsets such as the starting of large electrical drives. It is essential that such studies are carried out prior to specifying main generation and switchgear ratings in order to ensure the system is going to be stable.

If it is expected that a large proportion of the electrically driven machinery is of variable frequency or variable speed, harmonic analysis is also essential in order to calculate the percentage Total Harmonic Distortion (%THD). It is important that the grouped capacitance of lighting and any other large capacitive load is included in the model.

The following are examples of short circuit analysis and other outputs from three proprietary programs available. There are a many more programs available on the market, some of which are available free from switchgear manufacturers. The author is not recommending any particular software, and to some extent, suitability of the program will depend on:

- The particular engineer's experience
- What particular graphic outputs are required
- Whether other results such as harmonic analysis, arc-flash studies, relay discrimination graphs, etc., are required
- The hardware and software available for the program to run on
- Client and project preference

ETAP SHORT CIRCUIT ANALYSIS SOFTWARE

Etap.com Operation Technology, Inc. ©2017.

ETAP Short Circuit software program allows for fault calculations based on ANSI/IEEE, IEC and GOST standards. The user-friendly interface, combined with powerful analysis engine, saves hours of tedious hand calculations and takes the guesswork out of fault analysis by automating the process with multiple calculation and result analysis tools.

ETAP Short Circuit Software Key Features:

• IEC Standard 60909 • IEC Standard 61363	• Determine worst case device duty results
• ANSI/IEEE Standards C37 & UL 489	• Display critical and marginal alerts
• GOST (Russian) Standards R-52735	• Load terminal short circuit calculation
• Unbalanced L-G, L-L, and L-L-G faults analysis	• Generator circuit breaker evaluation
• Device evaluation for 3- and 1-phase systems	• Comprehensive short circuit plots & reports

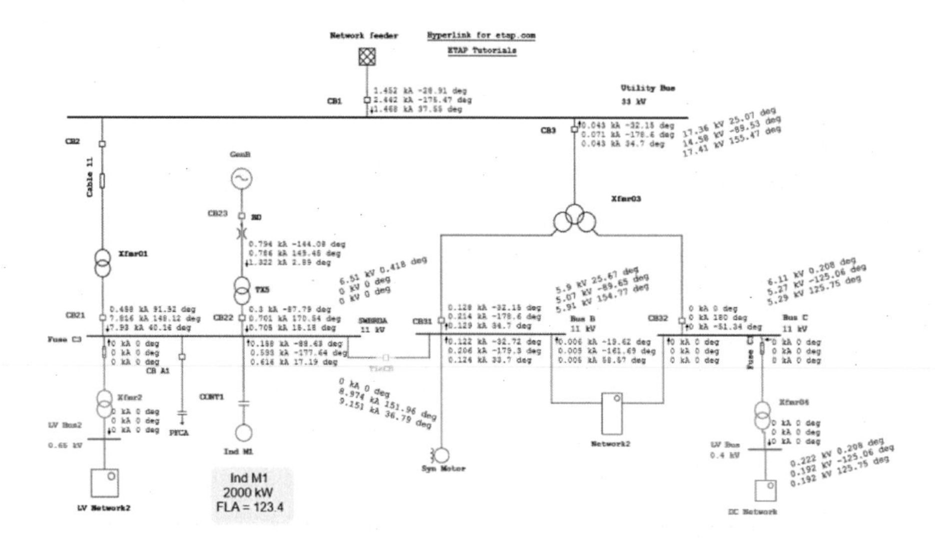

Built-in intelligence allows it to automatically apply all factors and ratios required for high- and low-voltage device duty evaluation. Overstressed device alarms are displayed on the one-line diagram and reports. The short circuit module seamlessly integrates with the ETAP Protection & Coordination and Arc Flash Analysis programs.

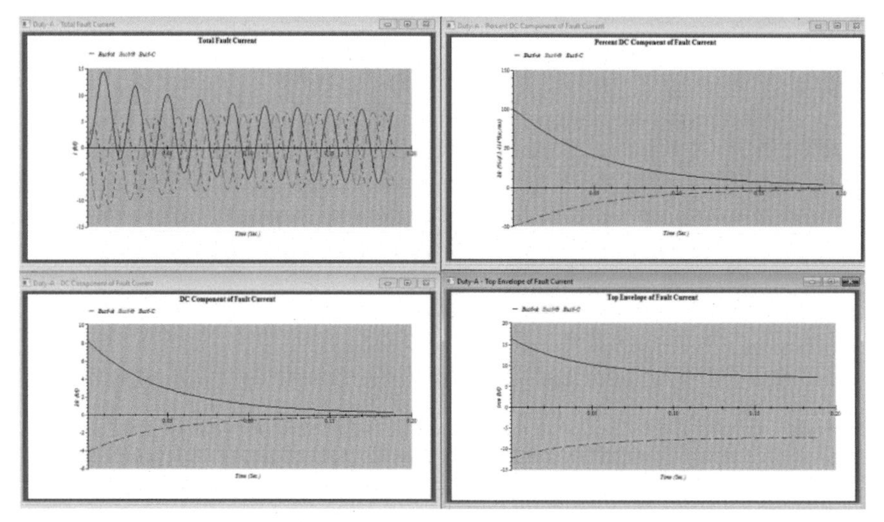

IEC 61363 individual phase fault current.

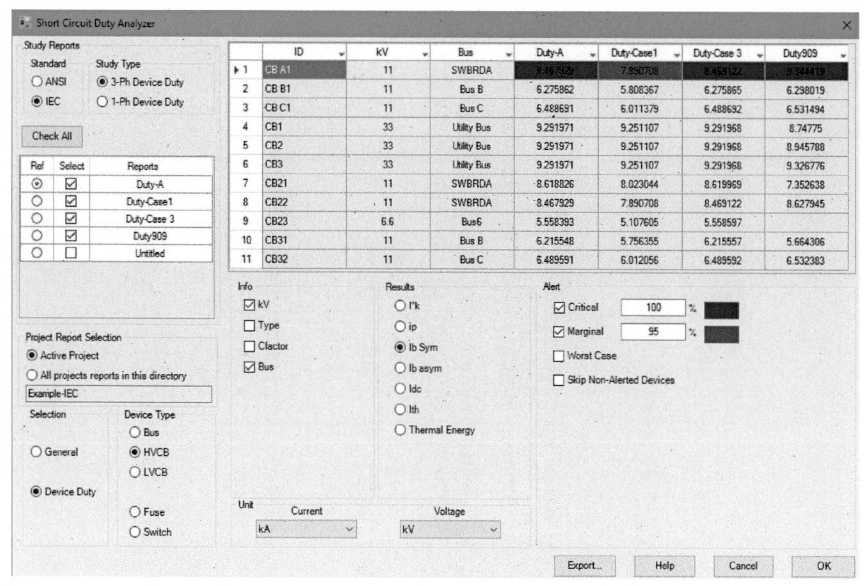

Report Analyser to Compare Studies and Identify Worst Case.

ETAP ANSI Short Circuit Features

- Determine maximum and minimum short circuit fault currents
- Calculate ½ cycle, 1.5–4 and 30 cycle balanced and unbalanced faults
- Check momentary and interrupting device capabilities
- Check closing & latching capabilities
- Evaluate symmetrical or total rated circuit breakers
- Special handling of generator circuit breakers
- Interrupting duty as a function breaker contact parting time
- Standard and user-definable contact parting time
- Automatically includes No AC Decay (NACD) ratio
- User options for automatic adjustment of HVCB rating

ETAP IEC Short Circuit Features

- Unbalanced L-G, L-L and L-L-G faults analysis
- Transient short circuit calculations to generate individual A, B and C phase currents
- Compares device ratings with calculated short circuit values
- User-definable voltage C factor
- Service or ultimate short circuit current ratings for LVCB breaking capability
- User-definable R/X adjustment methods for Ip (method A, B, or C)
- Negative or positive impedance adjustments for max/min I"k and Ik
- Automatic application of K correction factors (i.e., KT, KG, KSO)
- Automatically determines meshed and non-meshed networks for calculating Ib, Ik and Idc
- Considers both near and far from generator short circuits

Generator Circuit Breaker Evaluation per IEEE C37.013

Short Circuit Duty Summary Report										
Generator Circuit Breaker										
3-Phase Fault Currents: (Prefault Voltage = 102% of the Bus Nominal Voltage)										
Circuit Breaker			Momentary Duty (@ 1/2 Cycle)				Interrupting Duty (@ Contact Parting Time)			
ID	Type	Source and Generator PF	Symm. kA rms	Asymm. kA rms	Asymm. kA Peak	Symm. kA rms	Asymm. kA rms	Symm. kA Peak	DC Fault Current (kA)	Degree of Asymm. (%)
CB4	5 cy Sym CB	Gen-lagging PF	1.853	4.575	6.803	1.809	3.517	2.559	3.016	117.85
		Gen-leading PF	1.558	4.472	6.396	1.518	3.382	2.146	3.022	140.81
		Gen-no load	1.628	4.463	6.458	1.583	3.388	2.239	2.996	133.82
		Sys-lagging PF	8.660	14.155	23.444	8.625	11.228	12.198	7.188	58.93
		Sys-leading PF	8.969	14.660	24.280	8.935	11.631	12.635	7.446	58.93
		Sys-no load	8.867	14.492	24.003	8.828	11.492	12.485	7.358	58.93

UNBALANCED SHORT CIRCUIT ANALYSIS FOR MULTIPLE AND SINGLE PHASE SYSTEMS

The ETAP Unbalanced Short Circuit module applies 3-phase system modeling to represent unbalanced 3-phase systems, as well as 1-phase systems including panel, UPS and distribution transformers. It fully captures the effect of unbalanced systems, including unbalanced loads, transmission components, special distribution transformers and coupling between transmission lines.

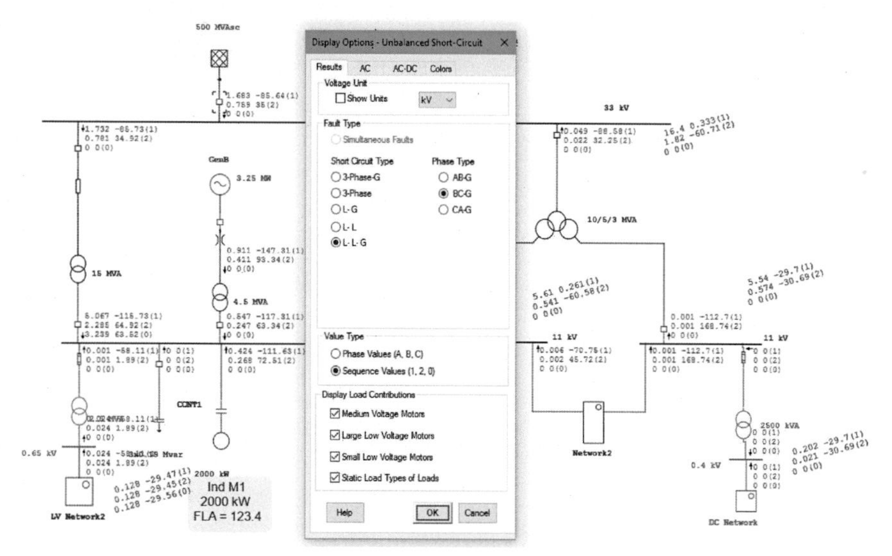

Display fault current in phase/sequence for any fault types.

IPSA SHORT CIRCUIT ANALYSIS SOFTWARE

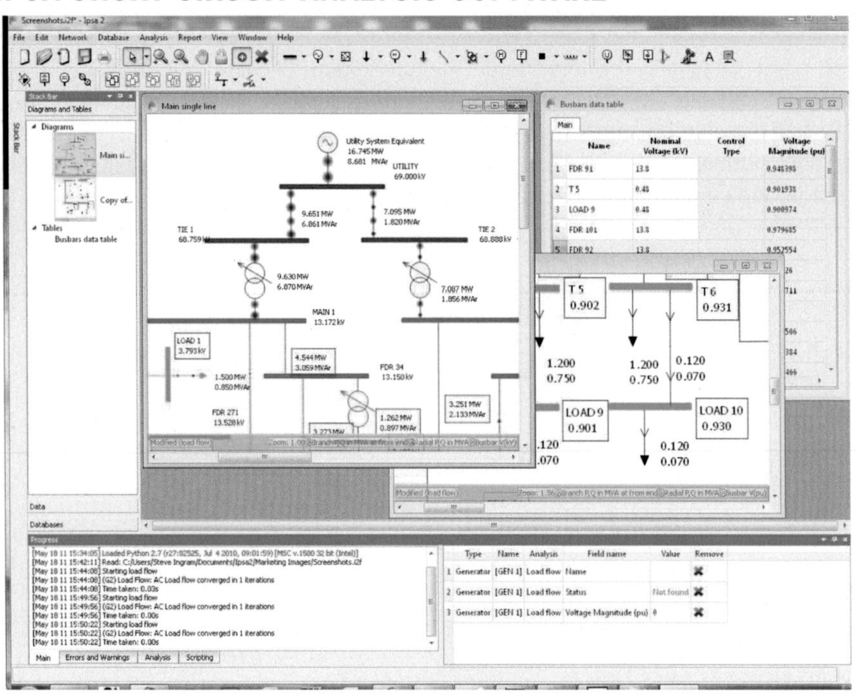

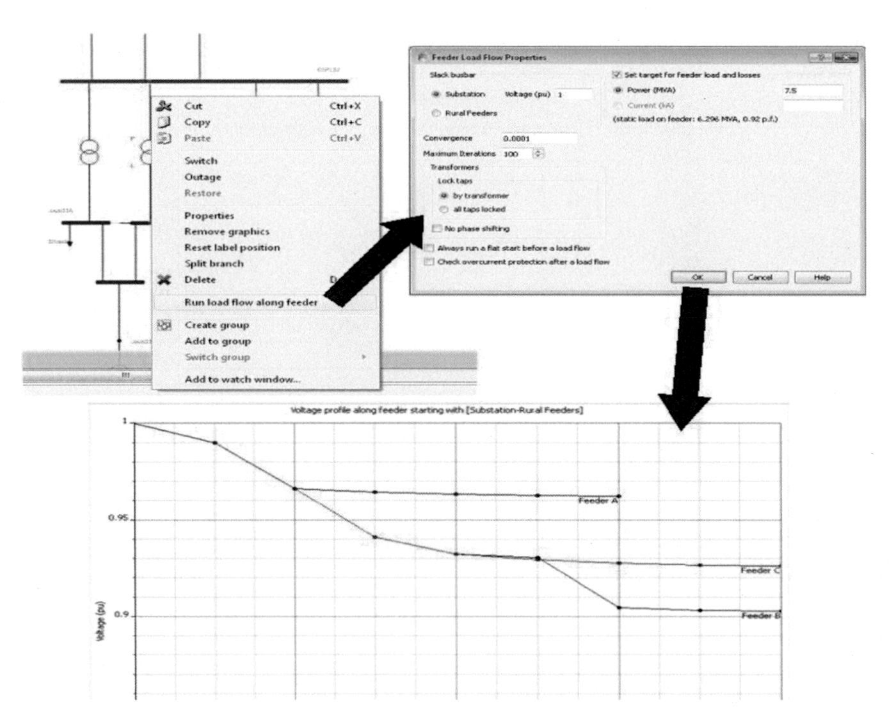

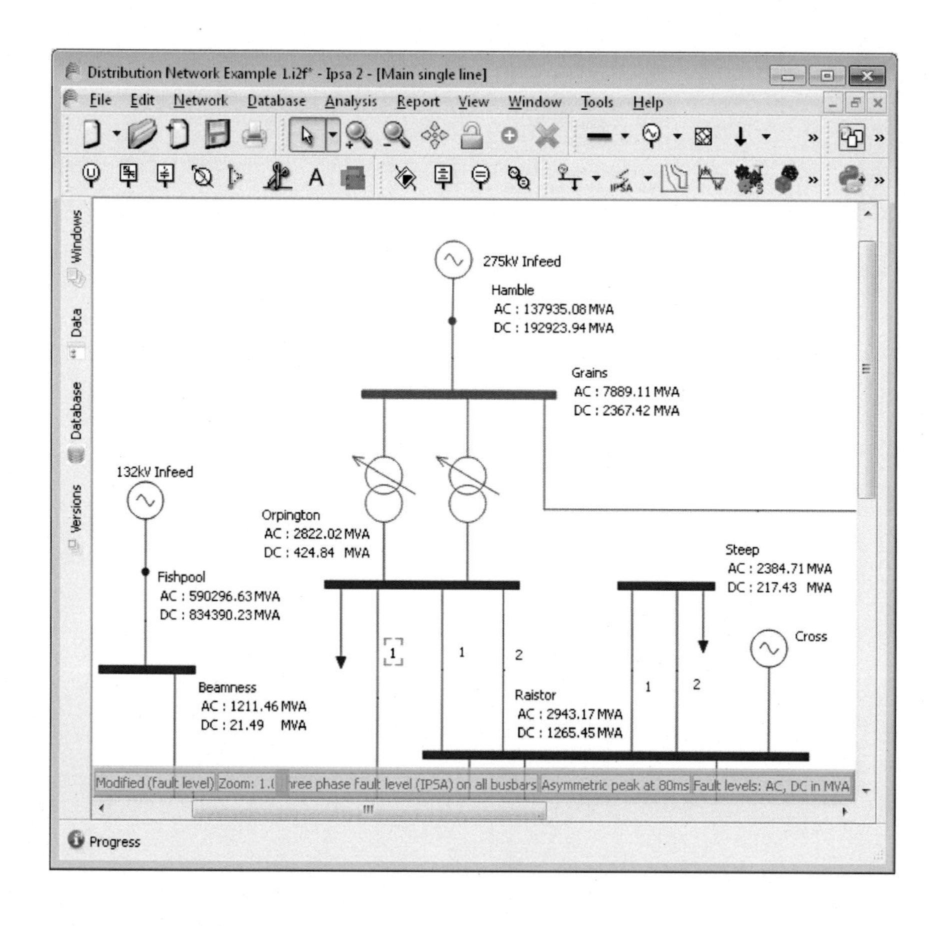

Distribution Network Example 1.i2f* - Ipsa 2 - [Main single line]
File Edit Network Database Analysis Report View Window Tools Help
Windows
Data
Database
Versions
275kV Infeed
Hamble
AC : 137935.08 MVA
DC : 192923.94 MVA
Grains
AC : 7889.11 MVA
DC : 2367.42 MVA
132kV Infeed
Orpington
AC : 2822.02 MVA
DC : 424.84 MVA
Fishpool
AC : 590296.63 MVA
DC : 834390.23 MVA
Steep
AC : 2384.71 MVA
DC : 217.43 MVA
Cross
Beamness
AC : 1211.46 MVA
DC : 21.49 MVA
Raistor
AC : 2943.17 MVA
DC : 1265.45 MVA
Modified (fault level)
Asymmetric peak at 80ms
Fault levels: AC, DC in MVA
Progress

POWER TOOLS FOR WINDOWS

Power*Tools for Windows (PTW) Software

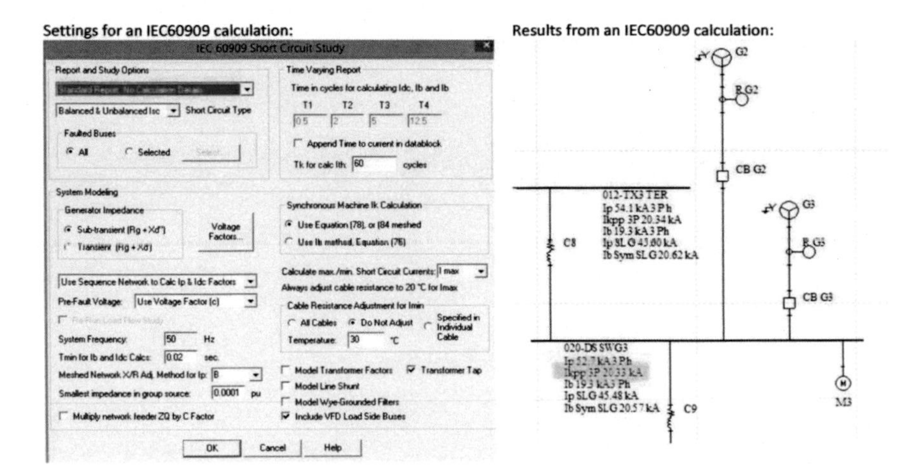

Results from the IEC60909 calculation automatically used in protection coordination:

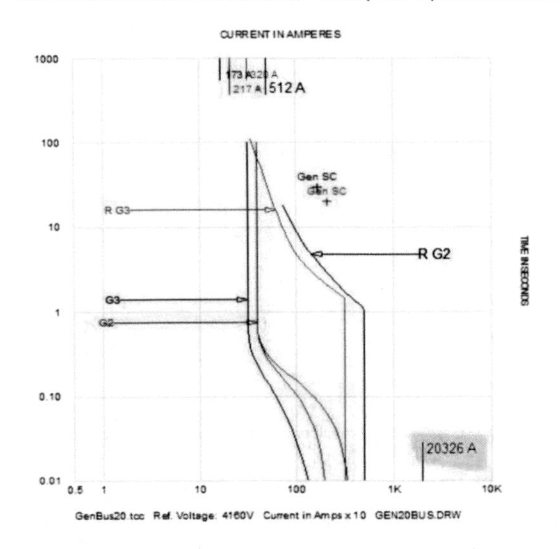

www.ceerelays.co.uk +44 1753 576477

Power*Tools for Windows (PTW) Software

A flashover study to IEEE1584 using results taken automatically from the IEC60909 calculation:

Bus Name	Protective Device Name	Bus kV	Bus Bolted Fault (kA)	Bus Arcing Fault (kA)	Prot Dev Bolted Fault (kA)	Prot Dev Arcing Fault (kA)	Trip/ Delay Time (sec.)	Breaker Opening Time/Tol (sec.)	Ground	Equip Type	Gap (mm)	Arc Flash Boundary (mm)	Working Distance (mm)	Incident Energy (cal/cm2)
020-DS SWG3	R6	4.00	20.33	19.49	10.54	10.11	0.015	0.0830	Yes	SWG	104	1784	910	2.3
020-DS SWG3	R G2	4.00	20.33	19.49	2.26	2.16	2.89	0.1000	Yes	SWG	104	10685	910	13
020-DS SWG3	R G3	4.00	20.33	19.49	1.02	0.98	5.81	0.1000	Yes	SWG	104	13898	910	17

Switchgear label created automatically by PTW:

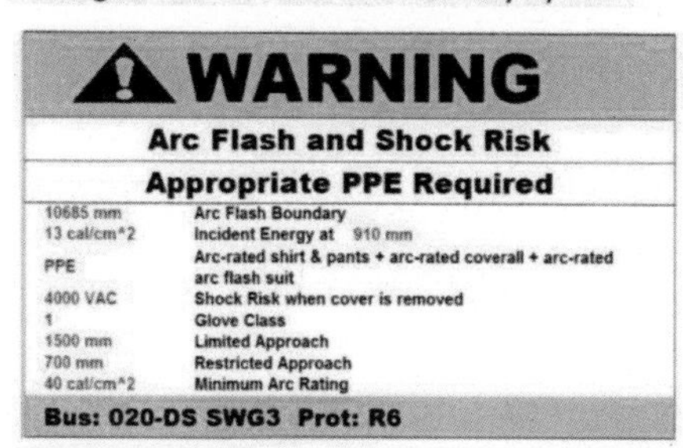

The Arc Flash Evaluation module in PTW automatically calculates the arcing fault current (normally using IEEE1584) and then uses the protection settings from CAPTOR to calculate the fault clearance time.

The incident energy is then calculated from these values, allowing PPE to be specified.

Dynamic generator simulation:

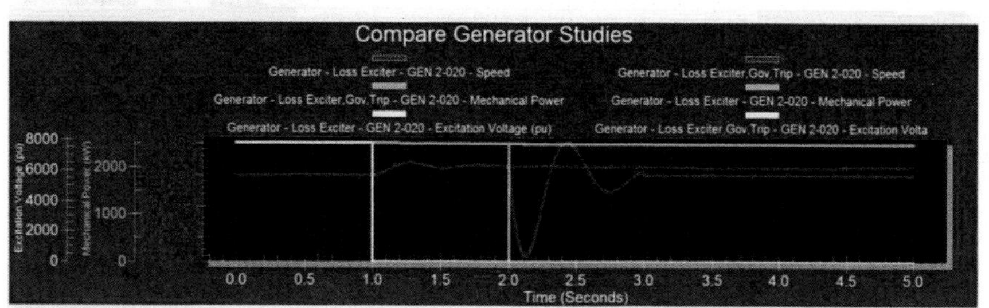

Harmonic Analysis:

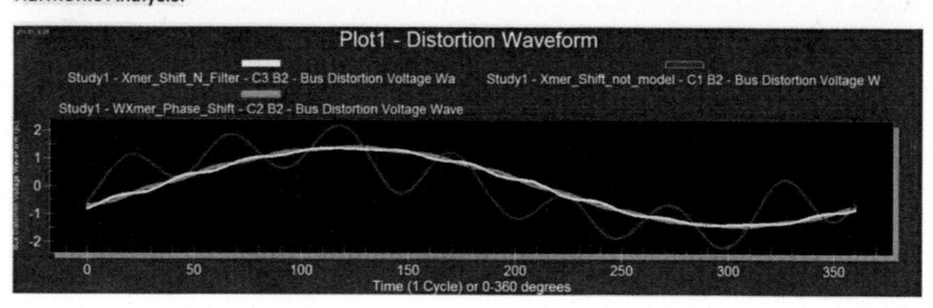

PROVING THE SOFTWARE

A set of example manual fault calculations with full workings should be available, for comparison with a set of input data (generated by the proposed software) and output results printout, (**using identical input data**) and carried out with the software, which is intended will be used for the proposed study.

This 'proving' example should be run on the hardware and software (e.g., Windows 10) intended to be used in the study, even if it has been run successfully before on other hardware.

Such calculations are essential to ensure that:

- The software is operating properly in the environment it is running on (e.g., Windows 10).
- The client can be confident of the software and consequently more confident of the results.

MANUAL CALCULATIONS

The engineer running the study needs to have a reasonable indication as to results expected and can check the input network and data before running the program, or rerunning the program if not happy with the initial result.

It is recommended that programs such as MatLab or Mathcad are used for the 'manual' calculation rather than a spreadsheet template, as spreadsheets may introduce rounding errors.

With such programs the workings can be demonstrated more easily by adding explanatory notes etc.

Please refer to PART 4 Chapter 5 for an example fault calculation.

Please refer to PART 4 Chapter 7 for example relay studies.

CHAPTER

Protection and Discrimination

7

INTRODUCTION

The worked example below is a continuation of the example in PART 4 Chapter 5 where Generator and Switchboard fault currents were manually calculated. Note that PART 4 Chapter 6 gives some guidance on calculation standards and commercial software available.

WORKED EXAMPLE (CONTINUED): RELAY SETTING OF TYPICAL MV PLATFORM SCHEME

The full load current of the largest motor on the system is 600 A. The starting current of this motor is 1300 A. (Note that this is with an autotransformer starter on 80% tap, and gives a starting time with this particular motor of 10 s). The maximum momentary fault current is:

$$17.7/(0.195 \times 3 \times 6.6 = 7.94\ \text{kA (rms)}.$$

The steady state fault current from manufacturer's test data is 3.54 kA (after 0.25 s).

Let us select in this instance a CEE ITG 7231 relay for three-phase duty, using a range of 0.7–2.0/n. A CT ratio of 2000/1 A is selected, with 15VA5P10 CTs. The relay settings must coordinate with the following:

1. The total standing load of 200 A on the switchboard that the generator is supplying, plus the starting current taken by the 600 A motor. Therefore the load current is 200 + 1300 = 1500 A for 10 s.
2. From simulation studies or voltage dip calculations, the maximum transient load for the maximum voltage dip of 80% (which must not be exceeded whilst this motor is being started) is obtained. This gives a load of 2010 A for an initially fully loaded generator.
3. The 600 A motor thermal overload relay. This is set to operate at 3.4 kA after 10 s.

Therefore the minimum relay setting current required is $I_r = 2010/2000$ or approximately 1 A.

To coordinate with point 2 above, the operating time should be approximately 10.0 + 0.5 = 10.5 s at 3.4 kA. The required plug setting multiple (PSM) is

Offshore Electrical Engineering Manual. https://doi.org/10.1016/B978-0-12-385499-5.00027-3

$3.4 \times 10^3/2010 = 1.69$ times setting. From the relay characteristic, at 1.69 times the relay setting and a time multiplier of 1.0, the relay operating time is 5 s. Hence, for a required operating time of 10.5 s, the required time multiplier T_m is $10.5/5 = 2.1$.

OVERCURRENT PROTECTION

To avoid repetition, only the coordination of the system operating in mode 1 will be considered. The relay and fuse coordination formulae have been obtained from the *GEC Measurements Protective Relay Application Guide* (PAG third Edition). Note that older GEC relays have been used in this illustration as resetting existing relays is a more likely task than setting new. Because of this, it is strongly recommended that data on old relays is retained. New (digital) relays are also usually easier to set.

DATA REQUIREMENTS

SYSTEM DATA

Before beginning work on such a scheme, ensure that you have at least the following information at hand:

1. Load flow study results for each operating configuration to be considered;
2. Short circuit study results for each operating configuration to be considered, including three-phase-faults and phase-to-earth faults;
3. single-line diagram of the system, identifying circuit breakers and their protective devices, fuses and any other means of fault interruption (see Fig. 4.7.1);
4. Fault decrement curves for any generator sets, and fault levels for any incoming subsea cable feeders.

BASE VALUES

The following base values will be assumed (Table 4.7.1):

$$\text{Base MVA} = 10\ \text{MVA}$$

$$\text{Base impedance at 440 V} = 0.44^2/10 = 0.0194\ \text{ohms}$$

$$\text{Base impedance at 6.6 kV} = 6.6^2/10 = 4.356\ \text{ohms}$$

$$\text{Base current at 440 V} = 10\ \text{MVA}/\left[\sqrt{3} \times 0.44\ \text{kv}\right] = 13.12\ \text{kA}$$

$$\text{Base current at 6.6 V} = 10\ \text{MVA}/\left[\sqrt{3} \times 6.6\ \text{kV}\right] = 0.875\ \text{kA}$$

$$\text{Z.pu for base 1} = \text{Z.pu for base } 2 \times \text{MVA}_{\text{base1}}/\text{MVA}_{\text{base2}}$$

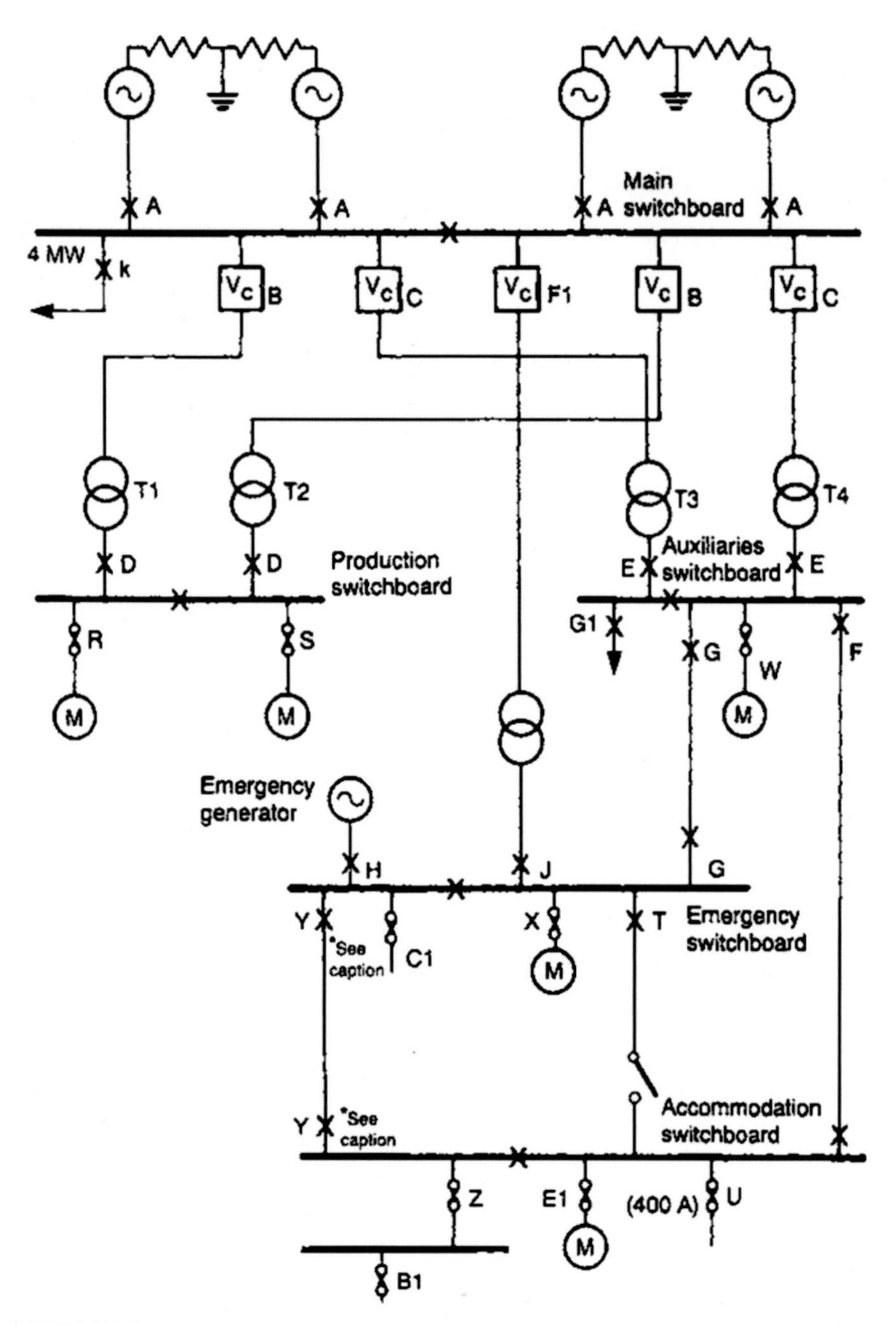

FIGURE 4.7.1

System diagram. Setting details given for relays T, H, G, F and Z only. Relay E is an earth fault only; overcurrent protection is provided by relay C. Circuit breakers either side of T3 are intertripped. Relays marked * are not considered, since this feeder is not used in operating mode1.

Table 4.7.1 Synchronous Machine Parameters

	Per Unit Value to Machine Base		Per Unit Value to 10 MVA Base	
Description	**Production Generator**	**Emergency Generator**	**Production Generator**	**Emergency Generator**
Rated MVA	4.729	1.313		
Rated voltage (kV)	6.6	0.44		
Subtransient reactance X''_d	0.181	0.233	0.383	1.775
Transient reactance X'_d	0.252	0.46	0.533	3.503
Synchronous reactance	1.95	2.45	4.123	18.66
Potier reactance	0.166	0.22	0.351	1.676
Saturation factor	1.45	1.562	1.45	1.562
Negative-sequence reactance	0.255	0.27	0.539	2.056
Zero-sequence reactance	0.103	0.14	0.218	1.066

OPERATING CONDITIONS

Mode 1: Maximum generation: four production generators, maximum standing load and all motors running. Main switchboard bus section closed; other switchboard bus sections open.

Mode 2: Minimum generation: one production generator running with no motors operating.

Mode 3: Emergency generation only.

The low-voltage switchboards are provided with two feeders. The normal configuration is with low-voltage switchboard bus section switches open, with each side fed by a separate feeder.

From Fig. 4.7.1 an overcurrent discrimination flow chart can be constructed, as in Fig. 4.7.2. All the devices including fuses should be identified, and the flow is downstream from left to right. The coordination exercise can now begin with the relays furthest downstream. It should not be necessary to start with the fuses U and Z as these obviously have fixed characteristics. However, it may still be necessary to change the fuses for ones with more suitable characteristics if grading problems are experienced.

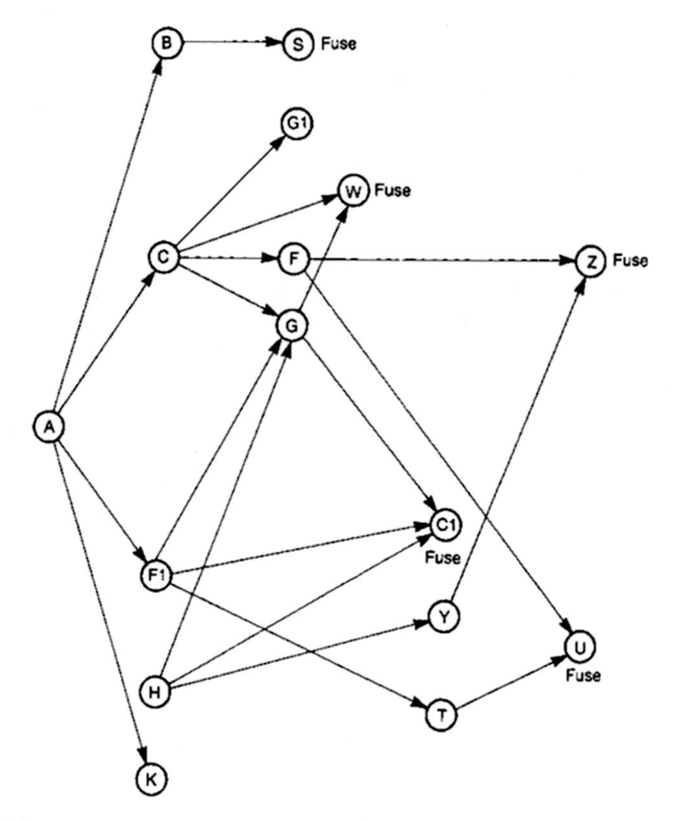

FIGURE 4.7.2

Overcurrent discrimination flowchart.

OVERCURRENT RELAY SETTING

RELAY F: RANGE 10%–200%, CT RATIO 1500/1

From the load flow study:

$$\text{Standing load} = 50\ \text{kW} + \text{j}20\ \text{kVA}_\text{r} \approx 71\ \text{A}$$

$$\text{Motor load (2 motors)} = 96\ \text{kW} + \text{j}40\ \text{kVA}_\text{r}$$

$$\text{Motor load (1 motor)} = 96/2 + \text{j}40/2 = 52\ \text{kVA} = 68\ \text{A}$$

$$\text{Motor starting current} \approx 6 \times 68\ \text{A} = 408\ \text{A}$$

Therefore the maximum load including motor starting, assuming that the currents are in phase, is

$$71 + 68 + 408 = 547\ \text{A}$$

The largest fuse on the accommodation switchboard is 400A (fuse U). The relay must discriminate with fuse U and permit the maximum load current of 547A to flow.

For discrimination with the fuse, the primary current setting should be approximately 3 times the fuse rating, i.e., $3 \times 400 = 1200$A. Therefore the nearest relay setting is $0.75 \times 1500 = 1125$A (0.75PSM).

The maximum fault level for fuse grading is 56.009kA (from computer science study). Therefore from the manufacturer's fuse characteristics, the fuse operating time t at 56kA is 0.01 s. From the GEC Guide, the minimum relay operating time is given by:

$$t + 0.4t + 0.15 = 0.164s$$

If the fault current from computer studies is 51.44kA, then the fault current as a multiple of the plug setting is 51,440/1125 = 46. The relay operating time at 46 times the plug setting with a time multiplier setting (TMS) of 1 is 0.255 s, from the relay characteristic. Therefore the required TMS is $0.164/0.255 = 0.64 \approx 0.7$.

RELAY T: RANGE 10%–200%, CT RATIO 1500/1

As the maximum loading of the accommodation switchboard is the same, the exercise is similar to that for relay F but using the fault currents that the computer has calculated for feeder T.

With the given fault current of 44,280A, the equivalent relay PSM is 44,280/1125 = 39.4. Therefore from the relay characteristic, using a PSM of 39.4 and a TMS of 1, the operating time is 0.255s. The required TMS is $0.164/0.255 = 0.64 \approx 0.7$.

RELAY G: RANGE 10%–200%, CT RATIO 2000/1

In mode 1, the utilities switchboard would be feeding the emergency switchboard, as the emergency generator would not be running. From the load flow data for the emergency switchboard:

$$\text{Load} = 455\ \text{kW} + j193\ \text{kVA}_r \approx 649\ \text{A at 0.92 power factor}$$

$$\text{Largest motor load} \approx 84\ \text{A at 0.8 power factor}$$

$$\text{Motor starting current} \approx 504\ \text{A at 0.2 power factor}$$

Therefore the maximum load during the starting of this motor is:

$$\text{Load} - \text{motor FLC} + \text{motor start}$$

$$(597 + j253) - (67.2 + j50.4) + (100.8 + j494) = 939.6\ \text{A at 0.67 power factor} \approx 1000\ \text{A}$$

Relay G must permit the maximum load without tripping and must also coordinate with the largest outgoing fuse on the emergency switchboard, i.e., fuse Cl

(160 A) must operate before relay G. Coordination with relay T is not considered necessary, since the accommodation switchboard would be supplied through circuit F. Relay Y does not need to coordinate with relay G, since feeder Y is only used in mode 3 operation (emergency generator only). However, the setting of relay G must also allow for back feeding of the utilities switchboard via the emergency switchboard and therefore coordination with fuse W, the largest utilities switchboard fuse.

The maximum fault current through feeder G is 56,000 A (from fault calculations). The operating time t of fuse Cl at 56 kA is 0.01 s, from the fuse characteristic. The setting of relay G for coordination with fuse Cl is three times the fuse rating, i.e., $3 \times 160\,A = 480\,A$, but the maximum load is 1000 A.

For back feeding of the utilities switchboard, relay G must coordinate with fuse W, which gives $3 \times 355\,A = 1065\,A$. From the load flow results, the maximum current flow through feeder G in the direction of the utilities switchboard is 1032 A at 0.8 power factor.

The starting current of the largest motor is 261 + j1281. Therefore the maximum load is:

$$(825.6 + j784.8) - (174.4 + j130.8) + (261 + j1281) = 912.2 + j1935$$
$$= 2139\text{ A at 0.43 power factor}$$

Therefore use the nearest setting of 2000 A. This is also the rating of the transformer which feeds the emergency switchboard.

From the SC calculations, the maximum SC current at fuse W is 44.3 kA.

From the fuse characteristics, the fuse operating time t is 0.01 s.

Therefore relay G operating time is:

$$t + 0.4t + 0.15 = 0.164\text{s}$$

The fault current as a multiple of the relay plug setting is 44,300/2000 = 22.

Therefore, from the relay characteristic the operating time of relay G at a TMS of 1 is 0.3 s. The required TMS is $0.164/0.3 = 0.54 \approx 0.6$ (nearest upward setting; using a setting of 0.5 would reduce the grading margin).

COORDINATION

The same methods may be adopted to obtain settings for the other overcurrent relays. The resulting coordination chart is shown in Fig. 4.7.3.

EARTH FAULT RELAY SETTING

Fig. 4.7.4 shows the earth fault discrimination flow chart for this system.

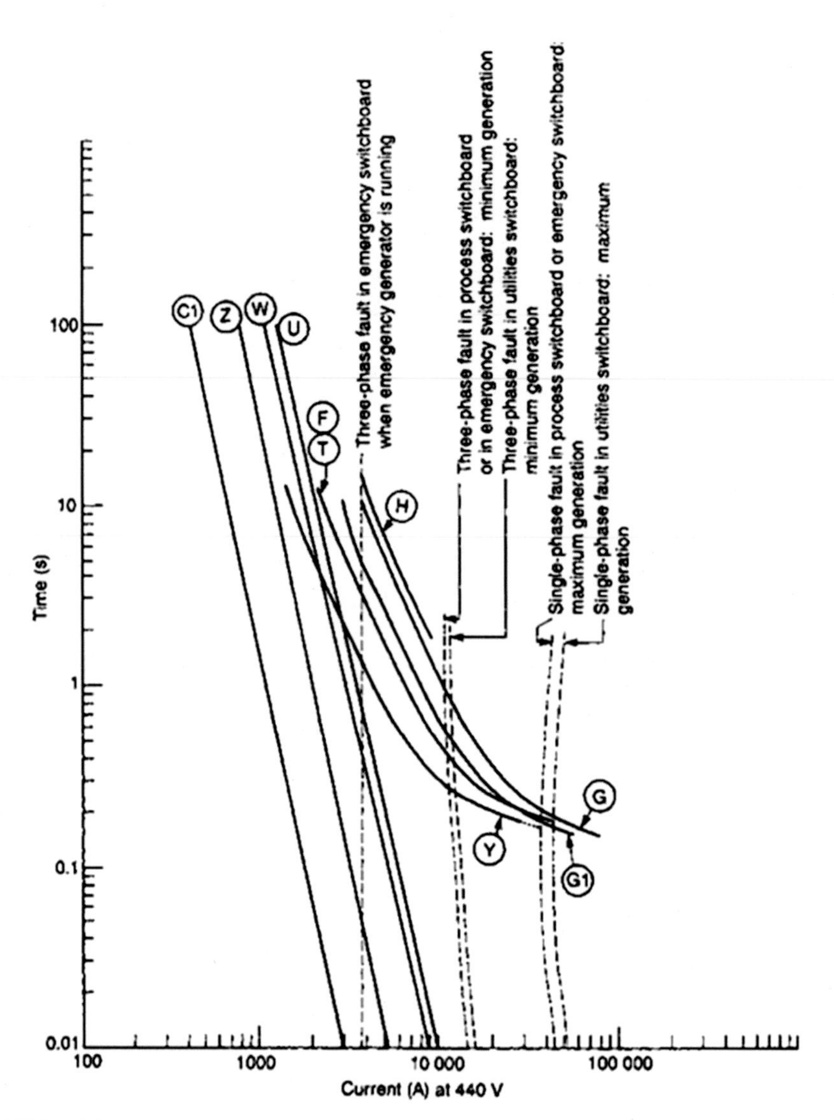

FIGURE 4.7.3

Overcurrent protection coordination chart – low voltage system.

CT SATURATION

It is necessary to check for CT saturation where it is considered to be a likely condition. The saturated CT current may be obtained as follows:

$$I_{\text{max sat}} = \frac{\text{CT rated VA} \times \text{Accuracy limit factor} \times \text{CT ratio}}{\text{Secondary burden at rated current}}$$

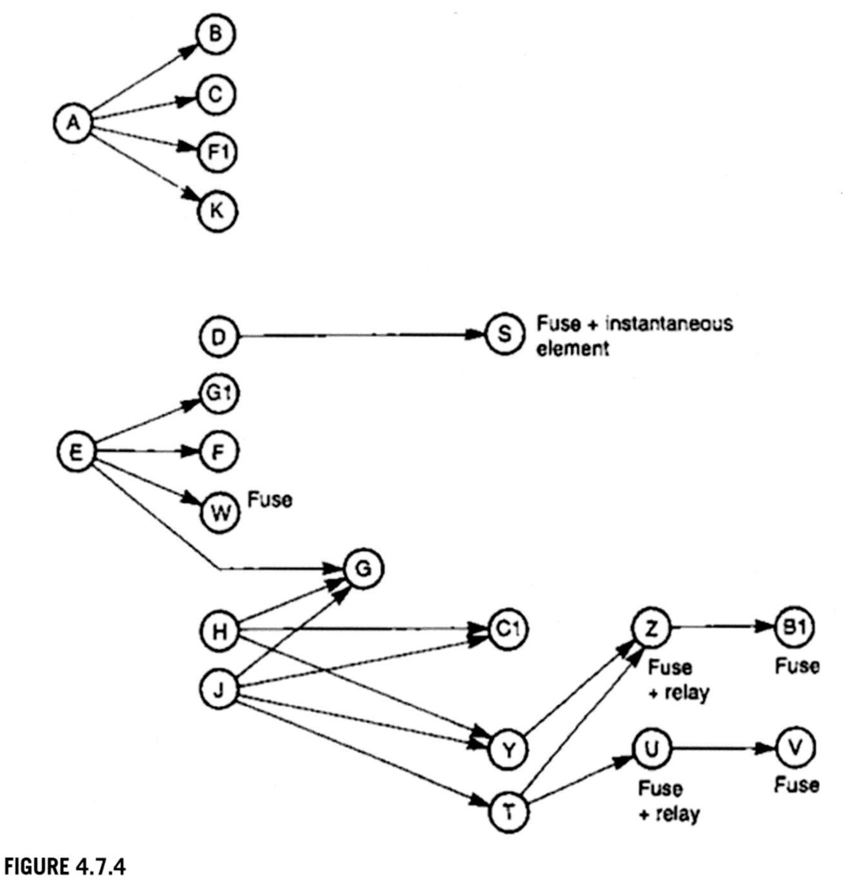

FIGURE 4.7.4

Earth fault discrimination flow chart.

If the maximum fault current available is less than this, coordination should not be lost owing to CT saturation.

RELAY Z: RANGE 10%–20%, CT RATIO 300/1. FUSE Z: 200 A

Relay Z and fuse Z must coordinate with the largest fuse in the next downstream switchboard, which is 125 A (B1 on the coordination chart).

From the fuse catalogue: I^2t (pre-arcing) of fuse Z is $2 \times 10^5 A^2 s$; I^2t (total operating) for fuse B1 is $9 \times 10^4 A^2 s$. Therefore as $I^2t_z \gg I^2t_{B1}$, coordination is assured.

Relay Z operates at lower fault currents while fuse Z operates at higher fault currents. Relay Z must also discriminate with fuse B1, however.

Therefore set relay Z at approximately three times the rating of fuse B1, i.e., $3 \times 125 = 375$ A; then the minimum relay setting is $1.5 \times 300 = 450$ A.

Fuse B1 operates at t=0.01 s at 2400 A, from the fuse characteristic, with fault current obtained from computer results.

From the GEC Guide, the relay Z operating time is:

$$t + 0.45\,t + 0.15 = 0.164\ \text{s}$$

The PSM is 2400/450=5.33. Therefore the relay operating time at a PSM of 5.33 and a TMS of 1 is 1.8 s, from the relay characteristic. The required TMS is 0.164/1.8=0.09≈0.1.

In this case, a check was made against the relay characteristic. It was found that the curve for TMS=0.1 gives an operating time of only 0.14 s at a PSM of 5.33. As this is too fast, the time setting was adjusted to 0.2. Relays U, T and Y may be set in a similar manner.

RELAY F: RANGE 10%–200%, CT RATIO 1500/1; 10 VA 5P10

This relay must coordinate with earth fault relays Z and U and with the largest outgoing fuse in the accommodation switchboard, fuse U, 400 A.

(This assumes that coordination with fuse U assures coordination with relay.)

Set relay F at three times fuse U rating, i.e., 3×400 A=1200 A. Then the nearest current setting is:

$$0.8 \times 1500 = 1200\ \text{A}$$

From the SC studies, the maximum earth fault current is 55.458 kA. At this current, fuses Z and U operate in 0.01 s, from the fuse characteristics. From the GEC Guide, the relay F operating time is:

$$t + 0.4 + 0.15 = 0.164\text{s}$$

From the SC studies, the average earth fault current at 0.164 s is 51.44 kA.

As a PSM, this is 51,440/1200=42.9.

From the relay characteristics, the relay operating time at a PSM of 42.9 and a TMS of 1 is 0.25 s. Therefore the required TMS is 0.164/0.25=0.66≪0.7.

It is necessary to carry out a relay saturation check. The relay impedance at the 0.8 setting is given by (relay VA)/$(0.8)^2$=0.39 Ω. Allowing 1 Ω for lead resistance gives 1.39 Ω.

Thus the secondary burden is 1.39×(1 A)=1.39 VA. Therefore:

$$\begin{aligned} I_{maxunsat} &= 10\ \text{VA} \times (10/1.39) \times (1500/1) \\ &= 107.9\ \text{kA (which is above the maximum fault current).} \end{aligned}$$

Note that if a saturation current significantly lower than the fault current is obtained, it is advisable to select a different CT. The burdens associated with electronic relays are much lower than those used above for induction disc types, and because of this are less likely to give rise to saturation problems.

COORDINATION

The same method may be used for the setting of the remaining relays. The resulting coordination chart is shown in Fig. 4.7.5.

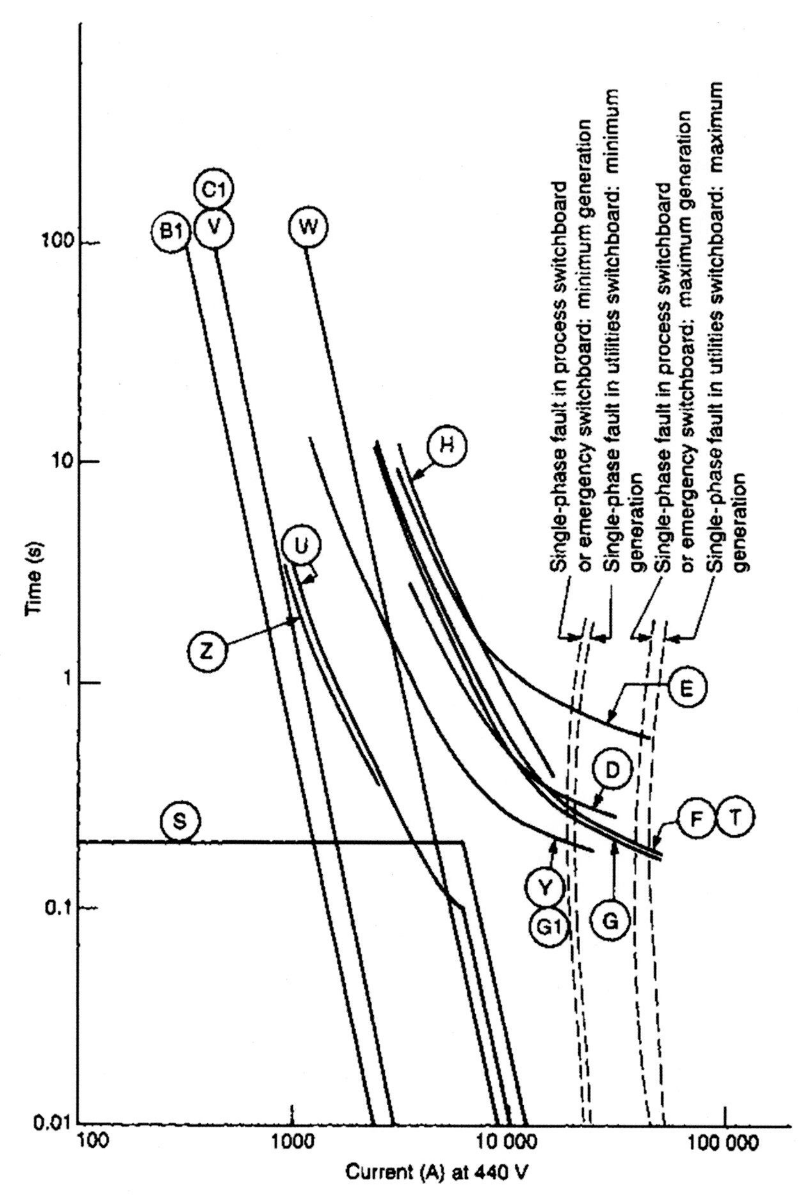

FIGURE 4.7.5

Earth fault protection coordination chart – low voltage system.

WORKED EXAMPLE: SETTING VOLTAGE CONTROLLED OVERCURRENT RELAYS

Some overcurrent relays provide dual inverse characteristics. The less sensitive characteristic gives a longer time interval for clearance of downstream faults, and the more sensitive characteristic removes close-up faults more quickly. Faults close to the generator have less impedance and hence produce greater voltage dips. At a set level of voltage dip, the relay will switch from one characteristic to the other.

In this case we will set the voltage to discriminate between a fault upstream (giving rise to a large voltage dip) and a fault downstream of a transformer (producing a smaller voltage dip). The inset on Fig. 4.7.6 shows the equivalent diagram. The maximum voltage dip occurs when the transformer reactance X_x is a

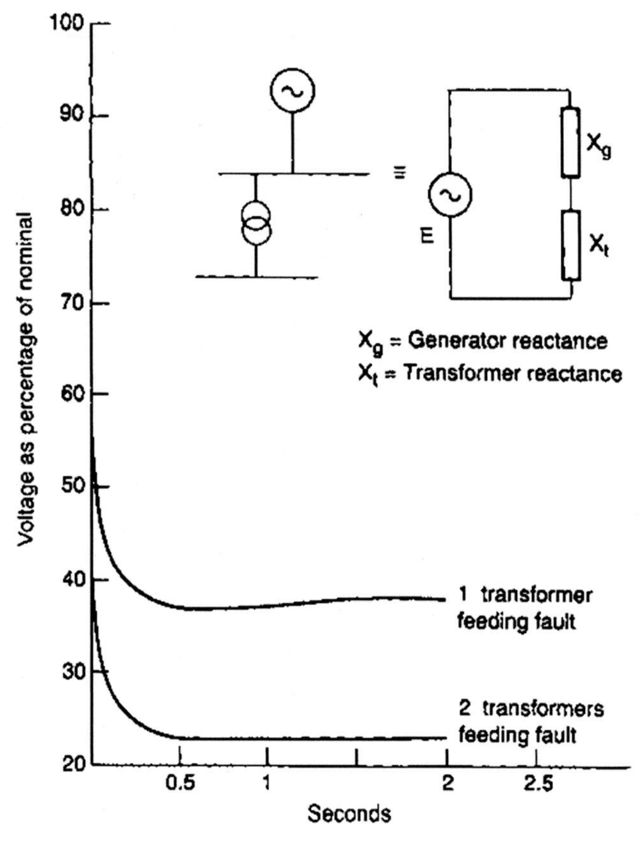

FIGURE 4.7.6

Graph of voltage downstream of transformers during system fault: Example for voltage controlled relays.

minimum compared with the generator reactance X_g. In this case we will assume this is when one generator is operating and two transformers are paralleled. The voltage at the generator (high-voltage) terminals is given by:

$$V = \frac{X_1}{(X_t + X_t)} - E \text{ (where E is the generated voltage)}.$$

The generator reactances from say 0–2 s can be calculated from the generator decrement curves or from values given by the manufacturer. In this case the generators are identical, each with reactances of $15.1\%_{t=0}$ and $33\%_{t=1}$, and each transformer has a reactance of 20.12% at the same MVA base.

Therefore with one transformer:

$$V_{t=0} = 20.12/(20.12 + 15.1) = 57\%$$
$$V_{t=1} = 20.12/(20.12 + 33) = 38\%$$

With two transformers in parallel:

$$V_{t=0} = 10.06/(10.06 + 15.1) = 40\%$$
$$V_{t=1} = 10.06/(10.06 + 33) = 23.4\%$$

Therefore a setting of 20% should prevent the relay from switching to the more sensitive characteristic on a low-voltage fault.

Two further points may be made:

First, this type of relay is known as a voltage controlled overcurrent relay, and should not be confused with the voltage restrained type where the sensitivity is continuously variable over a voltage range. The voltage restrained type is not recommended since sensitivity increases too quickly with dipping voltage so that discrimination with downstream fuses is lost.

Second, some American relays are also known as voltage controlled, but with these the inverse characteristic is inhibited at low levels of voltage dip, leaving only a definite time delay element to clear the fault (Fig. 4.7.7).

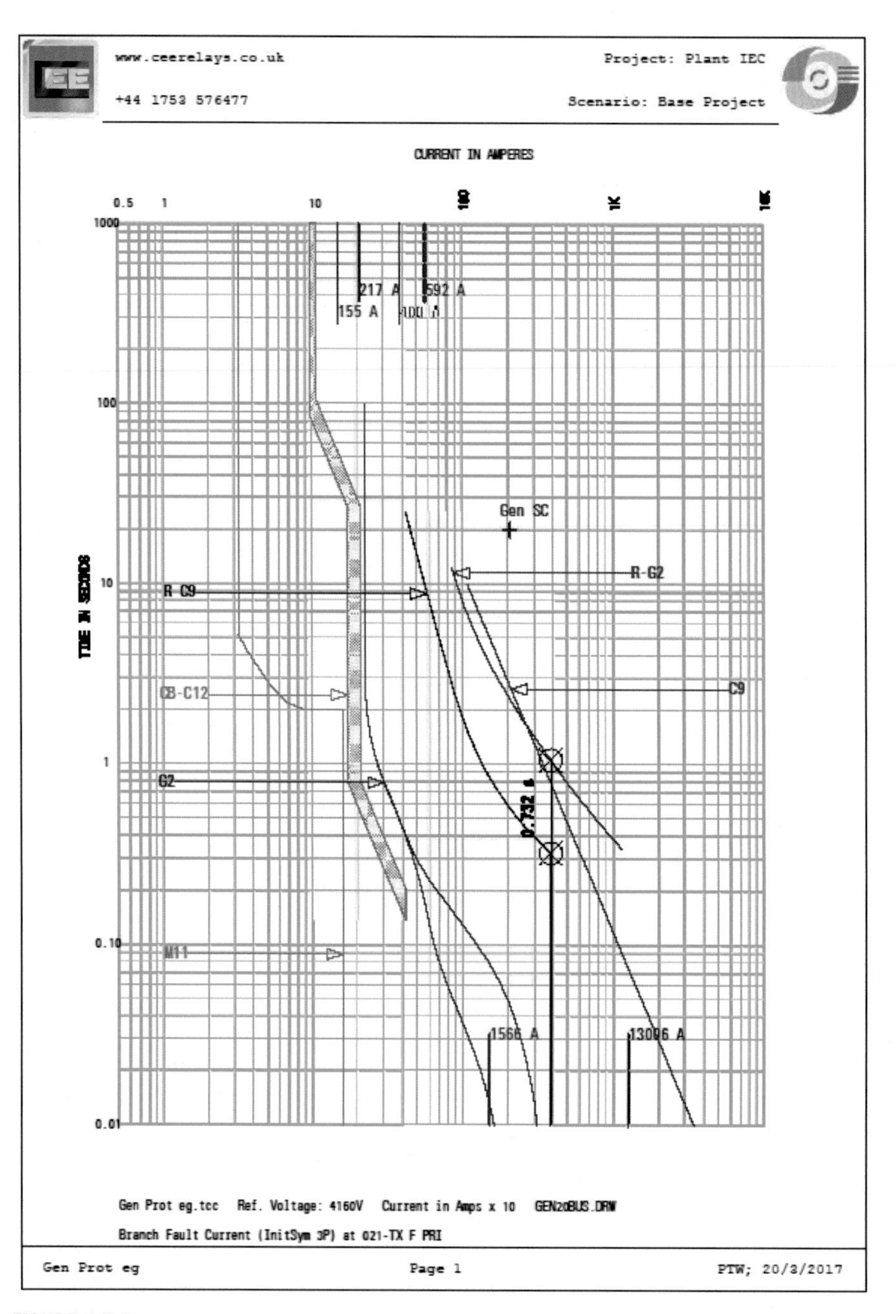

FIGURE 4.7.7

Typical computer-generated coordination chart using PTW captor.

Courtesy CEE Relays/SKM.

CHAPTER

Power Management

8

GENERATOR CONTROLS

With any electrical system, the importance of ergonomically designed controls cannot be overstressed. The following controls and instrumentation are considered to be the basis for a generator control panel. The logic of a typical generator control panel is shown in Fig. 4.8.1.

START AND STOP BUTTONS

The start control normally has the function of initiating the engine automatic start sequence and, depending on the control philosophy adopted, may automatically synchronise the generator with any generators already on load. An 'auto/manual' selector switch may also be provided. This is useful during commissioning to allow an individual to check each step in the sequence to be made. Two stop controls are often provided, a normal stop button which initiates a timed run down of the generator load and allows the engine to cool down before it is stopped. In the case of large gas turbines, a ratcheting sequence will also be required. This is a facility whereby the engine is rotated at intervals to prevent hot spots developing which may cause the misalignment of the main shaft.

AUTOMATIC VOLTAGE REGULATOR AND GOVERNOR RAISE/LOWER SWITCHES

These controls are used to set the voltage and frequency of the generator and, when in parallel with another generator, allow the sharing of reactive and real power to be adjusted, respectively. It is an advantage to group these controls and associated electrical metering such as voltmeters, kW and kVAr meters so that manual adjustment of real and reactive power can be carried out by one operator. If the controls are spread across four or five panels, each associated with an individual machine, adjustment can be difficult. If there are only two machines, then controls can be 'mirrored' so that raise/lower controls are located close together in the area on the panels near where they butt together. An auto/manual voltage control selector switch is also required, so that commissioning and routine checks may be carried out on the excitation system. In some systems, it may not be possible to switch easily from auto to manual voltage control or vice versa. As automatic voltage regulators (AVRs) are now solid-state devices and take up very little room in the control panel, a dual AVR system is

Offshore Electrical Engineering Manual. https://doi.org/10.1016/B978-0-12-385499-5.00028-5

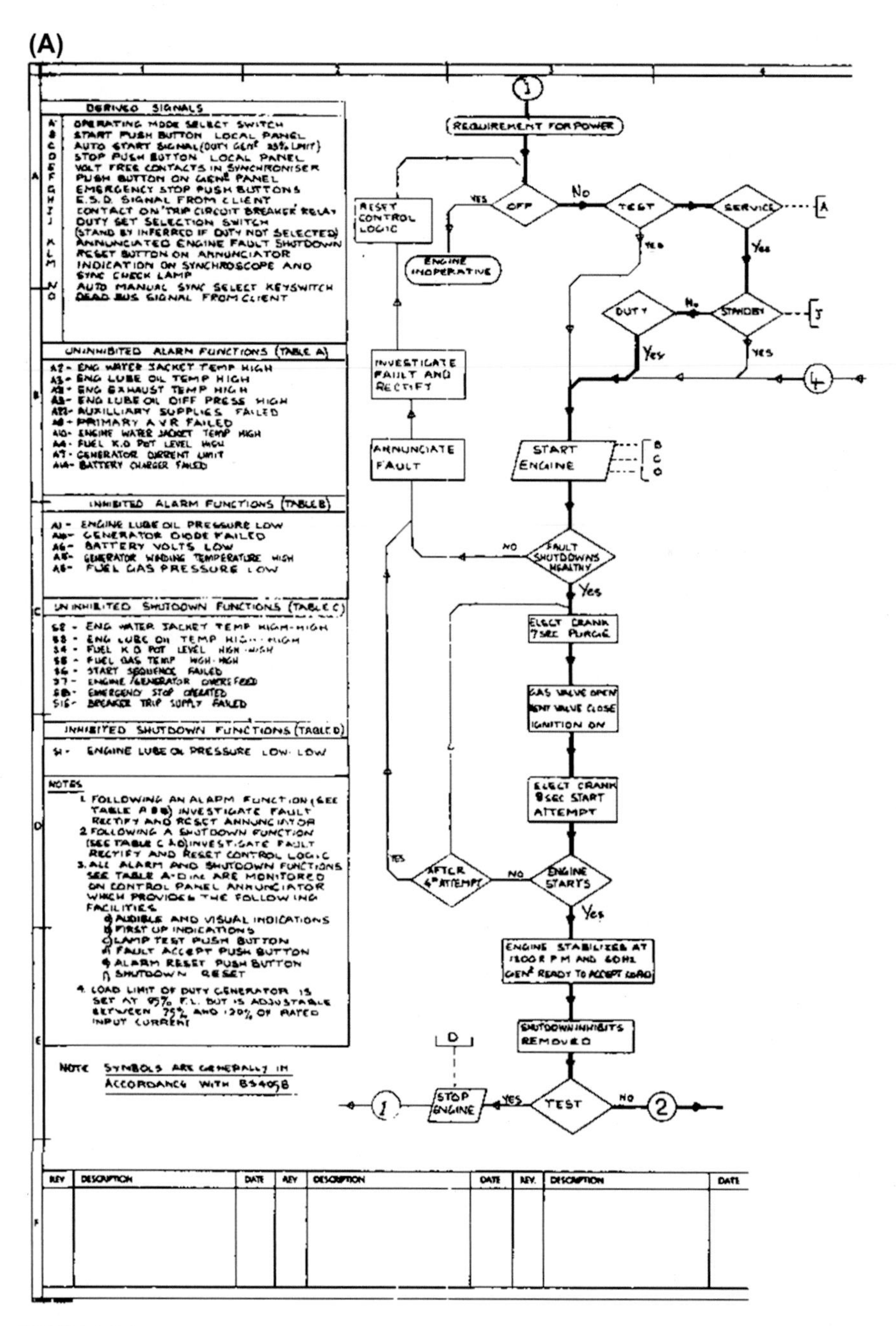

FIGURE 4.8.1

Logic diagram for a typical generator control panel: (A) left-hand side and (B) right-hand side.

(B)

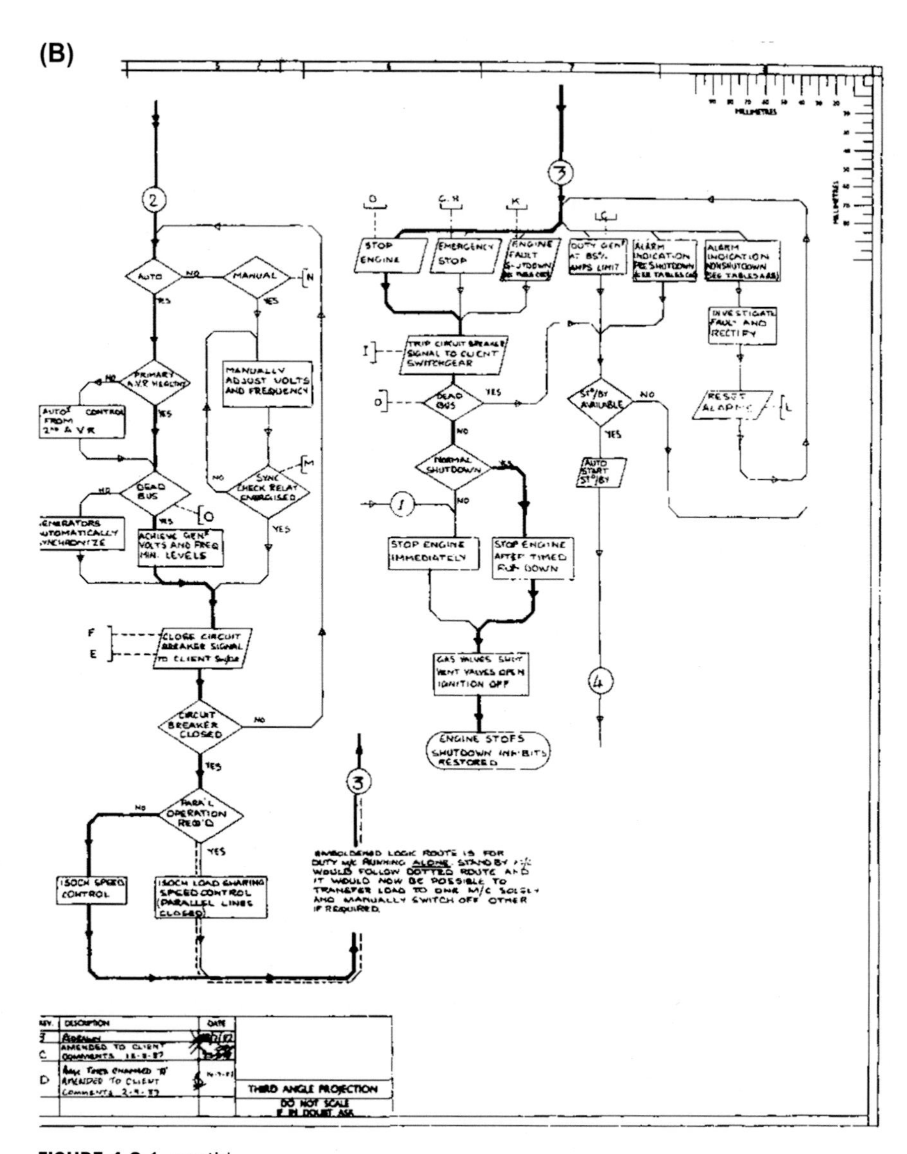

FIGURE 4.8.1, cont'd

recommended for all but the smallest machines. With a dual AVR system, a standby AVR follows the main AVR and automatically takes over if the main unit fails.

SYNCHRONISING EQUIPMENT

It is well worth providing a good selection of synchronising controls and indicators on the panel, as this not only provides for safer operation but also gives the operator more confidence during paralleling operations. An auto-synch unit and a check-synch relay should be provided; the auto-synch being used normally with automatic sequencing and the check synch when any paralleling is being carried out manually. The function of both devices is to provide a 'close permissive' signal to the generator circuit breaker when voltage and frequency conditions are suitable for paralleling. If the signal is not present, circuit breaker closing is prevented by the control logic.

INSTRUMENTATION

Metering

The control panel should provide at-a-glance information to allow the operator to carry out his work. Voltage and frequency displays should be digital and preferably to no more than two decimal places for fast reading. If the generator nominal voltage is more than 1 kV, the display should also be in kilovolts, again for fast reading. Although this means that the operator only sees changes in steps of 10 V, this is better than seeing in an equivalent analogue panel meter and is compatible with the accuracy of the transducer. Real and reactive power meters should preferably be analogue, as they are used mainly for balancing loads between machines and the pointer 'semaphore' effect is more quickly appreciated by the operator. A power factor meter should be provided of either digital or analogue type. However, in either case, the words 'leading' and 'lagging' should be displayed rather than '+' and '–'.

Synchronising Indicators

The indication provided by a particular instrument should preferably be duplicated by another for safety and operator confidence. For instance, a synchroscope should be supplemented by synchronising lamps (preferably three) for manual synchronising. Lamps should also be provided to indicate that check- and auto-synch permissive signals are present during circuit breaker closing. Where generator controls cannot be grouped conveniently, it will be necessary to provide busbar voltage and frequency meters as an aid to the operator during paralleling. If there are more than two generators in the system, a mimic panel is recommended. The mimic should show the electrical system in diagrammatic form. The various controls and indicators for each machine should be located on the mimic at positions where they are clearly associated with the machine whose parameters they are displaying or controlling.

Alarm Annunciator

As with any machinery, it is necessary to monitor each generator for faults and failures and bring them quickly to the attention of the operator. This is usually accomplished by using a matrix of windows, each engraved with a particular fault or failure. Each of these windows will be lit when a particular fault is sensed by a transducer located on the engine or generator. The transducers may be level switches, for example, monitoring lube oil level and operating on low oil level to light up the annunciator engraved 'LOW OIL LEVEL'. The logic of the annunciator should also indicate to the operator which fault occurred first. This 'first up' facility is usually provided by making the window of the first fault flash on and off or for it to remain steady and any later faults to flash. The occurrence of any alarm or shutdown will be accompanied by an audible alarm which will be silenced when an 'accept' button is pressed on the panel. There should also be a lamp test button which while pressed causes all the lamps on the matrix to light up. Once the fault has been cleared, the annunciator can be cleared of all fault and failure indications by pressing the reset button. Annunciator windows should be segregated into groups dealing with different types and severities of fault. Faults which cause an immediate shutdown must be indicated by windows of a different colour and be well segregated, usually by being placed on the lower lines of the matrix. Recommended colours are amber for faults which do not cause an immediate shutdown and red for those that do.

LOAD SHARING SYSTEMS

The operator's workload may be lightened by introducing facilities which automatically control the sharing of two or more generators running in parallel. The principle of a typical load sharing system is described in the following section.

An astatic control system is set up between the AVRs of each generator. This operates by comparing a direct current (DC) reference voltage with DC voltages derived from respective AVR control signals. The power sharing is accomplished by monitoring the output power of each machine and feeding this to an associated power comparator unit. Each machine power comparator also receives a signal from a frequency controller unit. This compares the actual supply frequency with a frequency reference so that the resulting comparator output signal is a function of the generator output modified by any frequency error. The outputs of each machine power comparator are linked to the next, forming a comparator loop which provides each associated governor with an appropriate power mismatch signal. The mismatch signal is used to drive governor raise/lower relays. The resulting system, once set up correctly, will provide good load sharing of both real and reactive power over a wide range of system loads.

POWER MANAGEMENT SYSTEMS

These are usually integrated with other systems in the installation or vessel in order to, for example,

1. avoid loss of production; by automatically tripping, nonessential large loads such as water injection pumps, which may be load shed automatically, both to maintain main generation stability and keep production up and running;
2. maintain dynamic positioning during the loss of a ship's generator.

With the advent of microprocessors, much more of the power generation system can be automated. To avoid embarrassing failures, microprocessors must be duplicated or triplicated so that errors may be detected and the system 'frozen' in a safe state so that operators may take over the function manually. The software used must also be extremely reliable, with built-in error trapping routines and means for ensuring a fast return to normal operation after a system crash. Memory devices should be read-only where possible; where read-write memory is necessary, this must be well protected from electromagnetic or other harmful effects. Any volatile memory must be permanently battery backed. Most microprocessor systems incorporate watchdog and visual indication in the occurrence of a fault.

Such a system, when properly programmed, can provide a whole range of useful facilities to make the operator's life easier, as follows:

1. Automatic starting and paralleling of generators as demanded by rising load and engine intake temperature according to a prearranged programme.
2. Automatic shutdown of surplus generators on falling load requirement or ambient temperature according to a prearranged programme.
3. Staged removal of noncritical loads from the system in the order of priority, the number of stages being dependent on the severity of the overload condition.
4. Automatic start-up and paralleling of another generator during serious fault or failure of a running generator. This facility can easily be provided in a system with only two machines but requires a programmable device where more than two machines are installed.
5. On a dynamically positioned vessel, backups to auxiliary machinery can also be programmed in; for example, additional cooling water pumps, lube oil pumps or fuel supply pumps can be started automatically.

POWER MANAGEMENT SYSTEM DESIGN

The principle documents involved in designing a suitable power management (PM) system are listed in the following.

1. Functional design specification
   This document sets out the detailed requirements of the system and has to be agreed between the manufacturer and the client usually before any design work starting.
2. Failure modes, effects and criticality analysis
   The purpose of the failure modes, effects and criticality analysis (FMECA) is to give a description of the different failure modes of the equipment when referred to its functional task. Most failure modes identified by this analysis will require to be covered by redundancy in some form if feasible and economic. If not, some operating procedure will need to be developed to provide a solution or mitigation.
3. Cause and effect matrix
   Once the detailed function of the system is fully described, the system logic can be developed in the form of a cause and effect matrix. It may also be necessary to use flow charts to clarify the logic in some cases.

4. Computer power simulation studies
Large nonessential loads must be tripped with alacrity by the PM system; otherwise generation will have already collapsed. Power simulation studies must be undertaken for the scenarios where the PM system functions include automatic load shedding or starting more generators, etc. Timers can then be set in the PM system based on the results of the studies. The studies should also determine whether the standby generators can be started and paralleled quickly enough to avoid a blackout. If not, more spinning reserve will need to be programmed in.

A failure within the PM system itself should not

1. cause a blackout,
2. cause alterations in the configuration of the main switchboard or generator controls but should detect the fault and initiate an alarm in the main control centre,
3. prevent or impede manual emergency intervention, i.e., the system should 'freeze' in a safe state so that operators may take over the function manually.

For more details on dynamic positioning, please refer to PART 6 Chapter 1.
For more details on FMECA, please refer to PART 8 Chapter 1.

CHAPTER

Harmonics

9

INTRODUCTION

Significant harmonics may be generated on oil installations by variable-speed drives and variable-frequency converters used in various process drives and particularly in drilling rigs. The harmonics are generated by solid-state switching devices which switch during the wave cycle, rather than at the voltage zero. Burst-fired or integral cycle fired heater controllers are not a source of harmonics but can generate low-frequency flicker or, at the right frequency, instability in generator automatic voltage regulators.

Type of Device	Number of Pulses	Harmonics Present
Half wave rectifier	1	2, 3, 4, 5, 6, 7,...
Full wave rectifier	2	3, 5, 7, 9,...
Three phase, full wave	6	5, 7, 11, 13, 17, 19,...
Three phase, full wave (transformer with zigzag secondary)	12	11, 13, 23, 25, 35, 37,...

OVERHEATING OF MOTORS

Motors running on supplies with nonsinusoidal waveforms will run hotter than those on sinusoidal waveforms because of the eddy currents generated. This is not usually a problem unless the motor is running close to its maximum loading, in high ambient temperatures or in hazardous areas where the certified T classification is compromised by a higher than normal casing surface temperature. Suitably derated motors should be specified for variable-speed drive or variable-frequency drive duty.

OVERHEATING OF TRANSFORMERS AND INCREASED ASSOCIATED LOSSES

For transformers feeding harmonic-producing loads, the eddy current loss in the windings is the most dominant loss component in the transformer. This eddy current loss increases proportionate to the square of the product's harmonic current and its

Offshore Electrical Engineering Manual. https://doi.org/10.1016/B978-0-12-385499-5.00029-7

corresponding frequency. The total transformer loss to a fully loaded transformer supplying to a nonlinear load is twice as high as for an equivalent linear load. This causes excessive transformer heating and degrades the insulation materials in the transformer, which eventually leads to transformer failure.

RESONANCE EFFECTS

Inductance and capacitance in the system, for example, capacitance in lighting, will intensify the waveform distortion by creating resonance at various points. This is why even lighting capacitance and the like is included in the model when harmonic analysis is undertaken, as well as the larger sources, such as capacitor banks for power factor correction. Filters may be utilised to minimise harmonics, tuned to the worst-case circuit condition. Alternatively, a full inverter can be included to handle variable levels of harmonics, but this is seldom an economic solution offshore.

UNINTERRUPTIBLE POWER SUPPLIES

Secure supplies for vital safety equipment such as emergency shutdown, telecommunications and fire and gas systems may be affected by the level of harmonic distortion in the installation supply, although they will also usually be contributors to the distortion themselves. For example, internal transformers and smoothing coils will run hotter because of additional hysteresis effects. Uninterruptible power supply purchase specifications should quote the maximum level of distortion in the supply in %Voltage total harmonic distortion (THD).

SWITCH MODE POWER SUPPLIES

Apart from the heating effects mentioned earlier, these devices are fairly immune to harmonics and tend to be small devices (e.g., power supplies in desktop computers). However, they do generate triplen harmonics, and if there are enough of these devices, they can overload the neutral.

PERCENTAGE DISTORTION DEFINITION

The equation for the harmonic expansion of a periodic function y(t) is

$$y(t) = Y_0 + \sum_{h=1}^{h=\infty} Y_h \sqrt{2} \sin(h\omega t - \varphi_h)$$

where Y_0, value of the direct current component, generally zero and considered as such hereinafter; Y_h, root mean square (rms) value of the harmonic of order h; ω,

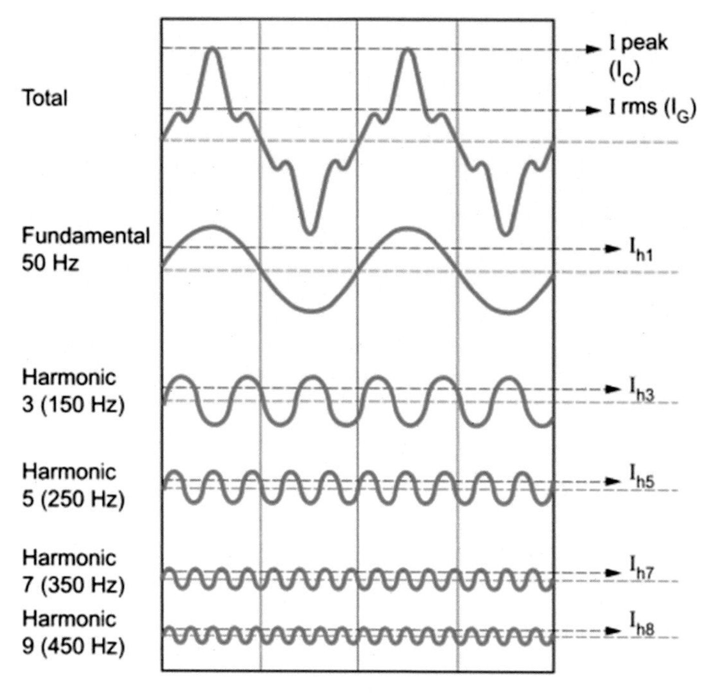

FIGURE 4.9.1

Example of a current containing harmonics and expansion of the overall current into its harmonic orders 1 (fundamental), 3, 5, 7 and 9.

Courtesy Schneider Electric/EIG_Wiki.

angular frequency of the fundamental frequency and φ_h, displacement of the harmonic component at $t=0$.

Fig. 4.9.1 shows an example of a current wave affected by harmonic distortion on a 50 Hz electrical distribution system. The distorted signal is the sum of a number of superimposed harmonics:

- the value of the fundamental frequency (or first-order harmonic) is 50 Hz,
- the third-order harmonic has a frequency of 150 Hz,
- the fifth-order harmonic has a frequency of 250 Hz, etc.

The THD is an indicator of the distortion of a signal. It is widely used in electrical engineering and harmonic management in particular.

For a signal y, the THD is defined as

$$THD = \sqrt{\sum_{h=2}^{h=H}\left(\frac{Y_h}{Y_1}\right)^2} = \frac{\sqrt{Y_2^2 + Y_3^2 + \cdots + Y_H^2}}{Y_1}$$

THD is the ratio of the rms value of all the harmonic components of the signal y to the fundamental Y_1.

H is generally taken equal to 50 but can be limited, in most cases, to 25. Note that THD can exceed 1 and is generally expressed as a percentage.

CURRENT OR VOLTAGE TOTAL HARMONIC DISTORTION

For *current harmonics* the equation is

$$THD_i = \sqrt{\sum_{h=2}^{h=H} \left(\frac{I_h}{I_1}\right)^2}$$

$$I_{rms} = \sqrt{\sum_{h=1}^{h=H} I_h^2}$$

By introducing the total rms value of the current, we obtain the following relation:

$$THD_i = \sqrt{\left(\frac{I_{rms}}{I_1}\right)^2 - 1}$$

equivalent to

$$I_{rms} = I_1\sqrt{1 + THD_i^2}$$

Example: For $THD_i = 40\%$, we get

$$I_{rms} = I_1\sqrt{1 + (0.4)^2} = I_1\sqrt{1 + 0.16} \approx I_1 \times 1.08$$

For *voltage harmonics*, the equation is

$$THD_u = \sqrt{\sum_{h=2}^{h=H} \left(\frac{U_h}{U_1}\right)^2}$$

PART

Installation

5

CHAPTER

Installation Practice

1

SWITCHGEAR AND MOTOR CONTROL CENTRES

Although the majority of the installation work is usually done by the offshore contractor, a technical representative from the switchgear manufacturer should be present to ensure that the work is done to the manufacturer's satisfaction and that any problems can be rectified quickly.

The installers should be completely familiar with the equipment and all the information regarding its installation. They should certainly not see it for the first time as it is removed from the packing case at the installation site. This is particularly important with regard to foundation and supporting steelwork arrangements. These must be in place, and of the correct dimensions, before installation of the equipment itself can be attempted. Once the framework is in place, correctly aligned, levelled and rigidly fixed, switchgear and control gear cubicles can be bolted together in the required order as shown in the design drawings provided. Identification labels should be checked to ensure that they correspond to the equipment and the associated drawings. All the copper connections for busbars and risers should then be made using torque limiting tools, under the supervision of the manufacturer's representative. Where the manufacturer recommends or provides special tools, only these should be used. 'Ductor' resistance testing of busbars can then be carried out. Any busducting between the switchgear and associated transformers or generators should be installed at this time, if this is possible. This should avoid any stressing of connections. If the switchgear is being installed in a module at a fabricator's yard before being shipped offshore in the module, then any transit packing removed should be retained, as this will be needed for repacking to prevent any shock or movement damage during the shipment offshore. All transit packing should be removed, and the equipment should be properly cleaned before testing.

A check for correct mechanical operation will then need to be made.

This should prove that

1. withdrawable trucks, cubicles and interlocking devices function correctly;
2. circuit breaker mechanisms, isolators, switches and relays are free to open and close properly.

By this time, the tripping battery and charger should have been installed and commissioned. Nevertheless, it is worth checking that the tripping batteries are fully charged before going any further.

Offshore Electrical Engineering Manual. https://doi.org/10.1016/B978-0-12-385499-5.00030-3

Having opened and closed all the circuit breakers electrically, a start can be made on primary injection testing of relays, and megger and high-voltage insulation checks completed, as these are more easily carried out before any cabling work commences.

Before any incoming or supply cables are connected, all the required earthing copperwork and cabling should be completed and tested.

The switchboard should now be ready for connection to incoming supplies. Cabling checks are discussed in a later section and commissioning is discussed in PART 7 Chapters 1–5. Having energised the switchboard, contactors may be checked for electrical operation. Supply circuits are then commissioned on a piecemeal basis as part of the commissioning procedure for the item of equipment being supplied.

DISTRIBUTION TRANSFORMERS

Distribution transformers are heavy devices and it may have been necessary to strengthen a particular section of the module floor or platform deck at the installation site. The installer should ensure that all transformers are placed on a flat, level, previously prepared base, in the location shown on the relevant design drawing. Once installed, it may be necessary to fill the transformer if it is a fluid-filled type. Some types of insulation fluid are toxic and therefore special precautions will need to be taken.

The manufacturer's representative should be called in after all the transformers have been located and filled, to assist and witness the final checking and testing. Insulation tests should then be completed, and after cabling, a second insulation test of all windings and connections should be carried out. On installations having two or more transformers connected to a single switchboard, a phase polarity check should be carried out across the associated bus section.

MOTORS AND GENERATORS

In most cases, rotating machinery will be installed as a complete skid-mounted package. Whether installed as separate items or not, great care must be taken to ensure that the motor and driven machinery are correctly located and aligned after installation. The equipment must be slung or lifted from the prescribed lifting positions such as eyebolts where these are available. A normal rule of thumb is that the lifting orientation of the machinery should be identical to its operating orientation, i.e., normally lift in the upright position.

During the entire installation procedure, it is important to use any tools or materials supplied specifically for the installation of the particular machine. It is not uncommon for such items to be left in the equipment packing case or otherwise overlooked and for installation to then go ahead using improvised methods which may jeopardise the success of the project and possibly the safety of offshore personnel.

Protective films on machined surfaces should be cleaned off with the recommended solvent. This cleaning should take place just before installation, where plinths are in contact with baseplates etc., and before commissioning for items such as motor shafts. Shafts and bearings need to be kept clean and covered whenever work is not proceeding.

Shims supplied with the equipment should be used wherever possible for levelling. Shims should be of similar area and shape as the machined surfaces of equipment footings they are used for, and the maximum thickness of shims should always be used so that the minimum number of shims is positioned under each footing. Where separate bearing pedestals are supplied, these should be positioned, levelled and lined up before the installation of other machinery components.

White metal bearings should be checked to ensure adequate bedding using engineer's blue or some other suitable method of indication.

Coupling faces should be truly parallel and level. This will need to be checked with each shaft rotated to several different positions.

Coupling bolts or flexible connections should be properly fitted without damage and all nuts tightened and locked before running the machinery.

When larger machines, particularly synchronous machines, are installed, the manufacturer's instructions concerning insulation must be carefully followed to prevent circulating currents flowing in the machine frame. Where the machine frame is installed in sections, conductive bonding between each section will need to be established, if necessary, by separate bonding conductors. This is particularly important when the motor is sited in a hazardous area, otherwise sparking may occur between sections.

The larger heavier machines should be provided with permanent runway beams and horizontal and vertical screw jacks to facilitate alignment.

Alternatively, suitable jacking points can be provided for temporary jacks. The use of block and tackle arrangements or tirfors is time consuming and likely to increase the risk of injury to installers.

Before commissioning, machines should be cleaned, and if of the open type, dust and dirt should be blown out. Commutators and slip rings will need to be checked for deterioration during transit and storage and, if necessary, cleaned. Carbon brushes will need to be checked for freedom of movement, and springs checked for correct pressure.

If machines have been stored for a long period, the bearing grease should be inspected and if deterioration has occurred, the bearings should be thoroughly washed, dried and regreased using methods recommended by the particular bearing manufacturer. Where oil lubricating systems are employed, bearing oil rings, flow switches, pressure switches, pumps, etc. must be checked to ensure they are fully operative before any running of the machine is attempted. Machine shafts will require to be rotated by hand or barring gear to ascertain that no foreign body is either inside the machine or between the external fan and its protecting cowl. Checks should also be made to ensure that ventilating air ducts are clean and clear of obstruction.

Where safety guards are provided with the machinery, these should be fitted before any rotation of the equipment under power. The guards should be constructed so as to facilitate removal. Location of both equipment and the guards on the equipment should be such that access for inspection and maintenance is practicable and safe. Both guards and couplings between motors and machinery should be removable without requiring removal of the motor or the driven equipment. Where guards are over belts and pulleys, a check should be made that the belt does not slap the guard at any speed. This is particularly important with variable-speed drives. In hazardous areas, nonsparking, antistatic belts must be used. Before any rotating equipment is commissioned offshore, the following precautions should be taken.

1. All necessary tools and ancillary equipment should be convenient to hand but not placed so as to pose a safety hazard.
2. Notices relating to firefighting procedures and treatment of electrical shock and burns should be displayed in switchrooms and control rooms.
3. All normal working and warning notices for the equipment should be if they are clearly visible.

LIGHTING AND SMALL POWER

During the design phase, calculations should have been carried out in order to fix types, light outputs and locations of all the luminaires to be installed, and if the required lighting levels are to be obtained, the installers must, wherever possible, adhere to the design details provided. This will not always be possible, particularly in congested areas, and to avoid shadow or obstruction of walkways, etc., an optimum location should be chosen by the installer.

Being, for obvious reasons, conspicuous, the acceptable appearance of luminaires should be preserved by careful installation, especially in accommodation areas. Rows of fittings should be installed accurately in a straight line and fastenings and suspensions set up rigidly so as to avoid distortion by handling during normal maintenance. The colour rendering of tubes and lamps should be consistent and suitable for the area where installed.

It is important both for balanced loading of phases and for ease of future identification that luminaires, power sockets and distribution boards are wired in accordance with design circuit schedules provided.

Care should be taken to ensure that polarities are correct when making connections to switches, convenience sockets, lamp holders and similar items.

When cabling up to swivelling floodlights, enough cable should be provided to allow the floodlight to be swivelled a full 360 degrees.

In some cases, luminaires with integral emergency batteries are provided with batteries unfitted. The batteries should be fitted as soon as possible to avoid deterioration, especially if the battery housing is required to seal the luminaire enclosure. Immediately after installation, the luminaires should be provided with a suitable electrical supply (even if this supply is temporary) for a continuous period of 80–100 h to ensure their batteries are fully charged.

SECURE POWER SUPPLY SYSTEMS

BATTERIES

Under normal circumstances, batteries will be shipped filled and discharged. Sealed type batteries will be shipped filled and charged. If batteries are shipped unfilled, appropriate safe filling facilities will need to be made available offshore.

Battery racking, cell units, connecting conductor links and associated nut, bolt and washer kits should be inspected for correct type and quantity on arrival. Any damaged components or shortages will need to be made up before installation can be completed.

Unsealed batteries will require either filling or topping up depending on the condition in which they were supplied.

It is important that batteries are assembled as shown in the drawings supplied. Incorrect configuration may provide the wrong output voltage and/or discharge duration and in any case, links will have been provided in the correct numbers and sizes for a particular configuration. Link and cable lug bolts should be tightened using a torque limiting tool, particularly with sealed type cells whose terminal posts snap off at very low torques.

In case of unsealed batteries particularly, all terminals and connections should be liberally greased with petroleum jelly or a similar substance according to the manufacturer's recommendations.

BATTERY CHARGERS AND INVERTORS

Cubicles should be positioned to ensure that a free flow of cooling air is available and that the ventilation entries and exits are clear of obstructions. An all-round clearance space of at least 150 mm is recommended.

Before commissioning, all cabling connections, circuit breakers and fuses should be checked for correct rating and operation. The manufacturer's representative should be present to inspect the equipment to his or her satisfaction before supervising its commissioning.

COMMUNICATIONS

PUBLIC ADDRESS SYSTEMS

The public address system is vital for communication of hazards and the calling of staff in the 'field', so it is often necessary to commission a part of the system as soon as it is installed and to minimise disruptions to its operation whilst installation of the rest of the system continues. This is quite possible provided a powered amplifier rack is available at an early stage in the proceedings. For reliability, it is a requirement that in every area, the public address circuits are duplicated. Therefore at least two circuits of loudspeakers will be required on any installation. In practice, more will be required because it will be necessary to mute speakers in sleeping areas during normal operating conditions and on large installations, more than two amplifier circuits will be required for reasons of loading and as a maintenance facility. As the system will be required to operate during abnormal conditions such as during serious gas leaks, junction boxes, changeover switches, loudspeakers and isolators should all be certified for use in zone one hazards, even when sited in areas classified as nonhazardous during normal conditions.

TELEPHONES

The installation of offshore private automatic branch exchange (PABX) systems is similar to that for onshore systems but with the addition of the following.

1. In hazardous areas, the equipment including the telephone itself must be certified for the zone of hazard concerned. Telephone instruments are normally of the flameproof type in such conditions.
2. In noisy areas such as machinery rooms, the telephone instrument will initiate, via a relay box, horn and light signals to indicate when the instrument is being called. The relay box, horn and signal lamp should be mounted near the associated telephone instrument. In some cases, a relay box may control more than one horn and lamp arrangement depending on the size and degree of congestion within the module concerned. Suitable acoustic hoods should be provided.

CABLE SUPPORT SYSTEMS

GENERAL

In the design phase, a great deal of care should be taken to ensure that the correct sizes, configuration and routes have been shown on the cable and racking arrangement drawings, and if route clashes with pipework or overloading of supports is to be avoided, it is vital to install both the support systems and cable strictly according to design drawings. However, there is always the possibility that a design error has been made, so the installer should carefully check the locations as shown on the drawings to determine if any conflict exists between the new cable route and any other equipment, steelwork, piping, ducting, etc., and whether, when installed, it will cause a hazard to personnel, obstruct accessways or prevent the installation or removal of equipment.

On minor routes, not detailed on design drawings, where it is necessary to site run cable support systems, it is important that separate racks/trays are used for the following categories to prevent electromagnetic interference between cables:

1. Medium-voltage alternating current (AC) (above 1 kV) with associated control cables.
2. Low-voltage AC (240–1000 V) with associated control cables. Direct current power cables, associated control cables and 110 VDC/254 VAC instrument cables.
3. Instrument, signal and alarm cables; telephone and communication cables and fire and gas cables.

Where practicable a minimum segregation distance of 1 m should be maintained. Crossovers should be kept to a minimum and as near perpendicular as possible. Some vital circuits associated with fire and gas, emergency shutdown or other safety related systems may have supply and/or signal cables duplicated. Such cables should be run on separate routes so that a single fire cannot destroy both cables.

SUPPORT STEELWORK

Drawings showing details of standard steel supports and brackets should be supplied as part of the design drawing package. If not, perhaps because a particular situation was not envisaged, then the installer will need to sketch a suitable arrangement and obtain approval for its use. The work will be made easier if one of the proprietary steel framing systems such as 'Unistrut' or 'Leprack' is used.

The steelwork is usually stainless or hot dip galvanised mild steel and should be free of sharp edges and burrs likely to damage cables. Nuts, bolts, etc. may be stainless or cadmium plated mild steel and ISO metric threads should be used throughout. Once installed, the whole arrangement may be given the standard paint finish before the cables are installed.

Supports for horizontal tray or rack should be spaced according to the type, width and estimated maximum loading, but should never exceed 3 m.

Supports for vertical tray or rack should be spaced at approximately 1 m intervals and should provide a clear space between the rack and the structure of at least 400 mm to allow for pipe lagging.

Steel or concrete members forming part of the module or installation structure must not be drilled or welded to provide a fixing point for supports, unless written permission has been obtained because such drilling or welding could weaken the structural integrity or reduce the seaworthiness of the installation.

CABLE TRAY

A variety of different types of cable trays are in use offshore and are made from a number of different materials. The following are two common types found offshore.

1. Heavy duty admiralty pattern: This may be stainless steel or high-quality 'Corten A' which has been hot dip galvanised several times.
2. Heavy duty reverse flange: This is much stronger mechanically than the equivalent admiralty pattern because of the doubling over at the edges (reverse flange). Again the material may be stainless steel or hot dip galvanised Corten A. Epoxy-coated mild steel types should be avoided, as they will deteriorate quickly, once any damage to the coating occurs.

TRAY INSTALLATION

The following points should be considered when specifying and installing trays.

1. Manufacturer's bend, tee and crossover sections should be used rather than site fabrication from straight sections. Site fabricated sections tend to be weaker mechanically and are more likely to damage cables by burring or unfinished metal. If it is necessary to cut a section of a tray, it should be cut along a line of plain metal rather than through the perforations.
2. Admiralty pattern and reverse flange trays should not be used in the same area, as the two types cannot be easily joined together. Reverse flange tray is normally used for external areas only.
3. At 25 m maximum intervals along the tray, it should be bonded to the platform structure.
4. With most types of trays, earth bonding continuity will be provided between sections by the tray itself, although some operators require a braided bonding connector to be used.

LADDER RACK

There is a variety of rack types used offshore, but all are based on a ladder design with various standard components which allow any three-dimensional configurations, including multitier, to be built up 'meccano fashion'. If stainless steel rack is to be used, great care will need to be exercised in quality control as there is a tendency for distortion to appear in some production batches. If reinforced glass fibre (grp) type ladder is used, the stronger 'pulltruded' type is recommended. As there is some fire risk associated with this type of tray, its use should be restricted to lower areas of the installation where sea spray is likely to cause corrosion with metal racking. For the same reason, long vertical runs of grp rack are not recommended.

THE DECOMMISSIONING AND REMOVAL/ABANDONMENT PHASES

It is vital that the same or better level of planning and monitoring of the work is maintained during these phases in the life of the installation. The organisation involved during decommissioning and removal is complex, as it is necessary to maintain services and utilities at every stage. This will include safe accommodation and provision of meals, etc. for the workforce. At some point, it will be necessary to provide an accommodation barge, i.e., before the platform accommodation can be decommissioned. In some cases the entire platform topsides is removed by a heavy-lift vessel, in which case the workforce can be accommodated on this vessel while preparations are made for the heavy lift.

If part of the structure is to be abandoned, some form of navigational aid package will need to be installed on the remaining structure, and possibly facilities for maintenance access should be provided.

WRITTEN SCHEMES OF EXAMINATION

The Safety Case Regulations apply to installation and decommissioning as well as design, so written schemes must be prepared and accepted before each of these phases in the life of the platform. These issues are covered in PART 9.

CHAPTER

Electrical System Earthing

2

INTRODUCTION

The earthing (grounding) of 'non-current-carrying parts' is essential for the safety of all those on the installation, to avoid both electrocution and voltage gradients between metal components, which can lead to sparking and hence risk of ignition of flammable gases and vapours.

PLATFORMS

It is necessary to have several main earth points on the platform, all of which will be of low resistance and with test facilities which allow the resistance to be measured without disconnection of the system.

ELECTRICAL EARTHS

This is required for safe operation of the electrical system. It consists of a network of protective conductors to ensure that fuses and protective relays operate effectively and that no dangerous voltage gradients exist. The resistance to earth of any part of this system should not exceed $1\,\Omega$. Some oil companies require this value to be $0.5\,\Omega$ to provide a safety margin against the effects of corrosion and dirt. At main earth bars, facilities such as duplicate connections and removable links should be provided and easily accessible to allow for testing and maintenance to be carried out regularly and safely. Conductor sizes should have been already calculated during the design phase.

A typical platform will have a main electrical earth connected to low-voltage (secondary transformer star points). The earth system will be in the form of a ring connecting to all low-voltage switchboards and connected to a main earth on the platform substructure (Fig. 5.2.1). Concrete leg platforms will be provided with main earthing points giving low resistance connections to the sea. Medium-voltage distribution transformers will be resistance-earthed at the star point to limit fault currents. For earth conductor sizes, please refer to

IEC 61892-4: Table 2	Mobile and fixed offshore units: cables
IEC 60092-352: Table 2	Ships: choice and installation of electrical cables

Offshore Electrical Engineering Manual. https://doi.org/10.1016/B978-0-12-385499-5.00031-5

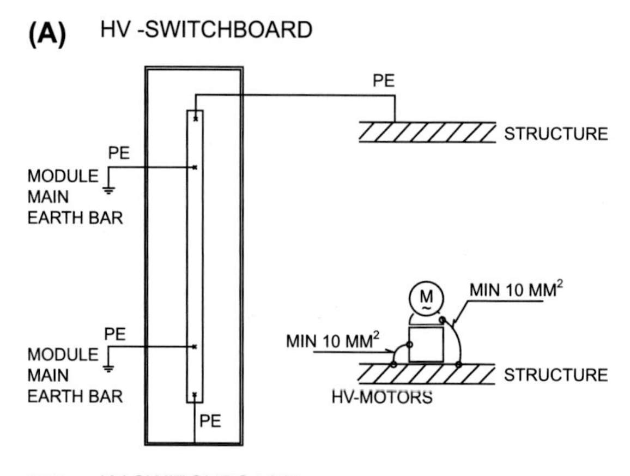

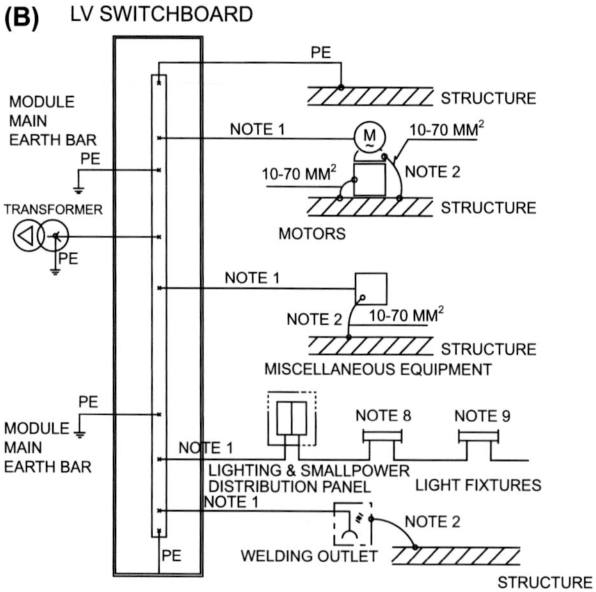

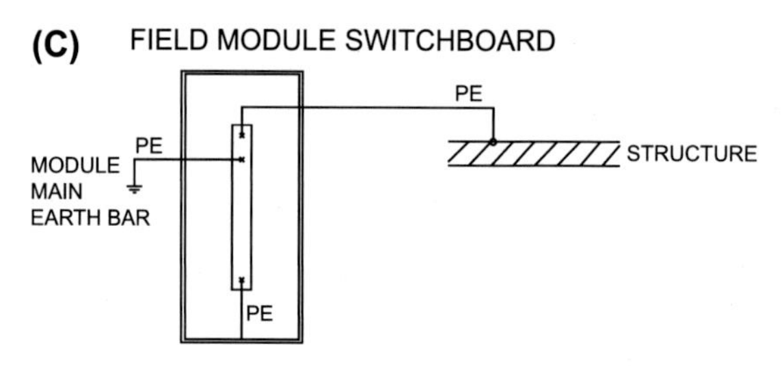

FIGURE 5.2.1

(A) Typical earthing for high-voltage (HV) switchboard. (B) Typical earthing for low-voltage (LV) switchboard. *Note 1*, earthed conductor may be in supply cable; *Note 2*, bonding conductors minimum 10 mm^2 copper; *Note 8*, bonding via fixing bolts or 10 mm^2 conductor. (C) Typical earthing: field. *PE*, protective earth.

INSTRUMENT/COMMUNICATIONS EARTHS

A separate instrument earth ring will be provided with a lower impedance than the electrical earth so that interference from the electrical system is minimised.

INTRINSICALLY SAFE EARTH

If intrinsically safe (IS) systems on the platform have the Zener barriers as well as optically isolated barriers, then an IS earth ring will also be required, in order to avoid ignition hazard if the Zener barrier fails (Figs 5.2.2 and 5.2.3).

EQUIPOTENTIAL BONDING

This is required to prevent any arcs or sparks occurring between adjacent metal sections, such as gasketed pipe sections, cable tray and steel tanks. On machinery bolted to a steel deck, separate bonding connections will be required. It is usual to provide the diagonal bonding mentioned in following list if resistance is increased by corrosion. Conducting copper grease can be used as a preventative measure. The resistance to earth and to adjacent metal parts should not exceed 10 Ω for any item, and this should be checked regularly. It is particularly vital that earth bonding is complete in hazardous areas in order to prevent ignition should a flammable gas leak occur.

A matrix of main earth bonding cables will be provided to bond

1. the casings of electrical and mechanical plant at diagonal corners of the bedplate and on the casings of electrical motors – cabins and superstructure need to be provided with earthing points at diagonal corners,
2. sections of cable tray with braided copper links between sections – gasketed pipe flanges should be provided with braided copper links.

The industry preferred cross-sectional areas are

- 70 mm^2 for main earth ring cables,
- 35 mm^2 for cables branching to individual earth bosses/lugs.

SHIPS

The major difference with ships is that the low-voltage system is usually an IT system, i.e., it is unearthed, and must be provided with earth monitoring, so that any earths can be easily detected and the low earth resistance cleared. Note that it is dangerous to leave low resistances in the system, as they may cause electrocution if the earth resistance is low enough. As such, it can contravene the statutory UK Electricity at Work Regulations on certain installations within British territorial waters.

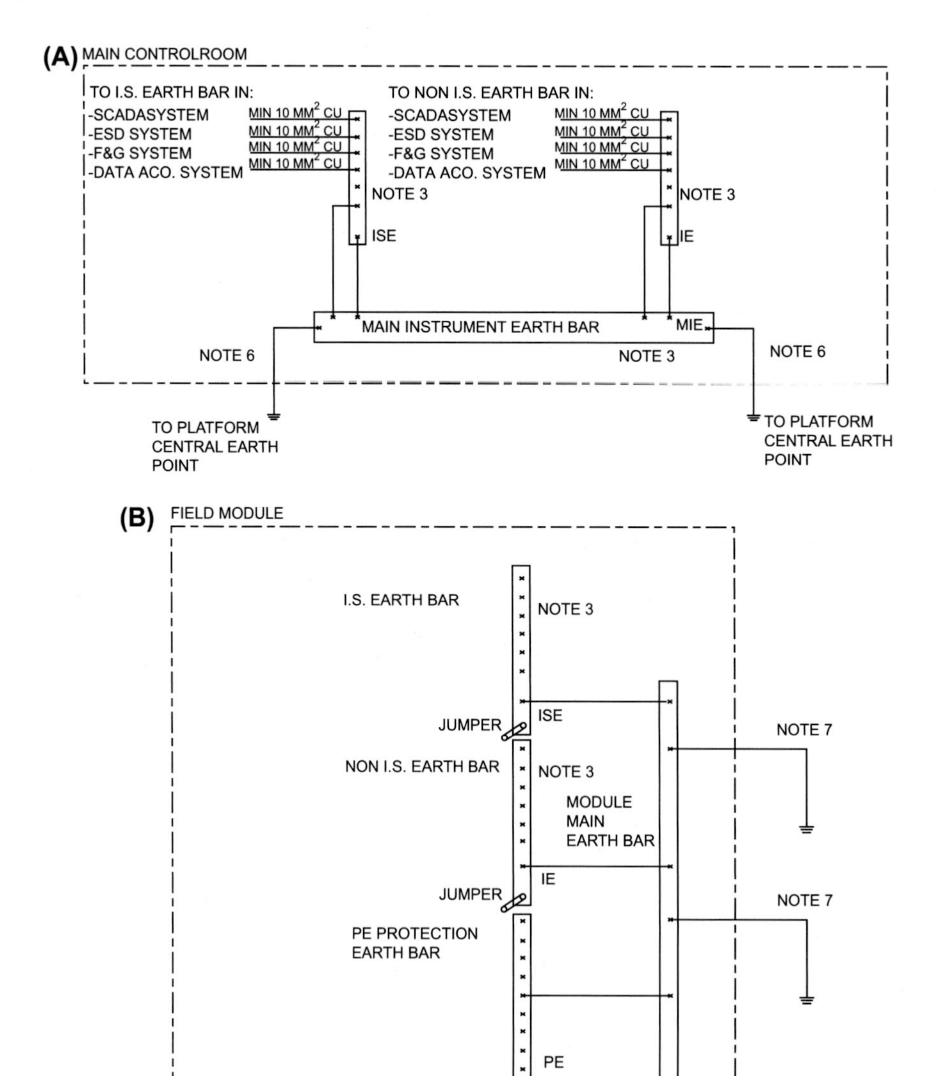

FIGURE 5.2.2

(A) Typical instrument earthing (control Room). *Note 3*, instrument earth bars to be mounted on insulators; *Note 6*, all main earth bars are 70 mm^2 (B) Typical instrument earthing (field). *Note 3*, instrument earth bars to be mounted on insulators; *Note 6*, all main earth bars are 70 mm^2; *Note 7*, protective earth (PE) terminals for components to be connected to the main structure by distributed local earth bus. *ESD*, emergency shutdown; *F&G*, fire and gas; *IS*, intrinsically safe; *SCADA*, supervisory control and data acquisition.

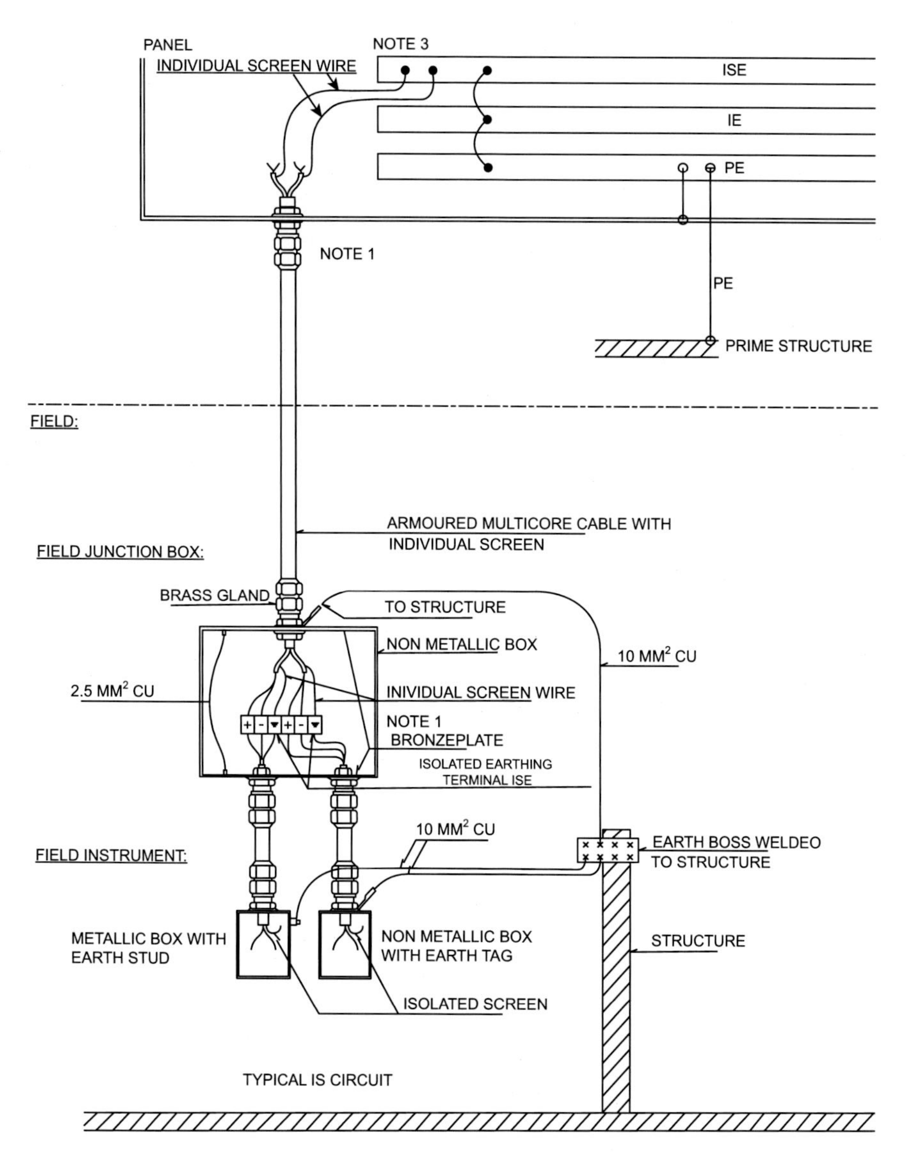

FIGURE 5.2.3

Typical instrument panel intrinsically safe (IS) earthing. *Note 1*, if the enclosure is made from metal <3 mm, a glandplate of minimum thickness 3 mm is to be provided; *Note 3*, IS instrument and electrical earth bars to be provided with linking bars. Instrument bar to be mounted on insulators.

LIGHTNING PROTECTION

Offshore structures and ships are not immune to lightning strikes just because they are located in the sea. Lightning tends to strike at the highest point, so any metal structures such as drilling derricks, telecommunications masts, flare stacks or crane booms may be vulnerable. Although lightning is less likely to strike over open water, there is still a risk to ships, rigs and platforms. In fact, changes in global weather patterns due to global warming may produce lightning strikes in areas where lightning has been historically rare. As offshore wind farms consist of a group of high towers, it is certainly necessary to provide lightning conductors and surge arrestors to protect equipment in the nacelles.

The main threats from lightning on both fixed and mobile (floating) oil installations are

1. direct ignition of explosive gases in hazardous zones,
2. current surges and voltage spikes causing damage to critical production and safety electrical safety equipment (telecommunications, process control and monitoring systems, fire and gas systems, etc.)

The main threat on wind turbine nacelles is damage to the control system due to current surges and voltage spikes.

Note: Danish and German offshore wind farm statistics indicate that the damages to control systems in the newer and larger wind turbines occur less frequently, probably as a consequence of improved lightning protection (see Reference 12).

CHAPTER

Environmental Protection

3

INTRODUCTION

The environmental conditions for most electrical equipment offshore are generally more onerous than those onshore. Environmental problems may be divided into three main categories:

1. those associated with sea and weather,
2. those associated with oil and gas,
3. those due to mechanical shock and vibration and the structural limitations of the installation.

WEATHER AND SEA PROTECTION

The salt-laden air and salt spray mists produced during bad weather have a searching corrosive effect on unprotected metal, as electrolytic cells are set up at these locations. Therefore, any protective coatings used on metal equipment offshore should remain impermeable, and preferably contain a sacrificial metal such as zinc.

A variety of different enclosure and support structure materials have been tried offshore, with varying results, but experience suggests that the following materials are suitable.

STAINLESS STEEL

This tends to be an expensive solution and needs to be carefully specified so as to avoid problems such as susceptibility to cracking. Stainless steel cable ladder and tray is more difficult to manufacture than that fabricated from the more common structural steels and may be delivered with some deformity. This should be checked by normal quality control procedures, as even a small out-of-true error will make the installer's life very difficult. Stress corrosion in stainless steel takes place in the presence of chlorides, and related failures occur well below the normal tensile strength of the metal involved. The stresses during the deformation processes involved in the manufacture may remain locked up until 'season cracking' occurs in the presence of sodium chloride (salt spray). However, good-quality low-temperature-annealed stainless steel products will provide long-term resistance to corrosion and are highly resistant to incidental damage.

Offshore Electrical Engineering Manual. https://doi.org/10.1016/B978-0-12-385499-5.00032-7

GREY CAST IRON

In grey cast iron, most of the carbon is present in the form of graphite flakes, which make the material softer, more machinable and less brittle than white cast iron. As the name suggests, cast iron is very fluid when molten and is therefore suitable for the manufacture of intricate castings. Its main use offshore is in the construction of flameproof enclosures. Its resistance to corrosion appears to be quite variable. Where it is exposed to salt spray, for example, in flameproof control stations on lower-deck handrails, a galvanised finish is advisable. The variation in the effects of corrosion is probably related to the method of producing cast iron, which involves the remelting of pig iron in cupolas. The qualities of the cast iron produced will depend on the selection of the pig iron, on the melting conditions in the cupola and on special alloying additions.

HOT DIPPED GALVANISED STEEL

This material is by far the most common for use in cable support systems offshore. The heavy-duty grade should provide a service life in excess of 20 years, particularly if Corten A steel is used.

POLYCARBONATE

This is a very tough plastic material used for junction box and similar electrical enclosures because of its corrosion-free property. It is particularly effective in areas close to the sea, where salt spray is common. It is resistant to mechanical damage and will deflect rather than break in most situations, but a heavy blow from a scaffolding pole, for example, is more likely to damage a polycarbonate enclosure than an equivalent steel one. Polycarbonate will tend to deform at elevated temperatures; it must therefore be shielded from the heat produced by flare stacks, etc. and must not be used with equipment which has to operate during firefighting operations.

MANGANESE BRONZE AND GUNMETAL

These very heavy corrosion resistant metals are sometimes used with success for the casings of floodlights and similar exposed electrical equipment.

WELDED AND CAST STRUCTURAL STEEL

Rotating machinery packages of all types are generally constructed of this material. It is not practicable to galvanise the whole package, and therefore, as with the structural steel of the module or platform jacket, a suitable offshore paint system must be applied by the fabricator. The integrity of this paint system must be preserved during the equipment's transit, installation and commissioning, if the package is to be presented to the operator in good condition.

GLASS FIBRE REINFORCED PLASTIC

Glass fibre reinforced plastic has been used successfully offshore for a number of years. It may be used in electrical equipment in the form of small junction boxes and for cable ladder and tray. Although fire resistant, it will burn when subjected to a gas flame but is more resistant to deformation, melting and fire than polycarbonate. Being strong and light and unaffected by seawater, it is useful particularly where installed in saltwater spray conditions. However, as it will burn, long unbroken vertical runs should be avoided.

ENCLOSURE INGRESS PROTECTION

All electrical equipment enclosures must be designed to prevent

1. the inadvertent contact by personnel of live parts inside the enclosure, when these are accessed for maintenance, etc.,
2. the ingress of dust and dirt,
3. the ingress of liquids, particularly conducting liquids such as seawater.

The degree of protection is specified in BS EN 60947-1 (and BS EN 60034-5 for motors). The degree of ingress protection (IP) rating is a two-digit number. The first digit refers to the degree of protection from foreign bodies and dust, and the second digit refers to the degree of protection from water. Some examples are given in Table 5.3.1. It should be noted that most hazardous area equipment enclosures must be capable of maintaining an IP rating of IP54.

STRUCTURAL CONSIDERATIONS

WEIGHT CONTROL

As in ship construction, the recording and control of weight are of vital importance to the structural integrity of the installation and the safety and survival of those onboard. As most equipment is shipped offshore in the form of prefabricated, precommissioned modules, weight control must start at the fabricator's yard and in the manufacturer's works. The selection of equipment may be restricted by maximum weight allowances. The power-to-weight ratio of a particular prime mover may be critical if the projected electrical load is to be met without the weight limit of the generator module being exceeded.

The weight of cabling and support systems, although more distributed, makes a significant contribution to the total and must be included in the weight control schedule. This schedule is usually compiled and updated regularly by structural engineers throughout the duration of a design contract.

The total weight of every module may be restricted, not only by the load-bearing capabilities of the offshore structure but also by the capabilities of offshore

Table 5.3.1 Ingress Protection Rating Examples

Code	First Digit	Second Digit
IP20	Contact with live or moving parts or ingress by objects with diameters of 12 mm and more.	
IP24S	As above-mentioned.	Harmful effects of water splashed against the motor from any direction.
IP44	Protection against contact with live or moving parts inside the enclosure by tools, wires or such objects of thickness greater than 1 mm. Protection against ingress of small solid foreign bodies (diameter greater than 1 mm) excluding the ventilation openings (intake and discharge of external fans) and the drain hole of enclosed machine, which may have degree 2 protection.	Harmful effects of water splashed against the motor from any direction.
IPW45S	Contact with live or moving parts by objects of 1 mm thickness. Ingress of foreign bodies with diameters of 1 mm and more, and rain, snow and airborne particles to an amount inconsistent with correct operation.	Harmful effects of water projected by nozzle against the motor from any direction.
IP54	Complete protection against contact with live or moving parts inside the enclosure. Protection against harmful deposits of dust.	Water splashed against the motor from any direction shall have no harmful effects.
IP55	Complete protection against contact with live or moving parts inside the enclosure. The ingress of dust is not totally prevented, but dust cannot enter in an amount sufficient to interfere with the satisfactory operation of the machine.	Water projected by a nozzle against the motor from any direction shall have no harmful effects.
IP56	As above-mentioned.	Motor protected against conditions on a ship's deck.
IP65	Complete protection against contact with live or moving parts inside the enclosure and against the ingress of dust.	Water projected by a nozzle against the machine from any direction shall have no harmful effects.
IP66	Complete protection against contact with live or moving parts inside the enclosure and against the ingress of dust.	Water from heavy seas or water projected from jets shall not enter the machine in any harmful quantity.

cranage available. The lifting capability of an offshore crane will vary depending on the operating radius of the jib required and the sea state, particularly the wave height, in which the crane is required to operate. Therefore, if a module weight is known to be critical, switchgear and other heavy equipment should be installed using bolted brackets rather than welding, as it may be necessary depending on the sea state, etc. to temporarily remove the equipment and reinstall it offshore.

SHOCK AND VIBRATION

With platform structures installed in sea depths of 150 m or more, rotating machinery rated at 5 MW or more may be operating near the top of a structure over 200 m high. The structure may resonate with vibrations produced by the machinery and, particularly if the machinery skid floor is cantilevered out beyond the main structure, a catastrophic failure may occur. To ensure that this is not the case, vibration analyses need to be carried out to establish structural resonance frequencies. If these are close to those of the machinery, modifications will need to be carried out to change the offending frequency to an acceptable value. At such heights, the swaying of the platform and the shock from drilling activities, etc. may produce damaging torques on the bearings and shafts of large motors and generators if these are not catered for in the machine design.

LOCATION OF ENGINE INTAKES AND EXHAUSTS

An offshore production platform is a compact three-dimensional arrangement of modules. Some modules contain prime movers and others, such as separator and gas compression areas, require to exhaust ventilation air from hazardous areas; all must take in fresh, uncontaminated air from the surrounding atmosphere. The prevention of cross-contamination between exhausts and intakes in all wind and weather conditions can be extremely difficult, and in some weather conditions, it may be necessary to accept a small drop-off in gas turbine performance due to the thermal contamination of other exhausts. Water curtains around turbine exhaust ducts are sometimes used to good effect in reducing contamination by taking exhaust gas downwards, away from intakes. Prime movers for emergency equipment such as emergency generators and fire pump alternators are not so critical because testing can be carried out in a favourable wind which blows the exhaust away from ventilation intakes.

With emergency equipment, it is more important that aspects of emergency scenarios such as gas cloud boundaries and likely flame passages are given consideration.

MECHANICAL PROTECTION

On any platform, routine production maintenance and drilling operations demand the movement of equipment, containers, scaffolding poles, drillpipe, etc. from the laydown area, at which it was received from the supply vessel, to its point of use, and then possibly the reverse journey back to the laydown areas before transferring back to the supply vessel. During these movements, there is a risk that the item will be dropped on to, or swung into, some exposed piece of electrical equipment. Worse still, an item may be dropped by a crane so that it pierces the roof or wall of an operating module. To avoid such occurrences the following should be considered.

First, exposed items of electrical equipment located in busy thoroughfares such as walkways, drill decks and laydown areas should be of steel construction or provided with steel impact protection of some form. No equipment should obstruct walkways. Overhead equipment such as luminaires and cable tray should not project lower than 2 m (safety helmets add about 100 mm to a person's height).

Second, the loci of crane loads should be carefully studied so that areas where the risk of dropped loads is high can be avoided as prospective critical equipment locations. If high-risk areas are unavoidable, then adequate mechanical protection will need to be provided, allowing for the shape and kinetic energy of potential dropped loads.

NOISE CONTROL

It is practically possible to run a platform unmanned and to shut down only for planned maintenance periods, which would then be the only time when the maintenance crew would be present. However, equipment reliabilities demand that unplanned outages be attended to, so noise levels must be limited throughout any installation if it is to be operated and manned at the same time. Sound levels less than 85 dB (A) should be the aim of any design, with much lower levels near accommodation areas.

PRIME MOVERS

Generator set procurement specifications must detail the maximum sound power level acceptable, with frequency spectra if possible. Equipment tests will need to be witnessed to ensure that the specifications are complied with. However, three main methods are available to the offshore design engineer to reduce the effects of prime mover noise offshore:

1. A close fitting acoustic hood can be installed over the engine, usually sized to fit round the skid on gas turbines. The enclosure sound insulation material must not absorb fuel, lubricant or coolant, as the enclosure itself must be designed to contain and extinguish engine fires with the aid of a halon or similar gas discharge.
2. It is difficult to enclose reciprocating engines in this way without impeding access for maintenance. Therefore the module housing the engine and generator may be soundproofed. If acoustic dampers are fitted to ventilation intakes and exhausts, this has the advantage that generator and radiator fan noise is also controlled.
3. Modules containing normally running prime movers should be located as far away from continuously manned areas as possible, although this distance is likely to be constrained by other limitations such as hazardous areas and air intake and exhaust positioning. In practice, a combination of all three methods will be used.

MOTORS

Unless the predominant noise is from the driven equipment, sound insulation of the module is not usually provided, as it should be possible for the motor manufacturer to control noise in the machine concerned. The specifier, however, may assist by avoiding the use of two-pole motors wherever practicable.

The sources of induction motor noise are as follows:

1. low-frequency magnetic noise from stator core laminations (100 Hz),
2. fan and airflow noise (400–2000 Hz),
3. core vibration noise (1000–2000 Hz),
4. slot tooth vibration noise (2500–4000 Hz),
5. bearing noise (6000–8000 Hz).

The magnetic noise can never be totally eliminated, as it is dependent on a large number of variables, which the machine designer must consider in order to trade off minimum noise against optimum performance and efficiency. Fan noise can be reduced by fitting a smaller fan on an oversized motor and also by employing closed air circuit, air cooled (CACA) cooling with an air-to-air heat exchanger. Silencing material may be fitted to the cooling air circuit, but this needs to be of a type suitable for the environment in that it should not absorb dirt, oil and sea spray as well as it does sound. The ingress of dirt, etc. is much less with CACA machines.

If a cooling water supply is available, then a substantial noise reduction may be obtained by replacing the air-to-air heat exchanger with an air-to-water type. However, the availability of the machine will then be dependent on the availability of the cooling water supply, which may itself be dependent on the satisfactory operation of a number of seawater lift and circulating pumps.

CHAPTER

Hazardous Area Installation

4

INTRODUCTION

This is not intended to be a comprehensive treatise on hazardous area installation, but more a consideration of the pitfalls often made in selecting and installing equipment suitable for locations where risks of explosive atmospheres are present. It is not known currently how Brexit will affect the ATEX regulations, but it can be assumed that the same explosive risks will be present, and that general trend is towards standardisation to facilitate world trade.

HAZARDOUS AREA APPLICATIONS

Much of the equipment on an offshore installation will be located in an area classified as potentially hazardous, because of the risk that flammable gases or vapours may be present in the atmosphere and could be ignited by equipment which creates electrical sparks or, during and perhaps for some time after operation, has an enclosure with a high enough surface temperature. This subject is covered exhaustively in a multitude of standards and codes of practice, many of which are listed in the Bibliography, and it is intended only to summarise the subject here. Those unfamiliar with the subject are recommended to study BS EN 60079 - Explosive atmospheres.

An explosive atmosphere is one where a mixture of air and flammable substances in the form of gas, vapour or mist exists in such proportions that it can be exploded by excessive temperature, arcs or sparks. The degree of danger varies with the probability of the presence of gas from location to location, and so hazardous areas are classified into three zones as follows:

Zone 0 (more than 100 h/year) –In which the explosive gas mixture is continuously present or present for long periods. The conditions in such zones are usually regarded as too dangerous for any electrical equipment to be located in.

Zone 1 (1–100 h/year) –In which an explosive gas/air mixture is likely to occur in normal operation.

Zone 2 (less than 1 h/year but more than 1 h/100 years) – In which an explosive gas/air mixture is not likely to occur in normal operation and, if it occurs, will exist for only a short time. By implication, an area that is not classified as zone 0, 1 or 2 is deemed to be a non-hazardous area.

The numerical values for exposure time shown above are for guidance only.

Offshore Electrical Engineering Manual. https://doi.org/10.1016/B978-0-12-385499-5.00033-9

TEMPERATURE CONSIDERATIONS

IGNITION TEMPERATURE

The minimum temperature at which a gas, vapour or mist ignites spontaneously at atmospheric pressure is known as the ignition temperature. To avoid the risk of explosion, the surface temperature of the equipment must always remain below the ignition temperature of the explosive mixture. Maximum permissible surface temperatures are classified as in Table 5.4.1. Classification applies to the equipment, not the gas; the actual classification designation can be taken as the next below the ignition temperature of the gas.

Motors to European standards conforming to BS EN 60079-0 and BS EN 60079-1 have temperatures 10°C lower than that shown in the table for classes Tl and T2, and 5°C lower for class T3 and below.

FLASHPOINT TEMPERATURE

The flashpoint of a liquid or solid is the minimum temperature at which the vapour above the material just ignites by application of an external flame or spark, in standard test conditions.

The flashpoint gives a very useful indication as to how hazardous a material is, and is used when drawing up a schedule of hazardous sources for a particular installation. As discussed later in this chapter, equipment designed for use in hazardous areas should never allow sparks or flames to come in contact with the external environment, although in the case of Ex 'd' equipment flames or sparks may occur within the enclosure. This is inherently the case with Ex 'd' switchgear.

If the flashpoint of a substance is higher than 38°C, it is not normally regarded as a source of explosive hazard on North Sea installations.

However, it should be noted that in hotter climates the storage area for the aviation kerosene (flashpoint 38°C) required for helicopters will have to be classified as hazardous, since there is a higher risk that ambient temperatures will exceed this temperature.

Table 5.4.1 Surface Temperature Classes

International (°C) IEC/BS EN 60079	US (°C) NFPA 70	
T1 – 450	T1 – 450	T3A – 180
T2 – 300	T2 – 300	T3B – 165
T3 – 200	T2A – 280	T3C – 160
T4 – 135	T2B - 260	T4 – 135
T5 – 100	T2C – 230	T4A – 120
T6 – 85	T2D – 215	T5 – 100
	T3 – 200	T6 – 85

EXPLOSION-PROOF EX 'D' EQUIPMENT (OLDER 'FLAMEPROOF' CERTIFICATION IS SIMILAR)

The principle of protection, as with the metal gauze on the Davy lamp, is to control the flow of and to cool any burning gas escaping from the motor casing, so that any gas that has escaped is no longer hot enough to ignite external gas. The size of gap or flamepath necessary to sufficiently cool the ignited gas on its way out of the enclosure varies according to the gas or vapour involved. Gases and vapours are subdivided according to experimental data which has been established to determine the maximum experimental safe gap (MESG). In the case of metal-to-metal joints in a flameproof motor, for example, that of the frame to the end shield, these will consist of a long metal spigot fitting into a long recess which will normally be clamped tightly to fixing bolts. A flamepath will always exist between the shaft and the motor interior.

For safety, all the flamepaths or gaps in the motor enclosure must never exceed mandatory dimensions, and the casing of the motor must be strong enough to withstand an explosion caused by the ignition of the maximum free volume of air/gas mixture it can contain, with all flamepaths at minimum production values. It should be noted that this could be exceeded in very cold regions (polar latitudes) owing to the increased density of the gas or at the elevated pressures used in diving. This type of protection is common in low-voltage motors up to a 315 frame size; on larger sizes, severe cost and weight penalties are likely owing to the cast method of construction.

The following criteria should be considered by the installer to ensure that certification is not invalidated and/or the risk of explosion increased:

1. Further entries should not be drilled out on flameproof enclosures. It is advisable to allow some margin in the number of entries specified to the manufacturer. Approved threaded plugs are available to plug unused entries. All unused entries must be plugged by such approved plugs.
2. Only tools suitable for use on flameproof equipment should be used.
3. Only flameproof cable glands should be used on flameproof enclosures.
4. The painting or obstruction of gaps between flanges must be avoided.
5. No flameproof equipment should be drilled, cut or welded.
6. Always check that all nuts, bolts, door hinges, isolators and interlocking devices are installed and are functioning as explained in the manufacturer's instructions.
7. All Ex 'd' equipment flanges should be greased with a certified grease such as Chemodex copper grease.

It is often operators' practice that explosion-proof equipment such as glands, luminaires and small motors are used throughout the installation in order to rationalise spares and simplify maintenance procedures. If this philosophy is in operation, the equipment sited in safe areas should be installed as if it were being installed in a hazardous area, with the correct seals, flange gaps, etc. The reason for this is that the equipment will be marked Ex 'd', and if hazardous zones are altered at some future date so that the equipment then comes within such a zone, it will not be obvious that internal seals or other components are missing or not correctly installed.

Table 5.4.2 Gas Group Examples

Material	Flashpoint (°C)	Ignition Temperature (°C)	Temperature Class	Gas Group
Ethylene	Gas	425	T2	IIB
Hydrogen	Gas	560	T1	IIC
Methane	Gas	538	T1	IIA
Kerosene	38	210	T3	IIA
Propane	−104	466	T1	IIA

EXPLOSION-PROOF EQUIPMENT GROUPS

The maximum dimensions for flamepaths in Ex 'd' equipment (see later), obtained from experiment, are used by the responsible standards authorities to subdivide flammable gases and vapours into three groups, A, B and C. The flamepath dimension is associated with the molecular size of the gas, the largest being in group A. Group C, which contains only hydrogen, is the most onerous as it requires the smallest flamepath cross-section and hence the finest machine tolerances.

Some examples of ignition temperatures and flashpoints are given in Table 5.4.2.

INCREASED SAFETY EX 'E' EQUIPMENT

This type of protection relies on the reduction or to make negligible the risk of an explosion by careful design of the motor and its control gear in order to eliminate any potential sparking which may come in contact with an air/gas flammable mixture, or an excessive temperature anywhere within or on the surface of the machine. Design criteria for this type of motor are given in BS 5501 and BS 5000.

This type of protection is not used in North American equipment, and the standards have therefore been written with European voltages only in mind, quoting a maximum nominal system voltage of 11 kV. Strictly speaking, equipment made for the North American standard voltage of 13.8 kV cannot comply with these standards. However, it is usual under these circumstances for such motors to be tested and then issued with a certificate of inspection rather than a certificate of authority. Acceptance of this method of protection must therefore be obtained from the platform certifying body and underwriters before any decision to use such a motor is made.

Ex 'e' motors offer increasing savings in weight and bulk as ratings increase. Beyond the 315 frame size, the only economic alternative is the Ex 'p' type of protection, as Ex 'd' would be too heavy or the castings too difficult to manufacture.

The following criteria should be considered by the installer to ensure certification is not invalidated and/or the risk of explosion increased.

TERMINALS

1. The terminals should be installed on 32-mm carrier rail to DIN 46 277/1, each group being completed by an end section and sandwiched between two end support brackets.
2. Any cross-connected terminal assembly fitted with jumper bar must be mounted on the 32-mm carrier rail between insulating partitions.
3. Except when shown in a certificate as being internal wiring of the apparatus, not more than one single-strand or multiple-strand wire or cable core should be connected into either side of any terminal.
4. Leads connected to terminals should be insulated for the appropriate voltage, and this insulation must extend to within 1 mm of the metal of the terminal throat.
5. All terminal screws, whether used or not, must be tightened down.
6. When used in general purpose Ex 'e' marshalling junction boxes, the circuits must be protected by close excess current protection which is designed to operate within 4 h at 1.5 times the designed load current.
7. The creepage and clearance distances between the installed terminals and adjacent equipment, enclosure walls and covers must be in accordance with BS EN 60079-7.

ENCLOSURE

1. Any cable gland or conduit entry into the enclosure must be capable of maintaining the degree of ingress protection IP54 and must be capable of passing the 7 Nm impact test required by BS EN 60079-7. If mineral insulated cable is used, the cable seal must be of a type having component approval for type 'e' applications.
2. The installer must ensure that the conductors are not cleated together in such a manner as to significantly increase the temperature of any individual conductor. A bunch of conductors at any point in the wiring loom should not contain any more conductors than are in the multicore cable or conduit from which the conductors originate.

Other non-hazardous related environmental design limitations are discussed in PART 5 Chapter 3.

EX 'NA' NON-SPARKING

An Ex 'n' motor is designed to prevent foreseeable ignition sources in situations, here the machine is operating within its design parameters in normal operation.

The protection principle limits the maximum temperature of all vital parts of the machine, and defines measures to prevent sparks. An explosive gas can thus penetrate the motor without the risk of explosion. Ex 'nA' motors are designed to be used in zone 2 locations.

NON-SPARKING: EX 'N' (NOW OBSOLESCENT)

The design concept for Ex 'N' machines is similar to Ex 'e', but among other minor differences lacks several important features. First, the Ex 'N' certification does not include the starting condition. Unlike an Ex 'e' machine, the surface temperature of the casing is allowed to exceed the maximum permitted by the temperature class during starting. Secondly, no special overload protection is called for. Because of these limitations, Ex 'N' machines are restricted to use in zone 2 or safe areas only.

Problems have been experienced with large Ex 'e' and Ex 'N' motors operating at voltages over 3.3 kV where the motor casing itself has exploded.

The ignition source is associated with partial discharges across the end windings in machines exposed to heavy salt or other contamination. Several operating conditions have been identified where gas could migrate into the motor casing to provide an explosive concentration.

First, where a gas compressor or crude oil pump shares a common lube oil system with its drive motor, gas can become entrained in the lube oil at the process end, and on reaching the motor bearing where the pressure is lower, leave the lube oil. Therefore common lube oil systems should not be used. Secondly, minor gas leaks often occur on gas compressor packages, and the cooling of the motor casing after a shutdown may draw gas in, so that a hazardous concentration exists within the motor when it is restarted. Existing motors considered to be an ignition risk are fitted with a prestart nitrogen purging kit. Type Ex 'N' high voltage motors are no longer specified.

PRESSURISED: EX 'P'

A pressurised motor relies on maintaining all internal parts of the motor enclosure at a greater pressure than atmospheric in order to prevent the ingress of any hazardous gases that may be present in the vicinity of the motor. No special precautions need to be taken with winding temperatures, but the motor interior, especially the stator windings, must be designed to allow a purging air flow to clear any pockets of flammable gas, within the required purging time. Be warned that this may not be the case with the standard industrial version of the machine.

To achieve pressurisation, an external source of dry air must be provided from a safe area. Although this type of motor is inherently cheaper than an Ex 'e' equivalent, the cost saving may be outweighed by the extra costs for the air supplies, pipework and control systems necessary. Another possible cause of delay and expense is the need for special certification to be obtained from an authorised certifying authority such as BASEEFA, UL or PTB.

INTRINSIC SAFETY – EX 'I'A AND EX 'I'B (REFER TO BS EN 60079-11:2012)

Where circuits are required purely for the transmission of signals, data, logic or instructions, the energy in the circuit can be limited to level below which ignition of explosive vapours/gases cannot occur. However, the circuits must also be provided with barriers which will prevent any faults from allowing these limits to be exceeded. The barriers can utilise zenor diode circuits to blow a fuse, or can convert the galvanic signal to an optical signal using 'opto-isolators'. The ignition curves for a resistive circuit are given in Fig. 5.4.1.

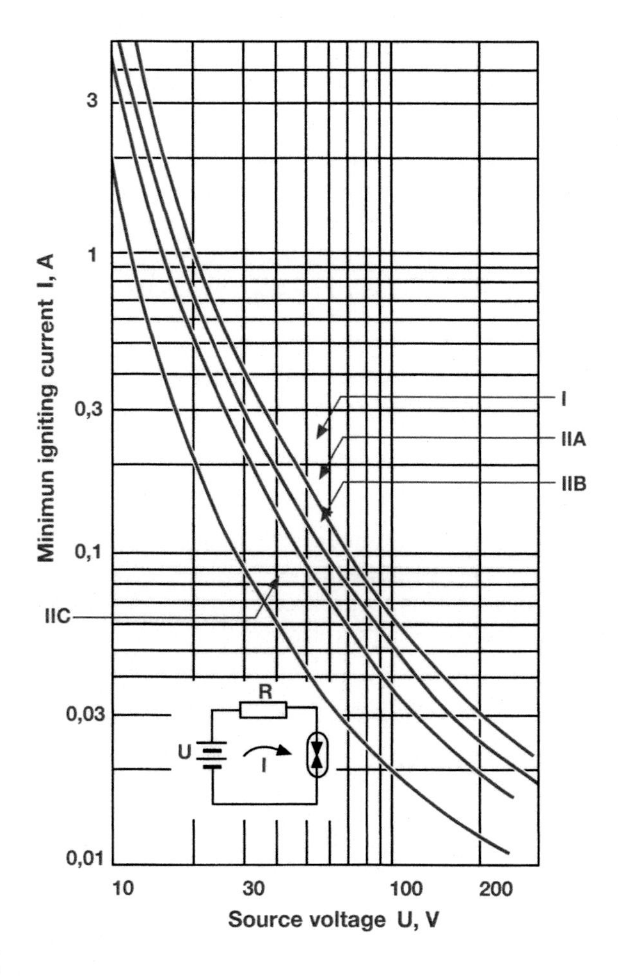

FIGURE 5.4.1

Ignition curves for a resistive circuit.

Table 5.4.3 Relationships Between Different Methods of Assessing Safety Levels

Level of Protection	Countable Faults	ATEX Category	IEC EPL	Normal Zone of Use
ia	2	1	0	0
ib	1	2	1	1
ic	0	3	2	2

LEVEL OF PROTECTION 'IA'

Intrinsically safe circuits in electrical apparatus of category 'ia' must not be capable of causing an ignition during normal operation when two faults occur. The different levels of protection and their interrelationships are given in Table 5.4.3.

1. in normal operation and with the application of those non-countable faults which give the most onerous condition;
2. in normal operation and with the application of one countable fault plus those non-countable faults which give the most onerous condition;
3. in normal operation and with the application of two countable faults plus those non-countable faults which give the most onerous condition.

LEVEL OF PROTECTION 'IB'

Intrinsically safe circuits in electrical apparatus of level of protection 'ib' shall not be capable of causing ignition in each of the following circumstances:

1. in normal operation and with the application of those non-countable faults which give the most onerous condition;
2. in normal operation and with the application of one countable fault plus the application of those non-countable faults which give the most onerous condition.

LEVEL OF PROTECTION 'IC'

Intrinsically safe circuits in electrical apparatus of level of protection 'ic' shall not be capable of causing ignition in normal operation. Where distances are critical for safety, they shall meet the requirements of IEC 60079-11.

(Note: A countable fault is a fault which occurs in parts of electrical apparatus conforming to the constructional requirements of IEC 60079-11). The manufacturer is required to issue a certificate of conformity as typically shown in Fig. 5.4.2.

(A)

IECEx Certificate of Conformity

INTERNATIONAL ELECTROTECHNICAL COMMISSION
IEC Certification Scheme for Explosive Atmospheres
for rules and details of the IECEx Scheme visit www.iecex.com

Certificate No.:	IECEx DEK 12.0070X issue No.:1	Certificate history: Issue No. 1 (2013-3-22) Issue No. 0 (2012-11-19)
Status:	**Current**	
Date of Issue:	**2013-03-22** Page 1 of 5	
Applicant:	**R.STAHL Schaltgeräte GmbH** Am Bahnhof 30 74638 Waldenburg **Germany**	
Electrical Apparatus: *Optional accessory:*	**Digital Output Module (DOM), Type 9475/3*-**-****	
Type of Protection:	**Ex ia, Ex nA**	
Marking:	Type 9475/32-**-e* (with e = 1-6): **Ex ia [ia Ga] IIC T4 Gb** **[Ex ia Da] IIIC** Type 9475/33-**-e* (with e = 1-6): **Ex nA ia [ia Ga] IIC T4 Gc** **[Ex ia Da] IIIC** Type 9475/32-**-e* (with e = 1-7): **Ex ia [ib Gb] IIC T4 Gb** **[Ex ib Db] IIIC** Type 9475/33-**-e* (with e = 1-7): **Ex nA ia [ib Gb] IIC T4 Gc** **[Ex ib Db] IIIC**	

Approved for issue on behalf of the IECEx Certification Body: R. Schuller

Position: Certification Manager

Signature:
(for printed version)

Date:

1. This certificate and schedule may only be reproduced in full.
2. This certificate is not transferable and remains the property of the issuing body.
3. The Status and authenticity of this certificate may be verified by visiting the Official IECEx Website.

Certificate issued by:

DEKRA Certification B.V.
Utrechtseweg 310
6812 AR Arnhem
The Netherlands

DEKRA

FIGURE 5.4.2

(A) Extract from IECEx certificate of conformity. (B) Extract from IECEx certificate of conformity – page showing limiting IS electrical parameters.

(B) Courtesy R. Stahl, Birmingham, UK.

(B)

Annex 1
To IECEx DEK 12.0070X, issue No.:1, and NL/DEK/ExTR12.0069/01.
Digital Output Module(DOM) Type 9475/3*-**-**

Electrical data

Circuit connecting to the IS1 or IS1+ System:

Power supply (input); Plug to BusRail V101/ Pin 7, 8, 9, 10 (+), Pin 27, 28, 29, 30 (–):
in type of protection intrinsic safety Ex ia IIC, with the following maximum values:
U_i = 26.2 V.
The circuit is equipped with an internal current limitation that limits the current to 450 mA.

Address- and Databus (communication); Plug to BusRail V101/ Pin: 4 (Bus Red.); 5 (Bus Prim.);
14, 15, 16, 24 (Bank 1-4); 1, 11, 21 (Mod. Select):
in type of protection intrinsic safety Ex ia IIC, only for connection to the internal Address- and
Databus of the IS1/IS1+ System with the following maximum values:
U_o = 6.6 V; I_o = 102 mA; P_o = 168 mW
U_i = 6.6 V; C_i = 0 nF; L_i = 0 mH

Electronic switch control (input); Plug to BusRail V101/ Pin: 18, 19:
in type of protection intrinsic safety Ex ia IIC, with the following maximum values:
U_o = 26.2 V; I_o = 5.4 mA.

Intrinsically safe field circuits:

8-Channel Devices Model 9475/3*-08-**:
X1 – Channel 0 (1+/2-); Channel 1 (3+/4-); up to; Channel 7 (15+/16-)

4 Channel Devices Model 9475/3*-04-**:
X1 – Channel 0 (1+/2-); Channel 1 (5+/6-); Channel 2 (9+/10-); Channel 3 (13+/14-)

The values of L_o and C_o in the following tables are the maximum values for combined inductance and capacitance (including cable inductance and capacitance). The values for L_o and C_o marked in grey are the values determined according to the curves and tables of EN 60079-11, Annex A. These grey marked values may be used for the assessment as per EN 60079-11, clause 10.1.5.2.

The internal capacitance per channel is already taken into account in the L_o and C_o values shown in the tables below. The internal inductance is negligibly small.

FIGURE 5.4.2, cont'd

'SIMPLE' APPARATUS AND COMPONENTS

Simple electrical apparatus and components (e.g., thermocouples, photocells, junction boxes, switches, plugs and sockets, resistors, LEDs) may be used in intrinsically safe systems without certification provided that **they do not generate more than 1.2V, 0.1A, and 25mW in the intrinsically safe system in the normal or fault conditions of the system.**

The 'Simple' apparatus should conform to all relevant requirements of the BS EN 60079-11. Simple apparatus intended for use in **Zone 0** requires specific protective measures. For such apparatus, EN 60079-26 and EN 1127-1 apply in addition to EN 60079-0 and 60079-11. Special requirements, e.g., for electrostatic charging of plastic materials, as well as partitioning walls and the mounting of the apparatus are specified in this standard. According to EN 1127-1, temperatures shall not exceed 80% of the limit temperature of the temperature class.

Junction boxes and switches, however, may be treated as T6 (85°C) at a maximum ambient temperature of 40°C because, by their nature, they do not contain heat-dissipating components.

A wide variety of 'Feed Through' and 'Disconnect' terminals can be fitted in simple apparatus enclosures. Disconnect terminals that do not require the conductors to be removed from the terminals for test and calibration purposes are particularly useful during operational conditions.

It is important that the external terminal connections maintain 3-mm clearance between bare metal parts of the same IS circuit and 6 mm between bare metal parts of different IS circuits. Some users prefer an Ex 'ia' certified enclosures for IS circuits.

Simple apparatus such as level switches and other safety switches should have gold-plated contacts due to the low voltages present in IS circuits, otherwise the change from closed to open circuit will not register.

The essential requirements of an intrinsically safe system are that:

- The system must work.
- The apparatus in the system must be 'certified' or 'simple'.
- The compatibility of the apparatus must be established.
- The required level of protection of the system is known.
- The required temperature classification and ambient temperature rating of each piece of apparatus is known.
- The parameters of the cable are known.

LOOP CALCULATIONS FOR GALVANIC BARRIERS

Every intrinsically safe loop will be made up of one or more field devices, at least one cable and the zenor barrier. The field devices and the cable types used will have known values of capacitance and inductance (capable of storing energy). In order to ensure that the barrier operates reliably, a loop calculation must be performed for each type of IS circuit, i.e., where the field components and the barrier type are identical, and the cable type is identical (see Fig. 5.4.3). The calculation will then be valid for those components, and for a maximum length of the cable type utilised.

Where the field device is a certified item (i.e., not simple apparatus, e.g., a temperature transmitter or a solenoid valve), extra checks are necessary. The certificate of the field device will include its maximum input parameters which will specify one or more of the values Ui, Li and Pi. Compatibility must be checked by ensuring that the maximum input figures of the field device are not exceeded by the maximum output values of the chosen barrier. If the system includes more than one item of certified apparatus, then compatibility with the barrier must be checked separately. The

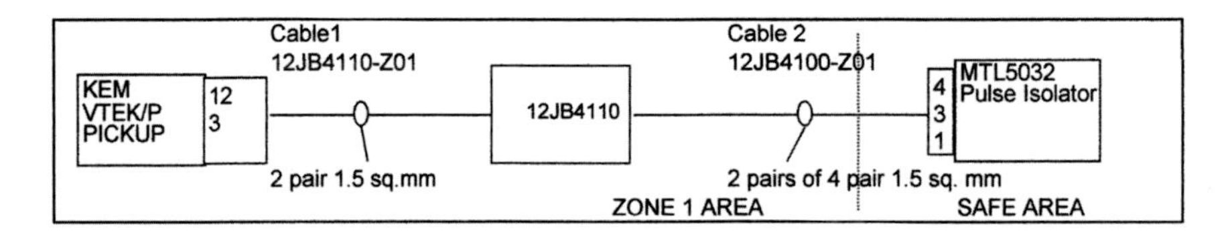

Field Apparatus	Zone 1			DMT BVS 99.E.2008
PICK UP				
U max (Va)		28.01	V	
I max (Ia)		0.1	A	
W max (Wa)	(at T5)	1	W	
Internal Capacitance (Ca)		0.18	uF	
Internal Inductance (La)		0.6	mH	

Hazardous Area Field Cable							
			Cable 1			Cable 2	
Cable parameters		per metre	length (m)	Total	per metre	length (m)*	Total
Capacitance - (Cc) (uF)		0.00014	30	0.0042			0
Inductance - (Lc) (mH)		0.00068	30	0.0204			0

Barrier Details MTL 5032 PULSE ISOLATOR			BASEEFA Ex 95D2416 / Ex 95D2417	
Open Circuit Voltage		Vb	28	V
Short Circuit Current		Ib	0.09	A
Max Allowable Power		Wb	0.63	W
Maximum Allowable Capacitance		Cb	0.65	uF
Maximum Allowable Inductance		Lb	14	mH

Safety Analysis			
Voltage	Va > Vb		Yes
Current	Ia > Ib		Yes
Power	Wa > Wb		Yes
Capacitance	Ca + Cc < Cb		Yes
Inductance	La + Lc < Lb		Yes

* ASSUMED CABLE LENGTH

FIGURE 5.4.3

Typical IS loop calculation.

addition of simple apparatus will not affect the compatibility, except that the system temperature might be derated.

The system will be categorised according to the least favourable components of the barrier category and the apparatus category. For example, a barrier of (Ex ia) IIC with a field device of Ex ia IIC T6 will categorise the system as Ex ia IIC T6. The addition of a field device of Ex ia IIC T4 will change the system category to Ex ia IIC T4.

SELECTION OF OTHER EQUIPMENT

Much of the explanation above regarding hazardous area motors may be applied to other hazardous area equipment. Switchgear, being inherently sparking equipment, must be enclosed in an explosion-proof or pressurised enclosure, as it cannot meet the criteria for increased safety Ex 'e' certification.

Enclosures used for housing cable terminals may be of explosion-proof or increased safety design.

Common types of enclosure and limitations affecting their use and those for cable glands and transits are discussed in PART 2 Chapter 8.

Hazardous area lighting is discussed in PART 2 Chapter 14.

AVOIDANCE OF IGNITION BY NON-ELECTRICAL EQUIPMENT (REFER BS EN ISO 80079-36 & 37)

There are many mechanical devices that operate with high surface temperatures or generate sparks, when running, or if they have developed a fault. The list below is not exhaustive:

- reciprocating engine turbo-chargers and exhausts
- gas-turbine casings
- pumps
- gearboxes
- Belt drives and continuously variable transmission (CVT) drives
- fan blades impinging on ducting
- process heaters

All of these devices will need to be risk-assessed at the very least, and mitigating devices fitted (e.g., spark arresters, water jackets, anti-static belts etc.). It is recommended that alternative equipment be considered if available, which does not need such protection, since mitigating devices will require critical safety maintenance. CVT drives are not recommended, because although anti-static belts will prevent static sparks, the belts are prone to becoming displaced and jammed against the driving pulley, generating a hotspot. It would be possible to pressurise or nitrogen purge the CVT housing, but an alternative electrical VSD would probably be a simpler solution.

AVOIDANCE OF IGNITION BY RADIO AND RADAR TRANSMISSIONS (REFER TO PD CLC/TR 50427)

Electromagnetic waves of sufficient power will generate sparks in metallic objects, as can be experienced with household microwave ovens. Radar and radio transmitters helicopter navigation beacons may generate high power, and therefore cause ignition should any gas be present close to the transmitter aerial (antenna). In any case, such high-powered transmitters should be shut down on a gas detection alarm. Guidance on which transmitters may produce an ignition risk is given in the above reference document (PD CLC/TR 50427).

Where ships are required to make a close approach to an installation (e.g., supply vessels), radar antenna on the bigger ships may be level with hazardous areas on the platform. Therefore, such vessel radar sets considered a risk should be temporarily shut down. GPS for dynamic positioning should not be affected.

HAZARD SOURCE SCHEDULES

The process design must be scoured for every possible point where flammable substances are likely to leak. Only continuously welded pipes and vessels are not regarded as potential points of release. This task is best done by a team of engineers which includes a chemical or process engineer, an instrument/systems engineer and an electrical engineer. It is recommended that a hazop study be carried out concurrently with this task.

The resulting hazardous source schedule should give details of each source of hazard, its location, risk of occurrence, extent and type of hazardous zone produced and remarks regarding any environmental aspects affecting the hazard such as dilution, ventilation and airlocks. The schedule should also refer to a relevant and up-to-date hazardous area boundary drawing. If the installation already exists and is due to be modified or expanded, it will require to be thoroughly surveyed to ensure that the existing source schedules reflect the current situation. A typical hazardous source schedule is shown in Fig. 5.4.4.

DEFINING BOUNDARIES

The extent of the hazardous zone round a source of release is usually determined in accordance with the British Institute of Petroleum Model Code of Safe Practice or equivalent American API Code. This gives details illustrated by diagrams of typical situations with distances of the zone boundary from the point of release.

VENTILATION

The degree of ventilation around the source of release will affect the classification of the hazard, unless one of the three following situations apply:

1. Open spaces with no structure or equipment restricting substantially free circulation of air, vertically and horizontally.
2. A module with a roof and not more than one side closed, free from obstruction to natural passage of air through it, vertically and horizontally.
3. Any enclosed or partly enclosed space provided with artificial ventilation to a degree equivalent to natural ventilation under low wind velocity conditions, and having adequate safeguards against failure of the ventilation equipment.

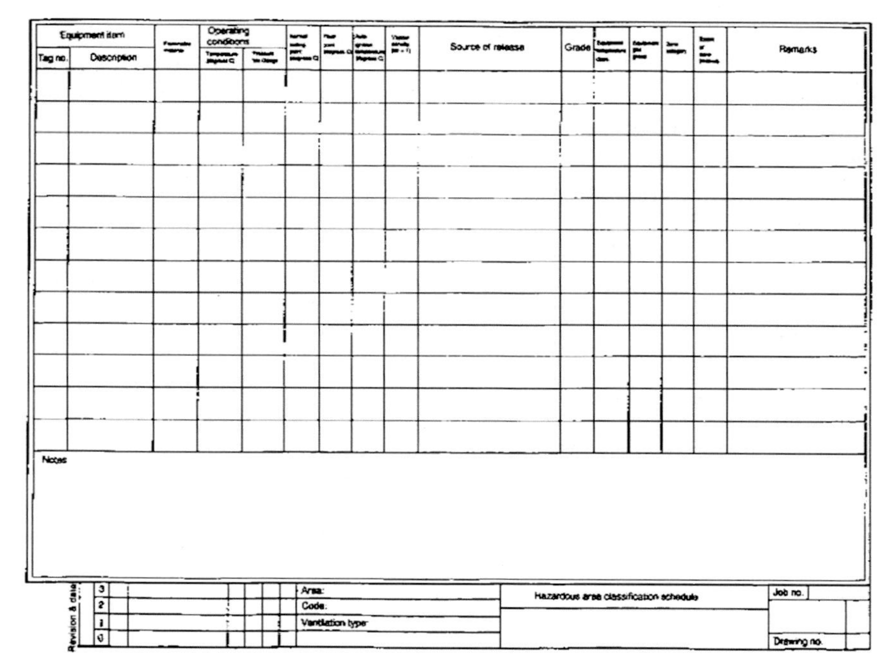

FIGURE 5.4.4

Typical sheet from hazardous source schedule.

The number of air changes per hour required to meet the natural ventilation equivalent condition in the third case will vary to some extent, depending on the degree of hazard, but a figure of 40 changes per hour is quoted by underwriters. The positioning of fan inlets and exhaust outlets must be such that no unventilated pockets are left. The adequate safeguard is usually an alarm followed, after a suitable time delay, by the shutdown and venting to the flarestack of all process equipment within the module.

Process modules are usually ventilated such that the internal air pressure is slightly negative with respect to the outside, in order to retain any minor gas leaks. Safe area modules are kept at a slightly positive pressure in order to prevent the ingress of gas. The ventilation systems of different modules must be well segregated to prevent cross-contamination of leaks. Fire and explosion dampers are installed to automatically seal ventilation systems if fire or gas is detected. Ventilation intakes for both safe and hazardous areas should be located well within safe areas to minimise the risk of drawing gas into the ventilation system.

A common problem on older installations is the location of secondary pressurised control rooms in the middle of a hazardous module. A typical example of this is a gas compression module with its own control room. If the control room pressurisation fails and remains failed for a timed period, the compression module should shut down and all non-certified electrical equipment in it should be isolated. Often, a loss of pressurisation will only give an alarm; the decision to shut down the plant will be left with the operator, and prevention of gas ingress will depend on an airlock door system. Apart from the serious ignition hazard that all the live non-certified equipment poses in this situation, the control operator's only means of escape is via the compression module where the gas leak has occurred. Owing to the amount of work involved in providing automatic isolation, with shutdown contactors for all the electrical supplies, and separating the instrument and control functions so that the main installation control room is unaffected, it could well be more beneficial to relocate the compressor control room to a safe area or possibly include its functions in an extended central installation control room. The last option would be more in line with the Cullen Report, as it would remove an extra manned control room. Both relocation options eliminate the need to provide an escape route through the compression module protected to some degree from fire and explosion.

The ventilation of engine enclosures is discussed in PART 5 Chapter 4.

A detailed discourse on the ventilation of offshore installations is beyond the scope of this book. However, the following is a list of basic criteria with which such ventilation systems should comply:

1. the positive pressurisation of non-hazardous areas with respect to adjacent hazardous areas or the external atmosphere;
2. the containment, dilution and removal of potentially hazardous concentrations of explosive gaseous mixtures in hazardous modules, and the adequate segregation of hazardous area ventilation exhausts from ventilation and engine intakes;
3. the provision of comfortable environmental conditions in accommodation and normally manned non-hazardous areas, and acceptable working conditions in normally unmanned areas;
4. the provision of acceptable working conditions within hazardous modules;
5. the isolation of individual areas and control of ventilation in emergency conditions, in accordance with the shutdown logic of the installation's ESD, fire, gas and alarm systems;
6. the provision of combustion air for essential prime movers, ventilation air for escaping, firefighting and rescue personnel and purging air services required to operate effectively during an emergency.

To avoid depressurisation and potential release of explosive mixtures, all exits to hazardous areas should be airlocked with self-closing doors. If single doors are used, a hazardous zone will extend outside the compartment.

Safe area control rooms may be located within hazardous modules provided they are kept pressurised with air obtained from a non-hazardous external area. Purging, alarm and timed shutdown facilities are required, similar to those for a pressurised motor. However, location of control rooms in safe areas is preferred for reasons given above.

LOGIC OF AREA CLASSIFICATION

Fig. 5.4.5 shows a typical flow diagram for selection of hazardous zones. The flow diagram is used in conjunction with a questionnaire, such as that shown in Fig. 5.4.6. The resulting zone boundaries are drawn on to a set of plans and elevations for the installation in order to produce the required hazardous area boundary drawings. Throughout the duration of a design project, regular meetings between discipline engineering heads and safety representatives are held to discuss and revise these drawings. Two typical hazardous area drawings are shown in Fig. 5.4.7.

SELECTION OF MOTORS FOR HAZARDOUS AREAS

It is common practice in offshore installations to require all motors for use in hazardous areas to be suitable for zone 1 areas, although in many cases the actual area classification might be zone 2. This allows

- areas to be reclassified from zone 2 to zone 1 without the need to replace motors;
- enables the safer transfer of motors from one hazardous area to another;
- tends to rationalise spare parts holdings.

Where a motor is used as part of a variable speed drive system, the hazardous area certification will not be valid unless the motor has been tested for safe operation over the whole speed range through which it is required to operate. At low speeds the motor will have different heat dissipation characteristics and, if dependent for cooling on a rotor driven fan, will have a reduced cooling air flow rate. Therefore there is a risk that the enclosure temperature may exceed the required maximum surface temperature requirement.

It is now known that large motors whose casings are made up of bolted sections may suffer sparking across the section joints due to voltages produced by stray currents induced in the sections. To avoid risk of ignition, all such motors should now be fitted with copper braid bonding straps across each section joint in order to prevent any potential build-up across the joint.

In general, there are four methods of motor design to achieve suitability for use in hazardous areas, as follows.

MIXING HAZARDOUS AREA CERTIFIED EQUIPMENT

VARIOUS GLOBAL AND NATIONAL STANDARDS

The certification concept, for example, Ex ‘d’ or Ex ‘e’ can be selected based on the most suitable and economic design. However, mixing between different national and international standards would be ill-advised and dangerous as they are not always compatible. This is also likely to lead to confusion in the maintenance workforce and will require separate storage of the two types of spares for certified equipment.

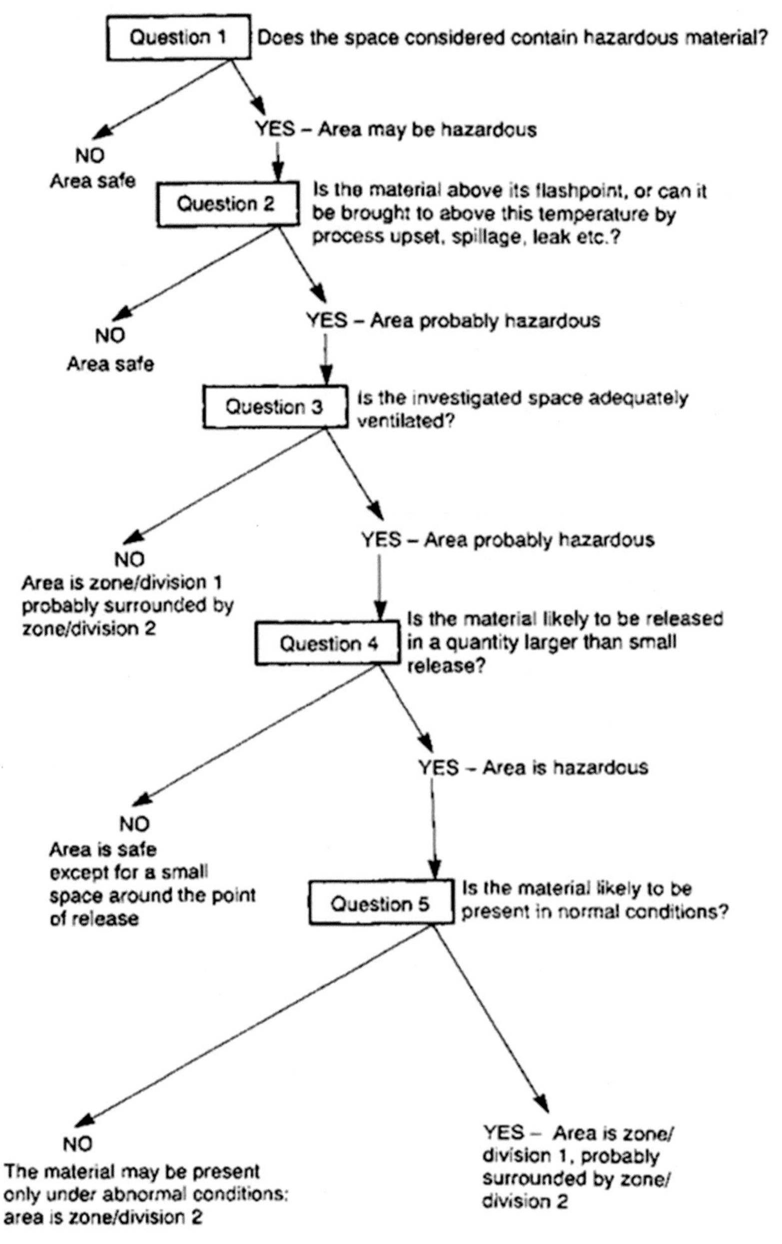

FIGURE 5.4.5

Logic flow selection of hazardous zones.

Hazard area classification sheet	Prepared by				
	Date				
	Issue	1	2	3	4
Client	Project no.				
Plant	Area		PDF no.		
Hazard point	Point no.		ELD no.		

Material / Vapour/gas / Liquid / Solid	Nature of hazard	Flash point	Material factor	Vapour density	Note reference
Component 1					
2					
3					
4					
Temperature of material	Flash point				

Density of the vapour/gas of the material relative to air at atmospheric pressure and ambient temperature, r =

1. **Does the investigated space contain potentially hazardous material**
   Otherwise than: (a) in pipes with no flanges, valves and fittings
   (b) stored in special containers approved for the material
   American (c) in pipes with flanges, valves and fittings located in space which is adequately ventilated

   If Yes – Area may be hazardous
   If No – Area is safe
   Yes ☐ No ☐

2. **Is the material at a temperature higher than its flash point, or is there:**
   (a) A source of heat at a temperature higher than the flash point of the material
   (b) An oxidising agent
   actually or potentially present within the distances from the point as indicated in the appropriate case in the examples of extent of hazard

   If Yes – Area probably hazardous
   If No – Area is safe
   Yes ☐ No ☐

   Note reference: **Typical examples of extent of hazard areas**

3. **Is the investigated space adequately ventilated**
   If No – Area is zone/division 1, probably surrounded by zone/division 2
   If Yes – Area probably hazardous
   Yes ☐ No ☐

4. **Is the hazardous material likely to be released in a quantity larger than small release**
   If Yes – Area is hazardous
   If No – Area is safe except for small space around the point of release
   Yes ☐ No ☐

5. **Is the hazardous material likely to come into contact with atmospheric air in the normal conditions**
   If Yes – Area is zone/division 1, probably surrounded by zone/division 2
   If No – Area is zone/division 2
   Yes ☐ No ☐

6. **Area and equipment classification** British ☐ American ☐ Other (specify) ☐
   A. Maximum extent from the point of zone/division 1 area
   B. Maximum extent from the point of zone/division 2 area
   C. Ignition temperature of hazardous material

7. Notes

FIGURE 5.4.6

Questionnaire form to aid hazardous zone selection.

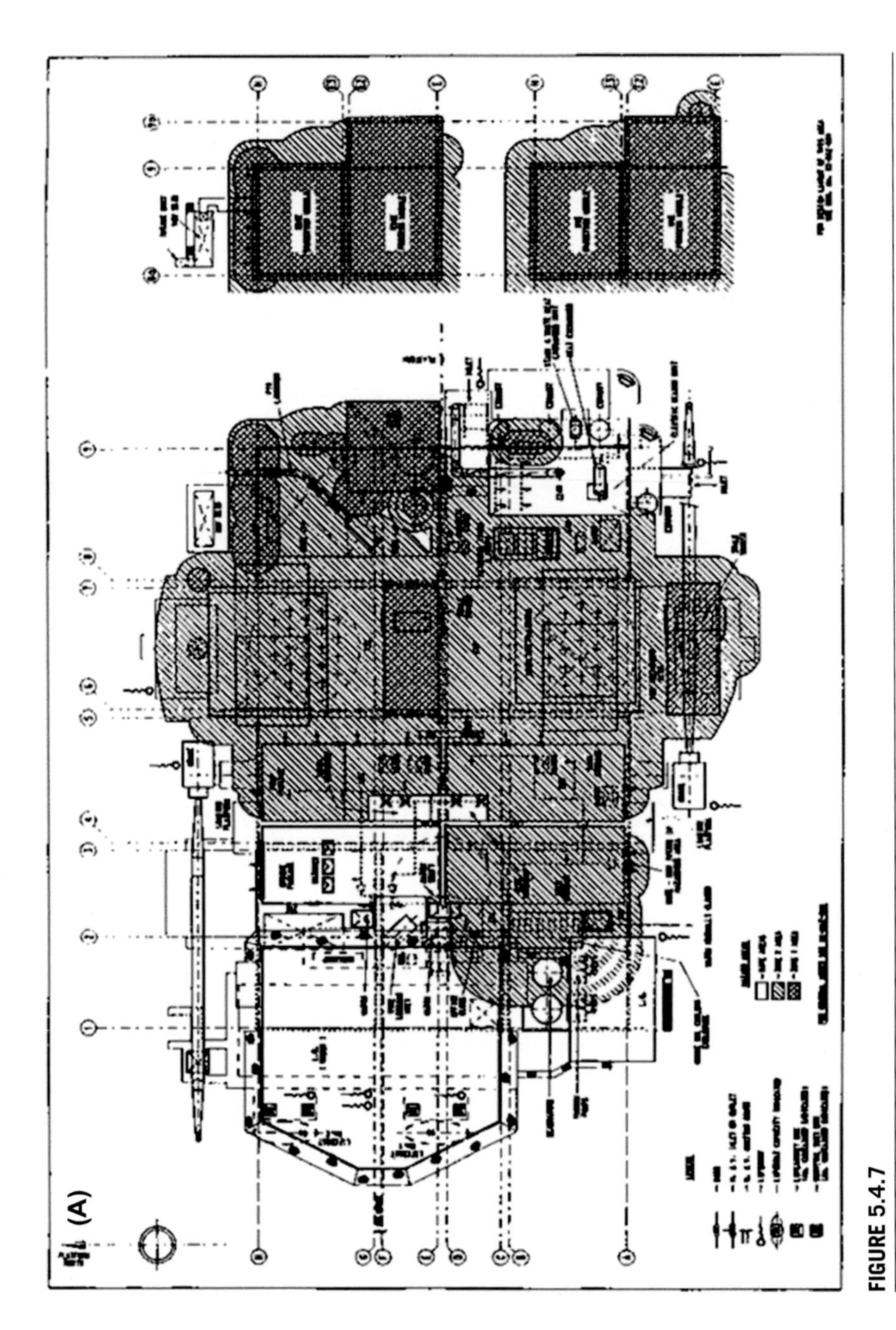

FIGURE 5.4.7
(A and B) Typical hazardous area boundary drawings.

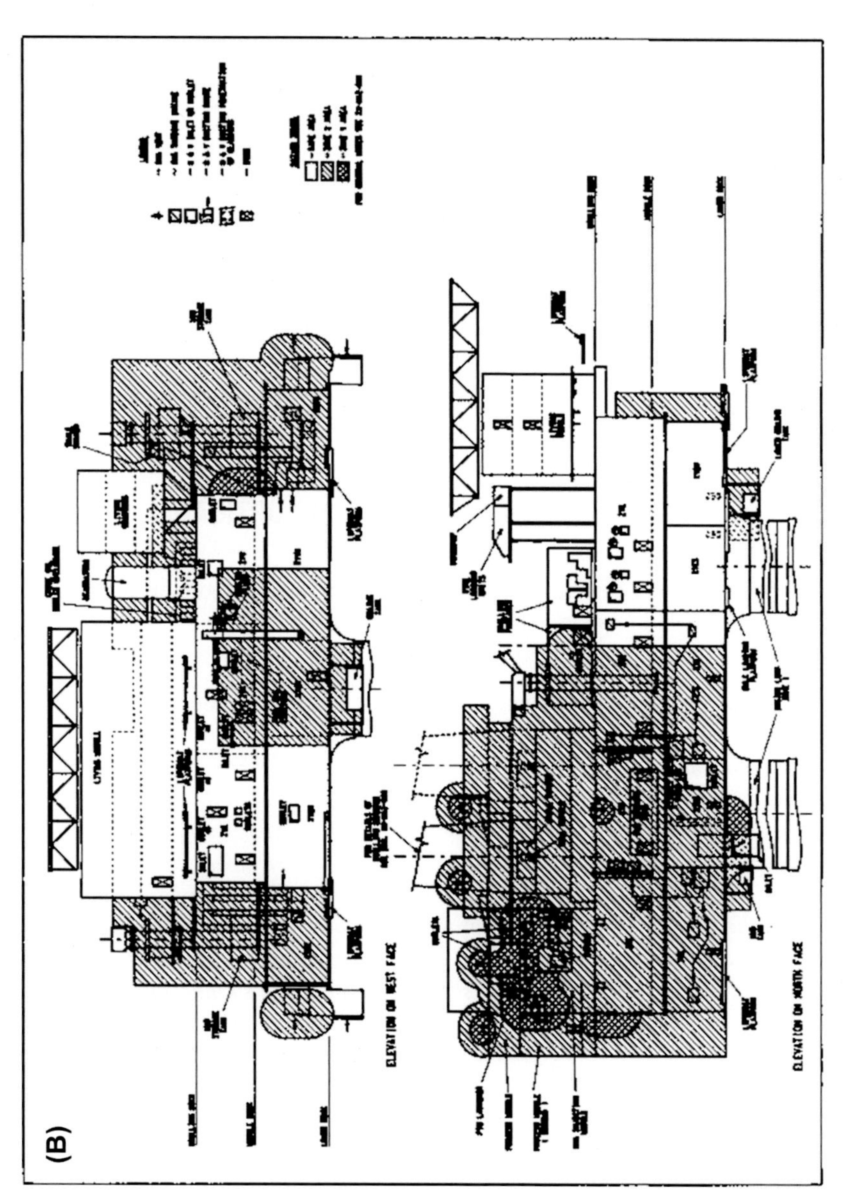

FIGURE 5.4.7, cont'd

The American standards allow defined, signposted area zoning of IEC standard areas, and this is more workable on a shore process plant, where say a new process plant certified to European standards is being built in the middle of an existing site. However, this is NOT recommended offshore and hopefully the Client Company and the Certification Authority involved would not accept it anyway.

EQUIPMENT FOR USE IN POLAR REGIONS

The very low temperatures involved may cause materials to behave very differently, and this must be taken into account when specifying, for example, cable insulation.

The certification for hazardous area Ex equipment, particularly Ex 'd' has a lower temperature limit, which must be observed. This is because if the explosive gas mixture is present, its density will be greater at low temperatures and hence the gas present in the Ex 'd' enclosure will have a higher calorific value, leading to a more powerful explosion. The 'explosion-proof' nature of the enclosure will then be compromised. Canadian Coast Guard standards are helpful with regard to this issue (See Canadian Marine Safety Guidelines Transport Canada).

INTRINSICALLY SAFE BARRIERS – GALVANIC OR OPTICAL ISOLATION?

In most situations, opto-isolation is preferred, although usually more expensive and this will add up if there are a large number of control loops.

Using opto-isolators may have the following benefits:

1. Prevents issues with earths (grounds) at different potential producing circulation currents.
   The MTL website states - '*The current flowing though the bonding conductor generates a potential difference between the IS earth point at the barrier and the neutral star point. The outside of the field-mounted instrument is bonded to the neutral star point, and the internal circuits of the instrument are connected to the barrier busbar. The potential difference between IS earth point at the barriers and the neutral star point is therefore transferred to the hazardous area. It is normally safe because these internal circuits are isolated from the instrument housing, but this potential difference should be minimised so that if there is a local insulation failure no danger can arise. The installation conditions of barriers, screened transformers etc. nearly all call for a return path impedance less than 1 ohm. A figure of 0.1 ohm is normally achievable and is more desirable.*
   It is important to remember that the resistance involved is that of the return conductor between IS earth points at the barriers and the neutral star point and the resistance of the earth mat is not important for this purpose. These principals are normally applied within the UK'.
2. Opto-isolators may be more reliable giving the control loop a higher SIL Level (see PART 9 Chapter 2).

PART

Specific Systems and Vessel Types

CHAPTER

Notes on Dynamic Positioning and Ballast Control for Floating Installations

1

DYNAMIC POSITIONING

Apart from the obvious advantages of docking a vessel more easily in crowded ports, dynamic positioning (DP) becomes vital to avoid collisions when it is necessary to hold position close to an offshore installation or maintain position over subsea equipment.

Examples of this are

1. a supply boat offloading containers to the laydown area on a platform,
2. a floating production, storage and offloading (FPSO) unit offloading crude to a dynamically positioned tanker,
3. a pipelaying vessel offloading pipe along a strictly contoured seabed route,
4. a diving vessel with divers deployed.

There are a number of defined grades or classes of DP depending on

1. the weather conditions that the ship can keep position in using its full propulsion capabilities,
2. the degree of redundancy in its machinery (e.g., standby generation),
3. the degree of redundancy in its position reference systems (e.g., GPS),
4. whether the back-up equipment is physically separated to avoid common-mode failure,
5. whether the DP system has an independent joystick control which the DP operator can use if the automated control system fails.

These classes have been developed by IMO MSC/Circ.645 and are listed in Table 6.1.1.

As can be seen from Table 6.1.1, for a dynamically positioned ship to hold position, there are a number of ship systems that are required to support the propulsion.

Class 2 and 3 vessels are normally utilised in offshore operations, with class 3 vessels reserved for the most onerous duties, for example, operation with divers deployed.

Offshore Electrical Engineering Manual. https://doi.org/10.1016/B978-0-12-385499-5.00034-0

Table 6.1.1 DP Class Notations: Comparison With DNV GL and ABS

IMO Equipment Class	Class Notations		LR	Additional Information
	DNV GL	ABS		
Not applicable	DYNPOS(AUTS)			DP system without redundancy.
		DPS-0		Manual position control system fitted with automatic heading control and with a freestanding position reference system.
			DP(CM)	No redundancy required, but secure UPS units provided for critical systems, and automatic starting of standby generator.
IMO equipment class 1	DPS(1)			DP system with an independent joystick system back-up and a position reference back-up.
	DYNPOS(AUT)			DP system with an independent joystick system back-up and a position reference back-up. Additional requirements to achieve higher availability and robustness as compared to DPS(1) will apply.
		DPS-1		In line with IMO equipment class 1.
			DP(AM)	An automatic and a manual control system are to be provided and arranged to operate independently. At least two position reference systems suitable for the intended service conditions and incorporating different measurement techniques are to be provided and arranged.
IMO equipment class 2	DPS(2)			DP system with redundancy in technical design and with an independent joystick system back-up. Verification by FMEA required.
	DYNPOS(AUTR)			DP system with redundancy in technical design and with an independent joystick system back-up. Additional requirements to achieve higher availability and robustness as compared to DPS(2) will apply. Verification by FMEA required.
		DPS-2		In line with IMO equipment class 2 • two main switchboards with bus-ties, in separate compartments • distribution system redundant, in separate compartments • power management system required • redundant thrusters in separate compartments • three position reference systems • wind sensors: 3 • MRU: 3 • gyro: 3 Verification by FMEA required. For other notes see ABS Guide For Dynamic Positioning Systems. 2013: Table 1, Summary of DP System Requirements for ABS DPS Notations.
			DP(AA)	Power, control and thruster systems and other systems necessary for, or which could affect, the correct functioning of the DP system are to be provided and configured such that a fault in any active component or system will not result in a loss of position. This is to be verified by means of an FMEA. At least three heading reference sensors are to be provided and arranged so that a failure of one sensor will not render the other sensors inoperative. Two automatic control systems are to be provided and arranged to operate independently, so that failure in one system will not render the other system inoperative. For other requirements see LR Rules PART 7 Chapter 4.

IMO equipment class 3	DPS(3)			DP system with redundancy in technical design and with an independent joystick system back-up. Plus a back-up DP control system in a back-up DP control centre, designed with physical separation for components that provide redundancy. Verification by FMEA required.
	DYNPOS(AUTRO)			DP system with redundancy in technical design and with an independent joystick system back-up. Plus a back-up DP control system in a back-up DP control centre, designed with physical separation for components that provide redundancy. Additional requirements to achieve higher availability and robustness as compared to DPS(3) will apply. Verification FMEA required. For other notes see DNV Rules PART 6 Additional Class Notations Chapter 3 Navigation, Manoeuvring and Position Keeping.
		DPS-3		In line with IMO equipment class 3, i.e., • two main switchboards with bus-ties, in separate compartments • distribution system redundant, in separate compartments • power management system required • redundant thrusters in separate compartments • three position reference systems • wind sensors: 2 + 1 in back-up control station • MRU: 2 + 1 in back-up control station • gyro: 2 + 1 in back-up control station Verification by FMEA required. For other notes see ABS Guide For Dynamic Positioning Systems. 2013: Table 1, Summary of DP System Requirements for ABS DPS Notations.
			DP(AAA)	The DP system is to be arranged such that failure of any component or system necessary for the continuing correct functioning of the DP system, or the loss of any one compartment as a result of fire or flooding, will not result in a loss of position. This is to be verified by means of an FMEA. Generating sets, switchboards and associated equipment are to be located in at least two compartments separated by an A–60 class division. An additional/emergency automatic control unit is to be provided at an emergency control station, in a compartment separate from that for the main control station, and is to be arranged to operate independently from the working and standby control units. For other requirements see LR Rules PART 7 Chapter 4.

DP, *dynamic positioning;* FMEA, *failure mode and effects analysis;* IMO, *International Maritime Organization;* MRU, *motion reference unit.*

INTEGRATED CONTROL SYSTEMS

Sophisticated computer-based Integrated Automation System (IAS) provide a supervisory control, alarm and data acquisition functions and integrate the control and monitoring systems into one single system. IAS is based on distributed processing units (DPUs) in which the various process parameters are controlled via input/output field stations, located in different places of the vessel. Man–machine interface for the vessel management system (VMS) are multiple redundant operator station that include the following functions:

1. power management system
2. alarm and monitoring system
3. auxiliary control systems
4. ballast automation systems
5. propulsion control systems including thruster control
6. DP systems including joystick system.

Operator stations allow control and monitoring vessel equipment including and are receiving process alarms. It is possible to change limits or parameters on one operator station and that information is automatically updated in all relevant operator stations. All data are displayed on custom-made process mimics. All monitoring and automation functions are carried out by the DPUs. Communication between operator stations is via redundant local area networks (Fig. 6.1.1).

PROPULSION REDUNDANCY

The automation should include an algorithm/truth table which decides whether the propulsion setup is still viable as a DP system when a propulsion device has failed. One bow thruster by itself is obviously not viable, but the system may be a mixture of conventional propeller and rudder, azimuth thrusters and tunnel bow thrusters, so depending on which device fails, and on its relative position, the DP system may or may not be viable.

BALLAST CONTROL SYSTEMS

The movement of heavy fluids aboard a floating vessel such as an FPSO or semi-submersible has a pronounced effect on the attitude of that vessel in the water, which must be counteracted by filling and emptying built-in ballast tanks with water (Figs 6.1.2 and 6.1.3).

FLOATING PRODUCTION, STORAGE AND OFFLOADING UNITS

If too much weight flows to one side or another, the vessel tends to list that way and, in extreme cases, may be in danger of capsizing, particularly if the vessel is lightly

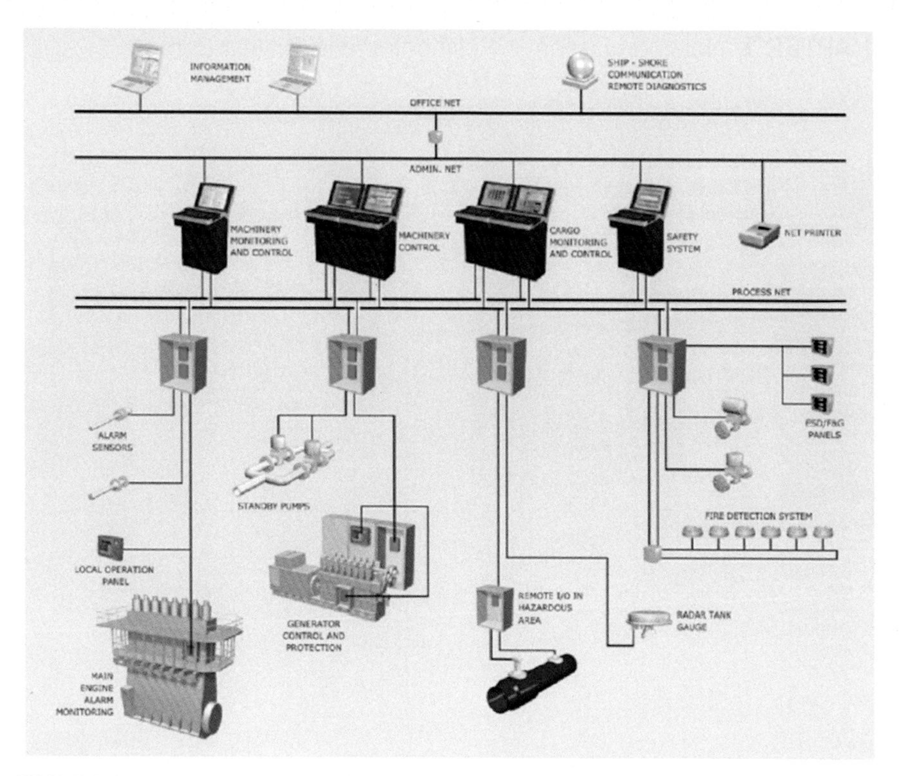

FIGURE 6.1.1

Integrated ship dynamic positioning control system.

Courtesy Kongsberg.

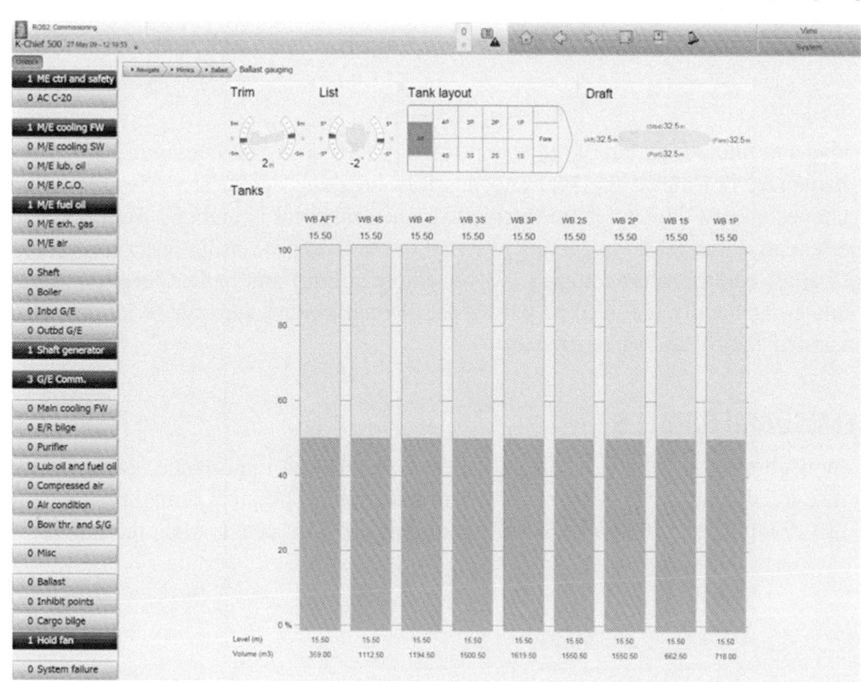

FIGURE 6.1.2

Typical tank level display.

Courtesy Kongsberg.

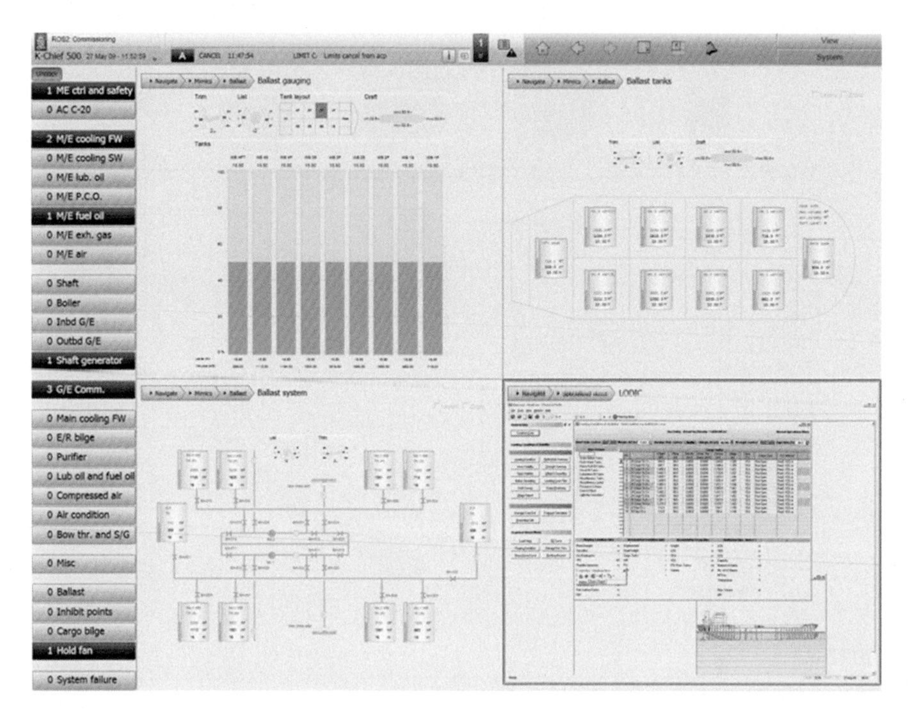

FIGURE 6.1.3

Typical multiscreen display. Top left, ballast gauging; top right, ballast tanks; bottom left, ballast system and bottom right, LODIC loading calculator.

Courtesy Kongsberg.

loaded and riding high out of the water. If too much cargo is stored amidships, there is danger of bowing at the keel and possibly breaking apart. The effects of fluid dynamics aboard the vessel are intensified when material is pumped from tanks in one section of the vessel to another as a part of onboard processing operations, making ballast control more complex. Valve remote control and ballast level measurements are generally included as part of the overall system and can be inputs to an integrated control and safety system.

SEMISUBMERSIBLES

A semisubmersible's ballast control systems are of vital importance, as they have to take in large amounts of ballast water to reach their operating draught from their transit draughts and discharge when operations are completed. Also, the nature of their work is such that they frequently require changes of draught, list and trim, often in a very few minutes' notice. They can be working in close proximity to other vessels, carrying out combined operations in the open sea in seriously bad weather conditions.

BALLAST CONTROL: BASIC CONTROL FEATURES

The central ballast control station should be located above the damage waterline and in a space not within the damage penetration zone, adequately protected from weather. The following control and indicating systems should be provided where applicable:

1. bilge and ballast pump control system
2. pump status indicating system
3. valve control system
4. valve position indicating system
5. tank level indicating system
6. draught indicating system
7. heel and trim indicators
8. power availability indicating system (main and emergency)
9. hydraulic or pneumatic pressure indicating system
10. monitoring systems, e.g., machinery alarm, fire and gas detection system.

To ensure that uncontrolled transfer of ballast water will not continue upon loss of power, ballast tank valves are to close automatically upon loss of power, or be provided with an equivalent (or better) facility.

The control and indicating systems should function independent of each other so that a failure in any one system will not affect the operation of the other systems. The ballast pump and ballast valve control systems should be arranged so that the loss of any one of their components will not cause the loss of operation in the other pumps or valves.

In addition, some facility to calculate the effects of tanks and flooded compartments is advised, especially in emergencies where the ballast control operator may need to act quickly to avoid serious listing or capsizing. As with any other safety critical system, the ballast control system will require to be fed from a secure electrical supply (uninterruptible power supply) and located in or near the main control room in a pressurised (nonhazardous) protected compartment.

It is vital that the ballast control panel is adequately illuminated after the failure of normal lighting (Fig. 6.1.4).

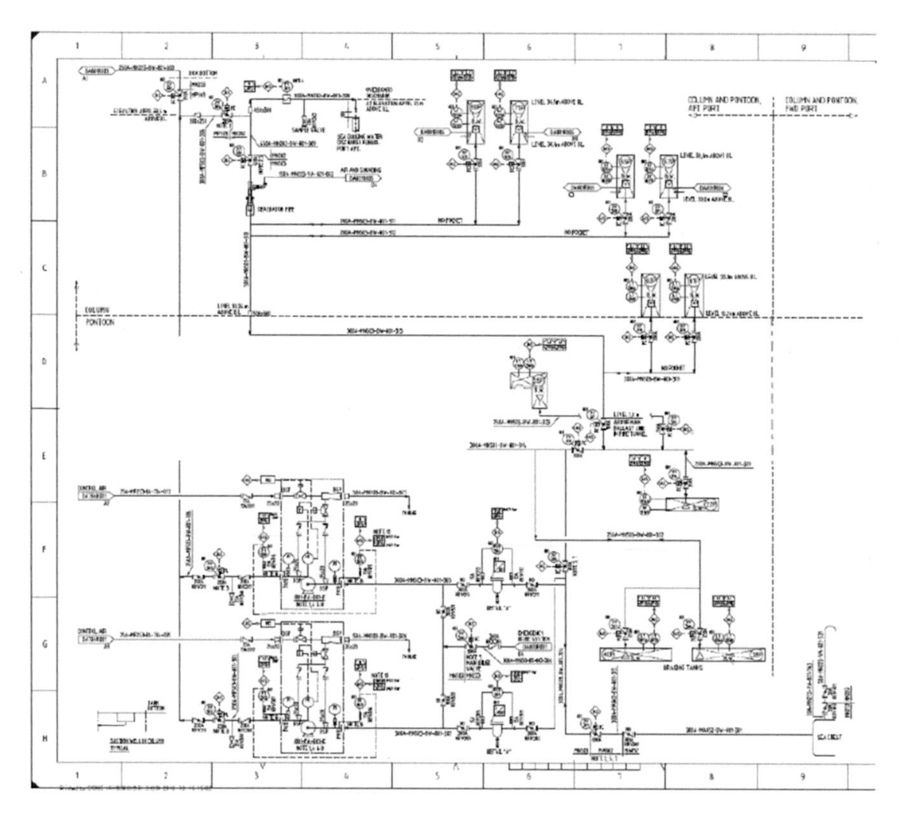

FIGURE 6.1.4

A typical ballast system piping and instrumentation diagram.

Courtesy Kongsberg.

CHAPTER

Floating Production, Storage and Offloading Units

2

INTRODUCTION

In deepwater where it is not economic to locate a platform, the floating production, storage and offloading (FPSO) facility comes into its own. Often based on a converted tanker hull, these vessels can accommodate the complete oil and gas recovery process on the main deck, whilst using the existing (ex-tanker) storage until the shuttle tanker arrives and offloads it. In order to achieve this, it is necessary to install some *extra* equipment not required on fixed platforms.

1. **Escape tunnel**
   Because of the length of the vessel, it is usually necessary to install a pressurised fire-protected tunnel for the length of the deck to allow personnel to reach the end of the ship where the accommodation and lifeboats are located. Usually, as with a tanker, this is the aft end of the vessel and the forward end is where the offloading facility is located.
2. **Offloading facility and ballast control**
   In order to offload the crude oil to a shuttle tanker, oil from the oil export pumps is fed to a hose reel which can be connected to the shuttle tanker using ropes fired by rockets. The shuttle tanker must be fitted with dynamic positioning (DP) to allow close approach to the FPSO. There are several safety systems forming part of the offloading facility:
   a. The green line system monitors the offloading hose for leaks by comparing the level of flow into the shuttle tanker with that flowing from the FPSO.
   b. The ballast control system must keep the FPSO stable and avoid stressing the hull structure by pumping water in to replace the oil, or vice versa, during oil production and storage.
3. **Differential Absolute and Relative Positioning Sensor**
   The *Differential Absolute and Relative Positioning Sensor* (DARPS) radar system is designed to detect possible collision threats whilst offloading is taking place. DARPS is a DP position reference system made for offshore loading operations. It combines sensors for reliable and accurate absolute and relative positioning of two vessels such as a shuttle tanker and an FPSO.
4. **Inert gas system**
   As with a tanker, it is necessary to maintain a blanket of deoxygenated air or nitrogen in the 'air' space above the crude in the hull storage tanks. This is usually generated with a main and standby nitrogen generator designed to combust

Offshore Electrical Engineering Manual. https://doi.org/10.1016/B978-0-12-385499-5.00035-2

the oxygen to a low level, monitored with an oxygen analyser. Deck water seals prevent the ingress of oxygen from above but allow the nitrogen to escape whilst the tank is filling.

5. **Heading control**
   It is not advisable to allow the accommodation/temporary refuge end of the ship to be downwind of the process areas, and therefore a limited DP system is provided just main to the vessel's heading so that the accommodation is always upwind of the process areas.
6. **Turret**
   Subsea wells on the seabed below the FPSO are connected to the ship via a turret. The turret is fixed relative to the heading of the ship and is where the flexible risers connect to the ship. Some turret systems use flexible hoses to connect between the turret and the ship, allowing the ship to change heading whilst the turret remains stationary. The other method is to use 'swivels'. These are rotation joints able to take the high oil pressures without leaking.
7. **Ground flare**
   If the vessel's movement is likely to be excessive (e.g., west of Shetland duty), it may be better to use a ground flare rather than the normal boom flare, which would be stressed by excessive rolling of the hull. The ground flare consists of a (usually) circular concrete open-topped tank, with the flare nozzles aimed down towards the centre of the tank. Above this is a stack to protect the nozzles and flame from the wind and funnel the heat and fumes upwards.
8. **Mooring system**
   It is usual for FPSOs to be 'permanently' moored (except for major refits, including the mandatory shallow water survey). Mooring, as such, is not an electrical issue, but the DP system, besides heading control, also has the duty of monitoring the tension on anchor chains and, if need be in bad weather, relieve chain stress by providing thruster assistance.

SAFETY CRITICAL SYSTEMS

Many of the above-mentioned systems may be classed as 'safety critical' by Safety Case Risk studies, in which case the system will have a 'Performance Standard' written for it, which defines its scope and various functions and how this is verified. These issues are covered in PART 9.

CHAPTER

Semisubmersibles and Mobile Offshore Drilling Units

3

INTRODUCTION

These are by far the most common type of installation and are used as mobile drilling platforms, accommodation, construction centres and fire tenders. They are sometimes modified for use as oil production platforms. With a displacement often exceeding 20,000 tons, these float on pontoons well below the zone of wave action, and with a draft of up to 90 ft, dampen the movement caused by the waves. The platforms are normally held in position by eight or more 15-ton anchors and like the fixed platforms are designed to withstand extreme weather conditions. There are obviously some differences to a fixed jacket or concrete platform which need to be borne in mind (see later discussion), but otherwise the requirements such as emergency shutdown, process control, fire and gas and power management are similar.

POWER GENERATION

The normal generation outfit of around six diesel generators is sufficient for all drilling operations, with some standby. However, if used for production, there may be some gas turbine sets.

PONTOON AND LEG EQUIPMENT INSTALLATION

It is usual to have mainly tanks and seawater lift and fire pumps in the pontoons and legs, but there is a design with diesel generators in the pontoons which makes maintenance difficult. In case of fire the legs can act like chimneys and so extinguishing systems are essential, especially if there are generator sets in the pontoons.

WATERTIGHT DOOR MONITORING AND BALLAST CONTROL SYSTEMS

Comprehensive monitoring and control of ballast tanks, fuel tanks, cement tanks, etc. is vital for control stability. Watertight door control and status is also essential.

Offshore Electrical Engineering Manual. https://doi.org/10.1016/B978-0-12-385499-5.00036-4

ELECTRICAL DRILLING DIRECT CURRENT AND ALTERNATING CURRENT SEMICONDUCTOR DRIVES

Refer PART 2 Chapter 11.

CEMENT KILL PUMPS AND CONTROLS

In addition to normal cementing facilities, emergency cement pumps are provided, usually consisting of one or more diesel-driven pumps, independent of the other rig systems, so that they can be used in emergencies to prevent well blowouts.

DRILLING MUD FACILITIES

It is also vital to provide a continuous supply of drilling mud of the correct weight and density to lubricate and cool the drill, to bring the cuttings to the surface and to control the well.

FIRE AND GAS SYSTEMS AND VENTILATION CONTROL ON RIGS

Good reliable ventilation is essential for the shale shaker mud pump and mud tank rooms to control the release of hazardous gases by the returning mud.

See also PART 2 Chapter 15.

THE UK OFFSHORE SAFETY CASE LEGISLATION

A semisubmersible operating 'permanently' as an oil and gas production unit will be classed as a 'fixed installation' and be required to meet the UK Offshore Safety Case Legislation (see PART 9 Chapter 1).

If the semisubmersible is working as a 'mobile' rig in the UK waters, the UK Safety Case Legislation does not normally apply. Such a rig will be subject to classification and survey by a certifying authority and be required to meet their rules. Other UK legislation may also apply, however.

CHAPTER

Self-Elevating Rigs

4

INTRODUCTION

A jack-up rig is a bargelike vessel with steel legs each having racking gear which allows it to be racked down to the seabed until the entire vessel is lifted out of water. These vessels are obviously limited by the length of the legs to water depths of around 300 ft. They are inherently a temporary facility, more suited to drilling and maintenance activities than production.

FREQUENT DUTIES

1. Drilling
2. Well intervention
3. Accommodation
4. Wind turbine installation
5. Heavy lifting
6. General offshore support (liftboat).

In UK waters, if the well intervention (Duty [2]) is over a fixed platform, a Safety Case must be written for the combined operation.

MAIN FACILITIES

1. Power generation
2. Position keeping (if applicable)
3. Propulsion (if applicable)
4. Steering (if applicable)
5. Fire and flammable gas detection, fire protection and extinguishing
6. Drainage and bilge pumping
7. Ballasting
8. Emergency shutdown systems (if applicable)
9. Jacking systems.

Offshore Electrical Engineering Manual. https://doi.org/10.1016/B978-0-12-385499-5.00037-6

CHAPTER

Tension Leg Platforms

5

INTRODUCTION

The tension leg system consists of a hull-like structure having positive buoyancy which is held to a deeper draught by means of tension legs at each corner.

The legs consist of flexible metal tubes anchored in foundations in the seabed and tensioned to hundreds of tons in the platform. A computer-controlled ballasting system prevents load movements on the platform deck causing large differences in tension and hence avoids dangerous stressing of the hull structure or the legs themselves. The tension leg mooring system allows for horizontal movement with wave disturbances, but does not permit vertical, or bobbing, movement, which makes tension leg platforms a popular choice for stability, such as in the hurricane-prone Gulf of Mexico. In a simpler form, it is also a useful concept for wind turbine structures, in locations where deepwater makes the usual tower design uneconomic.

TENSION LEG PLATFORM IN OIL AND GAS INSTALLATIONS

The tension leg monitoring, tensioning and ballasting system is the only major difference from the normal jacket or concrete platform installation. The safety facilities, such as hazardous area certification, emergency shutdown, fire and gas and process control, are similar, and in UK waters, a Safety Case would need to be prepared in the same way. As the tension leg monitoring, tensioning and ballasting system is vital to the safety of the platform, it would definitely be regarded as safety critical.

TENSION LEG PLATFORM AS A WIND TURBINE STRUCTURE

This would be a much simpler normally unmanned installation, and the flammable material inventory is very small.

(Note: As part of a wind farm, no oil-filled transformer should be required if the wind turbine generates at the export voltage (in the region of 20 kV). The transmission step-up transformer would normally be located on a separate substation platform, see PART 6 Chapter 6).

Offshore Electrical Engineering Manual. https://doi.org/10.1016/B978-0-12-385499-5.00038-8

CHAPTER

Notes on Offshore Renewable Energy Substation Platforms

6

INTRODUCTION

Large marine renewable energy projects such as wind and wave farms located offshore will generally require an offshore platform for power equipment to act as a collection node for the generator cables operating at medium (generating) voltage, the step-up transmission transformers and the necessary switchgear. The platform may not have to be continuously manned for operation of the renewable farm, but temporary manning will be required for maintenance purposes.

Where such platforms are close inshore, small service vessels can be used to ferry personnel and equipment out to them, and access is usually by ladder and harness.

However, if the platform is far offshore, boat access becomes time consuming and hazardous, and the platform will need to be equipped with a helideck.

DESCRIPTION

A typical wind farm in UK waters may consist of 200 wind turbines. For the purposes of the substation hub, the turbines would be grouped into 10 groups of 20, with a main 33-kV cable from each group connected to the hub platform. Each 50% capacity transformer outputs to metalclad SF6 outdoor type switchgear to a 132-kV subsea cable.

Main uninterruptible power supply (UPS) batteries would be typically 2 Ah × 200 Ah banks and would preferably be located in a separate battery room or enclosure.

The current UK Maritime and Coastguard Agency guidance document is MGN 543 (M + F) see UK Maritime and Coastguard Agency.

Substation platforms are usually unmanned except for normal summer maintenance campaigns and emergency visits.

The main transmission cables (both 33 and 132 kV) would normally have fibre-optic cores for both communications and telemetry.

The (exposed) transformer coolers should be protected from damage due to dropped objects, whilst maintaining cooling airflow.

A standby/emergency generator should be provided to supply power for lighting and utilities and to maintain UPS battery charge for telecommunications and control systems, when the utilities transformer is unavailable.

Offshore Electrical Engineering Manual. https://doi.org/10.1016/B978-0-12-385499-5.00039-X

Electric hoists/davits will be required for loading supplies from the small service boat but larger supply boats would need to use a crane. If the substation platform is well out to sea, consideration would need to be given to providing a helideck.

Sacrificial cathodic protection anodes should be fitted to the jacket structure and the splash-zone will be made of thicker metal with a special paint coating system as with any offshore structure.

HAZARDS

The biggest hazard is likely to be the two large 33/132 kV step-up transformers, which are currently oil-filled with separate coolers. There is a risk of explosion if an internal winding (interturn) fault causes an explosion though generation of gases from the hot oil. For environmental reasons, relief valves should be provided to discharge the oil pressurised by the hot gases into a dump tank. Assuming that transformers are geographically separated, the dump tank would only have capacity for one transformer failure. However, it is not considered acceptable to allow the transformer oil to be dumped into the sea apart from in the event of catastrophic loss of the substation platform. Therefore use of H120 fire barriers between the transformers, although beneficial, may not be sufficient without physical separation. Typical transmission transformers weigh 135 tons (108 tons dry) and installation of topsides would normally be complete with the main transformers in 'wet' condition.

PART

Commissioning

7

CHAPTER

Introduction to Commissioning

1

This chapter is intended as a general checklist for use when preparing commissioning workscopes and should not be treated as exhaustive. Every piece of equipment and the way that it is intended to be operated, will need to be thoroughly understood before writing such a document. Any document of this kind should also include reference to access and permit-to-work procedures specific to the installation where the commissioning is being carried out.

All test equipment used should have valid calibration certificates, as, apart from the results of tests being invalid, incorrect equipment control or protective device settings could render the equipment unsafe and, in some cases, create a fire or explosion hazard. The test equipment itself, when used in areas where flammable gas may be present, will also be an ignition hazard, and therefore, 'hot work' permits will be required for most of the tests described. It is important that all test results are recorded so that a complete maintenance history can be retained for reference, should a failure occur. Some typical test sheets are provided in Appendix B.

Lastly, all commissioning must accord with the client oil company's standard procedures, the relevant British Standards and the manufacturer's installation and operating manuals for the equipment concerned.

The commissioning details for the different types of equipment are covered in the following chapters.

Chapter 2	Generators
Chapter 3	Switchgear
Chapter 4	Electrical Protection, Monitoring and Control
Chapter 5	Large Motors
Chapter 6	Process Monitoring and Control

Offshore Electrical Engineering Manual. https://doi.org/10.1016/B978-0-12-385499-5.00040-6

CHAPTER

Generators

2

INTRODUCTION

At the time of testing, it is assumed that the manufacturer has already successfully conducted the following tests (please refer to PART 2, Chapter 1):

1. open circuit test
2. iron loss test
3. short-circuit test
4. transient performance tests (sudden loading and three-phase short circuit)
5. voltage balance
6. heat run at full power
7. waveform analysis
8. vibration analysis
9. noise test
10. friction and windage losses
11. automatic voltage regulator (AVR) and control panel testing and calibration.

It may be necessary to repeat some of these tests in the module yard and/or offshore if the machine is damaged or tampered with during transportation or installation.

Note that it is always essential to create a *detailed written commissioning procedure*.

Having finally reached the offshore commissioning stage, it will be necessary to make the generator safe for testing by carrying out the following:

1. Preparatory safety work
   Assuming the generator has already been taken out of service, the generator circuit breaker should be locked in the earth position and the associated AVR voltage transformers (VTs) should be isolated and padlocked. If the circuit breaker is not of the truck type, it must be padlocked off and a separate earth safely applied. VT fuses should be removed, if the VT is not withdrawable.
2. Work safety coordination
   It must be remembered that a large offshore generator will develop an appreciable voltage from its residual field if it is allowed to rotate. This is less unlikely than it sounds, especially if, at the same time, another commissioning team is anxiously attempting to complete the commissioning of the prime mover. Although barring gear is normally provided on the larger machines and may be used to prevent rotation, it may be necessary to provide some other

Offshore Electrical Engineering Manual. https://doi.org/10.1016/B978-0-12-385499-5.00041-8

temporary means of preventing rotation. The start permissive key should be removed from the prime mover control panel. Having made certain that everything is electrically and mechanically safe, work can then begin on testing the generator windings.

3. Cold resistance of main stator windings
   The method involves injecting a known direct current through each winding and measuring the voltage drop through the winding at the terminals. Before applying the current, the earth must be removed from the winding. Readings of voltage and current should be recorded for all three phases. The three resistances calculated from these readings after compensating for ambient temperature should not deviate from each other by more than 2% or from the manufacturer's design figure by more than 5%. For compensation purposes, the ambient air temperature will need to be measured at shaft height and 1 m from the alternator frame.
4. Rotor and exciter field windings
   The resistance of these windings should be measured in the same way as the stator windings. Using an alternating current high-voltage test unit, the generator windings should be tested to twice their working voltage plus 1000V, with a minimum of 1500V as required in BS 4999. The windings should withstand this voltage for 1 min. The voltage should be gradually applied, and the minute count started once full test voltage has been reached.
5. Exciter
   The exciter field should be tested in the same way and using the same criteria after temporarily removing the AVR connections. With a brushless machine, the rotor and exciter armature should be checked for continuity only, in order to avoid damage to the diode bridge.
6. Anticondensation heaters
   Anticondensation heaters should be high voltage tested to about 1500V to the machine frame, with adjacent windings earthed. A continuity check should also be carried out.
7. Insulation resistance
   Using a 'Megger' of suitable voltage rating (5000V for medium-voltage and 1000V for low-voltage machines), the insulation resistance should be measured. Minimum values acceptable are as follows:

Stators	50 MΩ
Exciter field	10 MΩ
Exciter armature	10 MΩ
Rotor	10 MΩ

Note that during any of the high-voltage or Megger tests, all voltage-sensitive auxiliaries, thermistors, auxiliary windings, etc. should be temporarily earthed to avoid damaging them. No part of the AVR or rotating diode bridge should be meggered, if damage is to be avoided.

8. Interface with mechanical components
   A check must be made that all nonelectrical parts of the generation system are complete and functioning, including
   a. the fuel system,
   b. the engine controls including the governor and overspeed protection devices,
   c. the engine and alternator cooling system,
   d. the ventilation system,
   e. the air intake and exhaust systems,
   f. the engine starting equipment,
   g. the generator package fire and gas detection and protection systems.
9. Associated electrical power connections
   Assuming the associated switchgear has already been commissioned, phase and earth cabling and conductors, including neutral earthing resistors, should be checked for continuity and insulation.
10. Before starting the prime mover
   Before starting the prime mover, the generator circuit breaker should be racked into the test position, the associated VTs racked in, fuses replaced and the means of preventing generator rotation removed.

RUNNING THE PRIME MOVER

With the prime mover running, the following equipment may be commissioned:

1. Under-/overvoltage protection check (note that this should have been checked beforehand by primary injection).
2. Automatic synchronising unit (note that with the circuit breaker in the test position, this consists of observing if the circuit breaker closes at the correct phase angle and slip frequency limits). A second generator is required for a reference.
3. Manual synchronising equipment and check-synch relay.
4. The excitation system including AVR setting.
5. Over-/underfrequency protection check.
6. A general check should be made to ensure that control panel instrument readings are normal for the generator on no-load. For generator load testing, it is advisable to connect a suitable load bank to the associated switchboard (or direct to the alternator if the switchboard is already in use), so that the electrical and mechanical stability of the package may be checked before it supplies load to platform consumers. The load bank must include both resistive and reactive switchable elements.

CHAPTER

Switchgear

3

AIR BREAK SWITCHING DEVICES AND FUSES RATED FOR USE AT VOLTAGES BELOW 1000 V

First, the equipment should be given a thorough mechanical and visual inspection. This includes the following:

1. The equipment should be clean and dry and free from extraneous loose items or tools.
2. Insulators and insulation should be clean and dry.
3. Indicating, protective and interlocking devices should operate satisfactorily.
4. All conducting connections should be secure and intact.
5. The ratings of fuses, relays and other devices should be as shown on design drawings.
6. Cables should be of adequate cross section and correctly glanded and terminated.
7. Internal wiring should be correctly secured.
8. Switching contacts should normally be sparingly applied with a suitable contact lubricant, but the manufacturer's manual (or the manufacturer's advice) should be consulted, as, in some conditions, this may lead to contact sticking or reduced fault clearing capability.
9. The mechanisms of switching and interlocking mechanisms should be checked for smooth and correct function.

Electrical tests will then need to be completed as follows:

1. A primary insulation test should be carried out on each phase to ground and from each phase to phase using the test voltages shown in Table 7.3.1.
2. The insulation resistance of small wiring and ancillary components should be tested using a Megger or similar device, where the test voltage is not greater than 500 V direct current (DC).
3. The overloads of motor control gear should be injection-tested at 100%, 125% and 600% setting values and checked against the manufacturer's tripping times.
4. Each thermal or magnetic element should be tested individually. Tests made on thermal elements should include for both 'hot' and 'cold' operating conditions.
5. The switchboard metering, i.e., all ammeters, voltmeters, current transformers (CTs) and voltage transformers (VTs), etc., should be checked for correct calibration.

Offshore Electrical Engineering Manual. https://doi.org/10.1016/B978-0-12-385499-5.00042-X

Table 7.3.1 Recommended Insulation Resistance Test Voltages (BS 6423)

Three-phase voltage rating test voltage recommended for primary insulation for the insulation resistance test (to earth and between phases) in kilovolt DC (1 min)	
Up to 1 kV	1 kV
Above 1 kV and up to 3.6 kV	2 kV
Above 3.6 kV and up to 12 kV	5 kV
Above 12 kV	10 kV

6. A DC overvoltage test of 2000 V should be applied for 1 min to all load-carrying busbars and circuit breakers. The test should be applied between phases with the circuit breaker closed, and both between phases and phases-to-earth with the circuit breaker open. Any auxiliary equipment which inherently cannot withstand the 2000 V DC test voltage (e.g., semiconductor devices) should be disconnected before the start of the test. VTs should be isolated by removing both primary and secondary fuses. CT secondaries should be fitted with short-circuiting links and earthed before starting the test.
7. When the electrical testing is complete, all protective interlocking and tripping devices (such as overcurrent, earth fault and undervoltage releases) should each be operated to its full extent to prove the function of each device before the switchboard is brought into service.

AIR CIRCUIT BREAKERS/SWITCHGEAR RATED FOR USE AT VOLTAGES ABOVE 1000 V

The equipment should first be inspected visually and for correct mechanical operation as follows:

1. Inspect for physical damage and compare nameplate details with platform design document requirements.
2. Make sure that the switchgear cubicle has been correctly installed and adequately anchored to the switchroom floor to withstand shock loadings imposed by operation of circuit breakers.
3. Refer to the manufacturer's manuals and perform all the mechanical operator and contact alignment tests on both the circuit breaker and its operating and interlocking mechanisms in accordance with the manufacturer's instructions.
4. Check that the operation of all safety shutters is satisfactory by removing circuit breaker trucks and physically checking that shutters are free to move easily.
5. Check that all insulation and insulators are clean and dry.
6. Check that main contacts, secondary contacts, auxiliary switches and earthing contacts are correctly fitted and aligned.

Table 7.3.2 Recommended DC Test Voltages on Metal-Enclosed Switchgear and Control gear After Erection on Site

(BS EN 62271-200 and BS 6423)
Rated voltage (U) site test voltage (kV RMS) in kV DC (15 min)

Up to 1 kV	1 kV
Above 1 kV and up to 3.6 kV	2 kV
Above 3.6 kV and up to 7.2 kV	7.5 kV
Above 7.2 kV and up to 12 kV	15 kV
Above 12 kV and up to 17.5 kV	25 kV
Above 17.5 kV and up to 24 kV	32 kV
Above 24 kV and up to 36 kV	45 kV
Above 36 kV	66 kV

Please check the current versions of the standards quoted to verify the correct values of these parameters.

Electrical tests will need to be carried out as follows:

1. Measure the resistance between pairs of contacts. This should be within the range given in the manufacturer's manual and not more than 500 μΩ if excessive contact heating is to be avoided.
2. Perform a voltage pick-up test on the tripping and closing coils.
3. Trip the circuit breaker several times by operating each protective device in turn.
4. Measure the primary insulation resistance for pole-to-ground, pole-to-pole and across open poles of the same phase.
5. Perform insulation resistance tests at 500V DC (using a 'Megger' or similar instrument) on all control wiring. Circuits containing semiconductor devices, such as solid-state protection relays, should not be tested. These tests should give results within the range given in Table 7.3.1, with a minimum value of 100 MΩ. The above-mentioned test should be performed both before and after a high-voltage DC test. The first application will determine if the insulation resistance is high enough for acceptance and whether the high-voltage test may be performed. The second test will verify that the application of the overpotential test voltage has not shown up any weakness in the insulation. In each case, the test voltage should be applied for 1 min.
6. Apply overpotential test voltages, as shown in Table 7.3.2, to phase conductors with circuit breakers in the open and closed positions. Arc chutes should be tested for watts loss within the manufacturer's allowable values. The test voltages in Table 7.3.2 apply to metal-enclosed switchgear and control gear (i.e., assemblies with external metal enclosures intended to be earthed and complete except for external connections). The switchgear under test should be free from all external cabling, VTs, CTs and other auxiliary equipment which cannot withstand the test voltage being applied.
7. If main cabling terminated at the switchgear is to be high voltage tested, then those parts of the equipment which cannot be readily isolated from the main cable terminals should be capable of withstanding the DC test voltage specified in BS EN 62271-200 for 15 min (i.e., the duration of the test).

When any overpotential testing is undertaken, the test voltage should be increased as rapidly as consistent, with its value being indicated by the test instrument. If the voltage breaks down the insulation as it is being increased, the test set operator should repeat the test, adjusting the voltage slowly enough to record the value at which the breakdown occurs.

Table 7.3.2 gives the maximum DC site test voltage, and this value should be maintained for 15 min. Test current values should be recorded at 30-s intervals for the first 2 min and at 1-minute intervals thereafter. The test potential should then be reduced to zero and earths applied for 10 min before continuing the next test. Although it is usually obvious if the equipment fails this test, values of leakage current equivalent to an insulation resistance greater than 100 MΩ are expected and leakage current at the full test voltage should remain steady. If the leakage current gradually increases during the test, this is likely to indicate an incipient failure or the connection of a semiconductor or similar component which should have been isolated before starting the test.

Commissioning checks for other forms of switchgear follow a similar pattern, but the exceptions are discussed in the following sections.

OIL SWITCHGEAR OVER 1 KV

The mechanical check should include a visual check of the circuit breaker tank oil levels and an examination of oil-filled bushings for any leakage. The electrical testing should include an oil sample insulation test in accordance with BS EN 60422.

VACUUM CIRCUIT BREAKERS

The mechanical check should include a visual inspection of the interrupters for damage or seal corrosion. If fitted, the wear gauge should be checked in each case.

The electrical testing should include vacuum integrity tests by applying a DC overpotential voltage (usually around 20 kV). The manufacturer should be consulted both as to the correct voltage to apply and to the safety precautions that should be taken because of the possibility of generating X-rays during this test (refer to BS 6626).

SULPHUR HEXAFLUORIDE SWITCHGEAR

The mechanical inspection should include the following:

1. A check that the gas system is operating at the correct pressure.
2. If the switchboard is of the sealed type, topping up of sulphur hexafluoride should be required infrequently by using a temporary connection to a gas

cylinder. If the gas system has a topping up compressor, the compressor gas supply, oil level, filters or desiccants should be checked for good condition. The permanently fitted gas pressure gauge should be checked against a recently calibrated test gauge.

As contact mechanisms tend to be inaccessible in this type of switchgear, electrical testing should include timing and travel tests, subject to the unit being properly equipped for this to be carried out. The results may then be compared with those provided by the manufacturer, with discrepancies indicating a fault in the operating mechanism or contact wear.

CHAPTER

Protection and Control

4

PRECOMMISSIONING

Before commissioning, it is essential that the installation work has been properly checked before any equipment or cables are energised. Cable connections are 'rang through' and cable insulation is checked where possible. Hard-wired logic and control and monitoring software must be checked, although this may be limited to critical circuits if the factory acceptance tests are comprehensive and well documented. The 'sense' of any current transformers (CTs) must be checked to ensure terminals are not reversed.

All such work is termed precommissioning.

PROTECTION SCHEME COMMISSIONING

The satisfactory operation of protection relays is important for the prevention of danger to equipment operators and local damage to equipment. By preventing an electrical fire, it may also have prevented growth of an incident to platform-threatening proportions. Therefore, great care should be exercised by the commissioning engineer to ensure that the protection devices will respond to electrical system disturbances as the designers intended. If the commissioning engineer considers that he or she has found design errors or omissions during commissioning, they must not be shrugged off but must be discussed with the designers and/or operators.

The testing of voltage transformers (VTs), CTs and protection relays is described in detail in the following sections.

The general procedure for the commissioning of protection schemes is to perform the following checks and tests:

1. obtain a full set of wiring diagrams, schematics and the protection relay setting schedules for the equipment to be tested;
2. ensure that you are in possession of the manufacturer's technical data and commissioning procedures;
3. carefully inspect the overall scheme, checking all connections and wires on CTs, VTs and protection relays;
4. measure the insulation resistance of all circuits;
5. carry out ratio, polarity and magnetisation curve tests on the CTs;
6. carefully inspect each relay before testing it by secondary injection in accordance with the manufacturer's installation instructions;

Offshore Electrical Engineering Manual. https://doi.org/10.1016/B978-0-12-385499-5.00043-1

7. check the stability of the protective relays for external faults and determine the effective current setting for internal faults, by means of primary injection tests;
8. perform closing, tripping and alarm checks for the equipment associated with protective relays.

Before commencing work on any equipment, the isolation and permit-to-work procedures must be strictly adhered to. In particular, CTs and VTs should be de-energised, isolated and discharged to earth before any commissioning operations are carried out.

VOLTAGE TRANSFORMERS (REFER TO BS EN 61869 AND BS EN 50482)

First the VT should be inspected both visually and mechanically as follows:

1. inspect for physical damage and for compliance with design documents,
2. check mechanical alignment, clearances and proper operation of any disconnecting and earthing devices associated with the VT,
3. verify proper operation of grounding or shorting devices.

After the initial examination has been carried out, the following electrical tests should be completed.

INSULATION RESISTANCE

1. Remove all deliberate earth connections from the VT windings.
2. Electrically isolate the VT primary and secondary windings by removing fuses, links or connections.
3. Perform insulation resistance tests on secondary windings to earth and between secondary windings at 500V direct current (DC) for 1 min.
4. Having obtained approval from the VT manufacturer, perform a high-voltage DC insulation test on the primary winding insulation.
   Recommended test voltages are given in Table 7.4.1, which is for fully insulated windings. This test is only to be used on new VTs which have yet to be placed in service and is for 15 min duration.
   Secondary windings should be shorted together and to earth for the high-voltage DC test, so that the insulation is tested with a potential from primary to earth and primary to (earthed) secondary. The test voltage should be gradually increased up to the full value and maintained for the duration of the test.
5. An insulation test of 500V DC should be applied to all windings to earth, and between windings, for a duration of 1 min. This test should be performed both before and after the high-voltage DC test described earlier and should result in a minimum resistance reading of 100 MΩ.

Table 7.4.1 Recommended Test Voltages for Voltage Transformers (Fully Insulated Primary) and Current Transformers

Rated Highest Equipment Voltage (kV AC)	Test Voltage (kV DC)
1. Primary insulation test	
Up to 0.66	1.5
Up to 3.6	7.5
Up to 7.2	15.0
Up to 12.0	25.0
Up to 17.5	32.0
Up to 24.0	45.0
Up to 36.0	**66.0**

1. The application time should be 15 min applied between the primary and earth, with secondary windings shorted and earthed.

2. Secondary insulation tests. All secondary insulation tests should be conducted using a test voltage of 500 V DC to earth for 1 min duration.

AC, *alternating current;* DC, *direct current.*

POLARITY CHECK

The VT polarity can be checked using the simple circuit shown in Fig. 7.4.1.

The battery must be connected to the primary winding with the polarity ammeter connected to the secondary winding. For the circuit shown, the ammeter will flick positive on making the switch and negative on opening the switch.

VOLTAGE RATIO TEST

The approximate voltage ratio of the transformer can be tested by a simple voltage injection test on the primary winding by measuring the primary and secondary voltages.

Extreme care must be exercised when carrying out this test to ensure that the test voltage is applied to the primary and not the secondary winding. Otherwise, very high voltages may be applied to the test instruments and create a hazard to the tester.

PHASING CHECK

It is essential to check the phasing of the secondary connections on three-phase VTs, or three single-phase VTs, when these are used for either metering or protection purposes.

The simplest method of testing the phasing is by the application of the system voltage on to the primary winding of the transformer, i.e., with the main busbars live. The secondary voltages can then be checked between phases and neutral, and the phase relationship is verified with a phase indication meter.

Open-delta secondary windings of three-phase transformers can also be checked to verify a balanced primary three-phase supply.

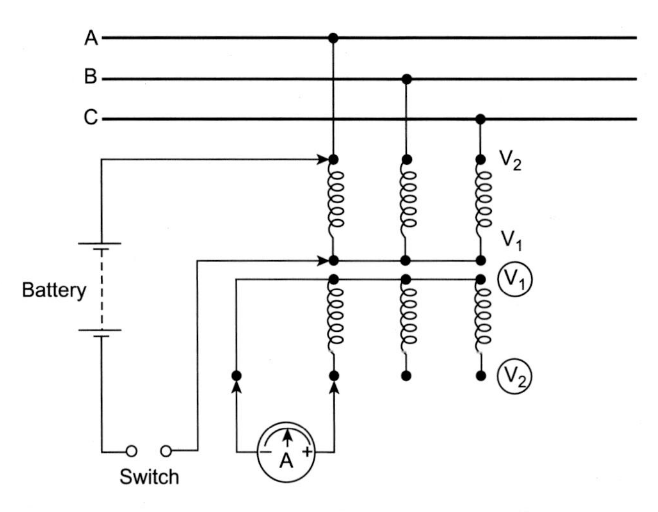

FIGURE 7.4.1

Circuit diagram for carrying out voltage transformer polarity checks: for the polarity shown, ammeter flicks positive on switch make and negative on switch break.

RESTORATION

1. Restore all electrical connections to their correct state after having removed any temporary test wiring.
2. Check all fuses and current limiting resistances, if provided, for continuity, correct rating and good condition. Replace any component which shows signs of thermal or mechanical damage.

CURRENT TRANSFORMERS

Visual, mechanical, insulation resistance and polarity checking are carried out similar to the methods for the VTs discussed earlier. Fig. 7.4.2 shows the connection arrangements for polarity checking.

The following tests should also be undertaken.

CURRENT RATIO CHECK

There are two methods to perform the check.

Method 1

This test is performed by secondary injection, as shown in Fig. 7.4.3.

Current is passed through two primary windings using a primary injection test set, and the primary and secondary currents are measured. A temporary short-circuit

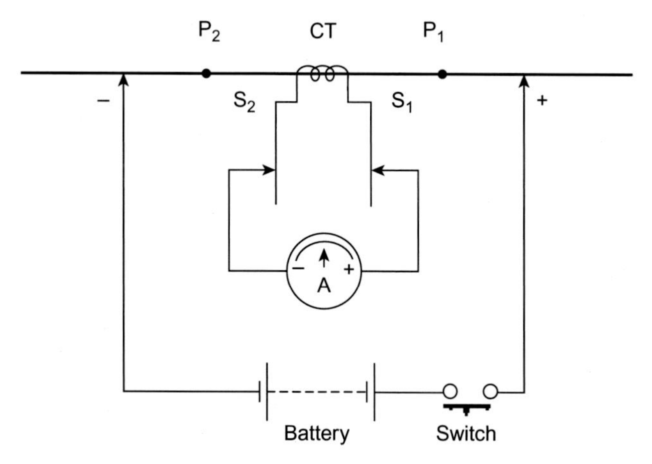

FIGURE 7.4.2

Circuit diagram for carrying out current transformer (CT) polarity checks: for the polarity shown, ammeter flicks positive on switch make and negative on switch break.

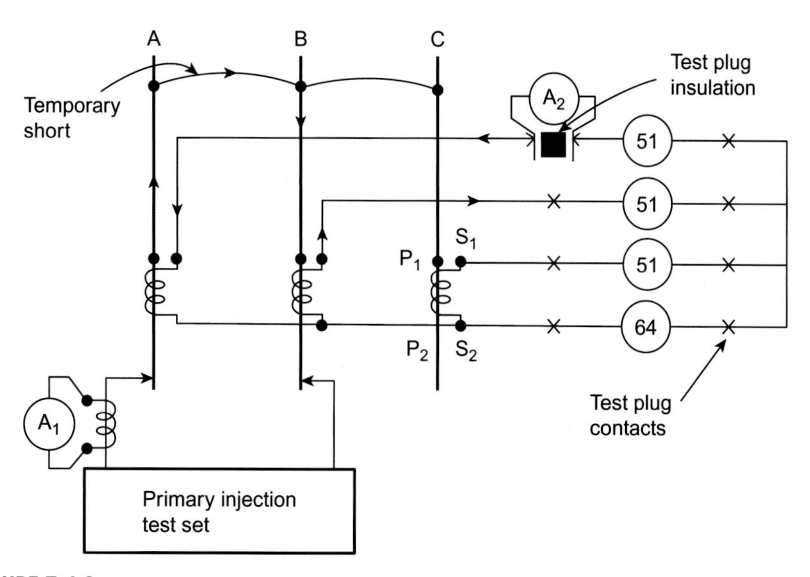

FIGURE 7.4.3

Circuit for current ratio check on current transformer by primary injection (method 1).

is placed across the primary windings at one end and the injection is applied at the other. The ratio of primary to secondary ammeter readings should be approximately equal to the marked ratio on the CT nameplate.

This method of current ratio test also provides a second check on polarity, as the residually connected secondary ammeter will read a few milliamperes if the CTs are of the correct polarity, but will read a value approximating to the CT ratio current if the polarity is incorrect. In some cases, such as in metal-clad switchgear, it may not be possible to access the transformers physically, in which case the protection circuit drawings should confirm that the polarity is correct.

Method 2

An alternative method is to perform a primary injection test on one CT only, as shown in Fig. 7.4.4. As with method 1, the ratio of primary to secondary test currents should approximate to the transformer ratio marked on the nameplate. For multitapped CTs, the ratio should be checked at each tap.

MAGNETISATION CURVE TESTS

The CT magnetisation curve should be checked at several points so as to establish the approximate position of the knee point of the transformer.

The test circuit for obtaining this point is shown in Fig. 7.4.5. A test voltage is applied to the CT secondary winding, and the resulting magnetising current is measured with the primary winding open circuit. The applied voltage is increased gradually until the magnetising current is seen to increase rapidly for small increases in voltage, indicating that the knee point has been reached. The test voltage should then be reduced carefully in stages. The voltage and current should be recorded at several points both above and below the knee point and down to zero test voltage.

The knee point is defined as the point on the CT magnetisation curve where a 10% increase in applied voltage results in a 50% increase in magnetising current.

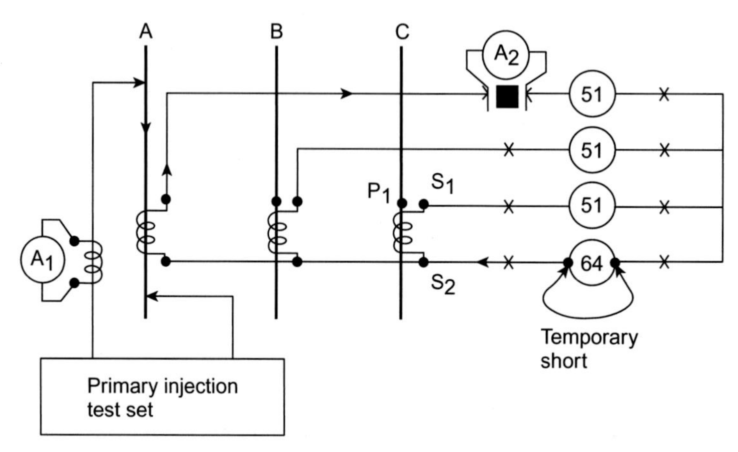

FIGURE 7.4.4

Circuit for current ratio check on current transformer by primary injection (method 2).

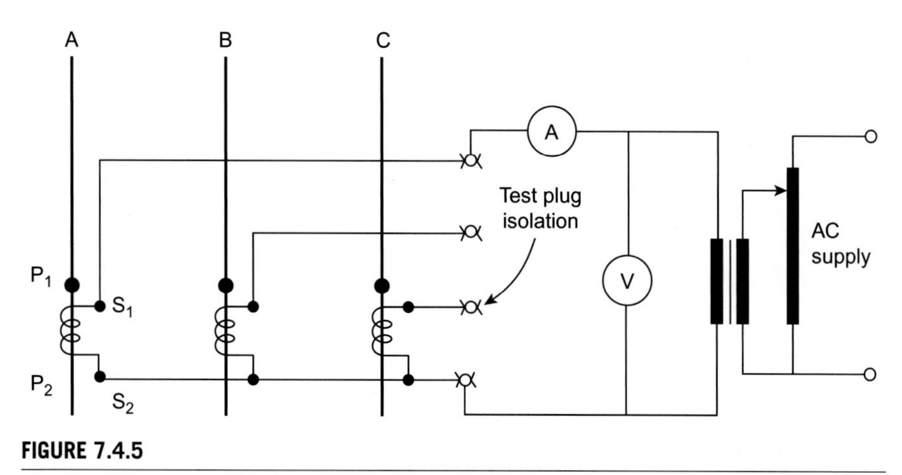

FIGURE 7.4.5

Test circuit for obtaining a current transformer magnetisation curve. *AC*, alternating current.

Generally, the knee-point level should be reached when the secondary voltage is raised until the magnetising current is equal to or less than the rated secondary current.

Typically, the magnetising current is about one-tenth of the secondary current rating. The test voltage required may be higher than 240 V, in which case a step-up transformer will need to be included in the testing kit.

After the magnetisation curve check has been carried out, the secondary winding resistance of the CT should be measured.

RESTORATION

When all the tests have been completed satisfactorily, the test connections should be removed and all permanent electrical connections should be restored to their correct state.

PROTECTION RELAYS

In general, secondary injection is used to check the operation of relays via test blocks. An overcurrent relay, for example, can be checked against the manufacturer's operating curve and tolerances for various current taps and time multiplier (TM) settings.

Once the relay is operating correctly by secondary injection, primary injection tests may then be performed to determine the effective settings for internal faults, whilst proving the stability of the protection for external faults.

The most common element types used in protection relays are

1. the induction disc,
2. the attracted armature,

3. the bimetallic thermal element,
4. electronic circuitry which replicates the characteristics of the above-mentioned electromechanical devices.

One relay case may contain a combination of different elements to meet the particular protection requirements of the equipment it serves, and therefore a number of different test procedures may be involved in the testing of one relay case. The following sections provide only the general principles on relay testing, and the manufacturers' literature and the references given in PART 4 Chapter 1 should be referred for a deeper appreciation of the subject. A guide to the principal applications of protection relays offshore is given in PART 4.

MECHANICAL CHECKS ON INDUCTION DISC AND ATTRACTED ARMATURE RELAYS

1. Check that all fixing and terminal screws are secure.
2. Check that air gaps are free of dirt and foreign matter.
3. Check that the hair spring on induction discs is reasonably spiral in shape, with no turns touching.
4. Check that turn multiplier dials have a smooth action and show no sign of sticking.
5. Ensure that all the transit packing has been removed.
6. On attracted armature relays check that, when the armature is closed with a finger, normally open contacts close and normally closed contacts are opened properly by their push rods.
7. Ensure that any trip indicator flags and their latching mechanisms operate satisfactorily and that they may be unlatched by pushing the reset rod.
8. With the aid of the system schematic drawing check that the relay wiring is as intended.

ELECTRICAL CHECKS ON ELECTRONIC (STATIC) AND INDUCTION DISC RELAYS

Insulation Tests

A cautious approach is required when applying insulation test voltages to protective relays, especially when they contain semiconductor devices.

However, secondary wiring insulation can usually be tested at 500 V DC, and tests should be made on relay coils, contacts and between AC and DC circuits, with secondary CT and VT connections removed. The resistance values obtained should exceed 10 MΩ.

Secondary Injection Tests

These tests are normally carried out using test plugs. However, for relays which cannot be withdrawn from their cases, the test connections must be made directly on to the relay terminals.

If secondary injection tests are to be carried out on relays whose CTs are energised, then care must be taken to ensure that the CT secondary connections on the test plug are shorted out by links, so as to prevent high voltages appearing across the CT secondary when the test plug is inserted.

Overcurrent induction disc relays used offshore usually have a current/time operating characteristic in accordance with BS EN 60255-151 for an inverse definite minimum time relay. There are, however, relays with other inverse time characteristics in use, and so it is important that the manufacturer's commissioning and operating documentation is immediately available during any testing. The arrangements for performing tests on the various elements are as follows. Figs 7.4.6 and 7.4.7 show how the test circuit is connected:

1. With the TM at 1.0% and 100% of CT secondary current apply two times current setting. The operating time should be checked and compared with the relay characteristic.

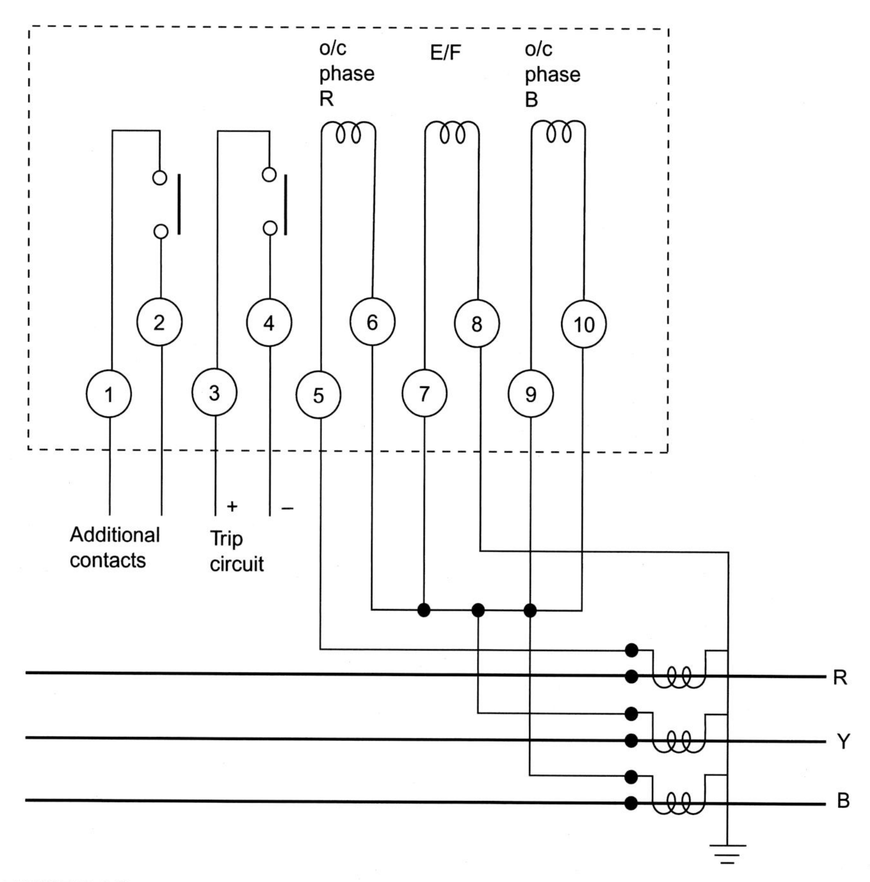

FIGURE 7.4.6

A typical triple-pole overcurrent and residually connected earth fault relay. *E/F*, earth fault; *o/c*, overcurrent.

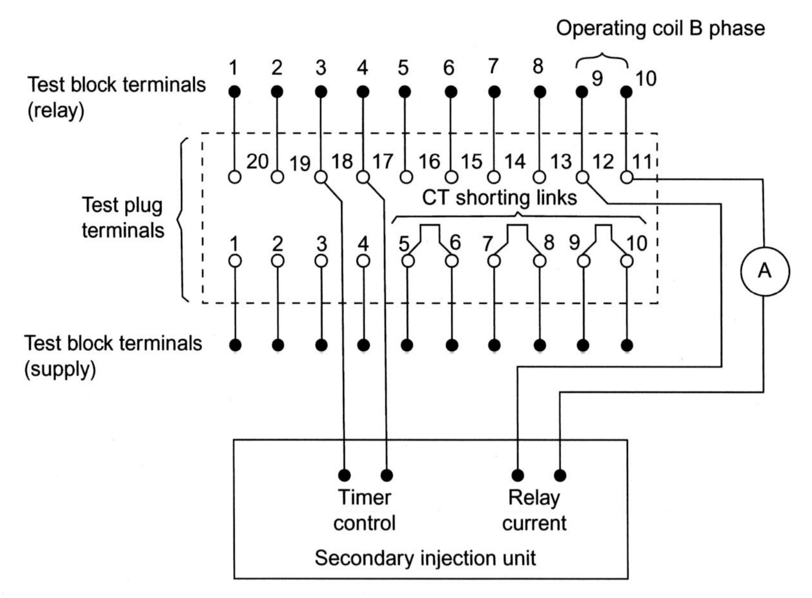

FIGURE 7.4.7

Secondary injection test circuit for triple-pole overcurrent and residually connected earth fault relay.

2. With the TM at 1.0 and 10 times current setting check the operating time against the characteristic once again. In each case, the relay operating time should be within the allowable tolerance band of –7.5% to +7.5%.
3. If the relay is within tolerance the characteristic may be checked over a range of values so that the actual characteristic may be checked against the manufacturer's standard.
4. The minimum operating current of the relay should be checked to ensure that it does not exceed the maximum tolerance given by the manufacturer.
5. Check that the maximum current at which the relay will reset immediately following operation (the resetting current) is as stated in the manufacturer's data.
6. By inspecting the measured characteristic check that the operating time required by the system design, shown in the relay setting schedule, is obtainable.
7. If there are any attracted armature and/or electronic instantaneous elements in the relay, these should be tested for pickup (i.e., operation at the minimum specified voltage or current). Attracted armature relays used for this purpose are designed so as not to chatter, and if chatter occurs, it is likely that the relay contacts need to be cleaned.
8. Auxiliary relays for operation of flag indicators, remote indication circuits, etc. may require a separate supply in order to be tested.

ELECTRICAL TESTS ON ATTRACTED ARMATURE RELAYS

Attracted armature relays are used as lockout, tripping, control, counting and indicating relays designed to sense variations in either system current or voltage. Such a relay may be located either as a single unit in its own case or as one of a number of relays within a switchboard cubicle.

Insulation Testing

With all secondary wiring disconnected from the relay, a 500V DC insulation test should be carried out on all terminals of the relay unit. The resistance value to earth should exceed 10MΩ.

Secondary Injection Testing

In general, voltage-operated relays should be tested to ensure that they operate satisfactorily at 75%–120% of normal coil rated voltages for DC operation and 80%–115% for AC operation. Where voltage measuring relays are used, the pickup and dropout voltages should be checked against calibrated values. (The dropout voltage is the voltage at which the relay coil is no longer able to hold the armature in its energised position, so that it opens or drops out.)

Current-operated relays should be tested to ensure that the minimum operating current does not exceed the manufacturer's recommended value.

In the case of auxiliary relays, tapped coils are provided and the pickup current should not exceed the value recommended for the particular tap. For current measuring relays, the pickup/dropout ratio should be checked against the manufacturer's reference value.

ELECTRICAL TESTS ON THERMAL RELAYS

Thermal relays are frequently used for motor overload protection and may be designed to provide protective sensing for single phasing, stalling and unbalanced loading. The P&B Golds relay is the most common electromechanical type, and the testing information given later reflects this. However, electronic relays, often with some data processing capability, are now available in production quantities, and these are designed to replicate the thermal characteristic of the motor in a similar manner. The following tests are usually required as a minimum.

1. Check that each heater element operates in accordance with the calibration curve, from cold, by applying a multiple of the rated current through the elements, having connected them all in series.
2. The time to indicate a given percentage (say 115%) of rated relay current whilst the above multiple of the rated current is applied should be within the range quoted in the manufacturer's calibration data.
3. With the heater elements still connected in series check the hot operating time by increasing the current from 100% running current to a multiple of

this representing the motor starting current (about 600% running current). Measuring the time for the relay to operate at about 115% running current will enable the hot characteristic time to be checked.

4. Check the operation of the relay at different tap positions by varying the injection current. This should confirm that the operation of each bimetal phase element is balanced with that of its neighbour.
5. Check that if the heater of each element is injected in turn, representing a single phasing situation, the relay becomes unbalanced and trips.
6. Test any attracted armature elements (instantaneous overcurrent, earth fault, flag auxiliary, etc.), if fitted, for operation at the correct settings.

Primary Injection Tests

Primary injection tests are carried out on the relays following the completion of all commissioning tests on CTs and VTs and relay calibration tests by secondary injection. The purpose of the primary injection test is to check the stability of the system for external through-faults and to check the entire system, including the circuit breaker control circuitry, for satisfactory operation when subjected to a fault current to which it is designed to react.

Transporting primary injection test sets offshore is difficult, as they tend to be too heavy to be flown out to the installation and have to be transported by sea. Having seen the treatment such equipment can be subjected to when lifted off a supply boat in heavy weather, the author advises that great care should be taken in packing the equipment to avoid damage and hence delays to the commissioning programme.

As switchgear, cabling and protection system designs can vary greatly, design documentation and the manufacturer's manual will need to be carefully considered for each part of the system to which primary injection currents are applied.

The most common types of protection are overcurrent, earth fault (residual) and restricted earth fault, for which typical test circuits are given in Figs 7.4.8–7.4.10.

Overcurrent Relays

The effective setting for overcurrent relays can be checked using the circuit in Fig. 7.4.8. The method shown checks the operating time of each element, and the residual current flowing in the common CT connection represents the imbalance between the phases under test. This residual current should be very low, normally less than 1 mA.

Residual Earth Fault Relays

The sensitivity of residually connected earth fault relays can be tested as shown in Fig. 7.4.9. Relays connected in this manner can be either instantaneous current operated elements or induction disc elements. The effective setting of the relay for primary faults is checked by single-phase injection.

Restricted Earth Fault Relays

For restricted earth fault relays, both sensitivity and stability tests are necessary, as these relay schemes may be subject to through-fault currents owing to faults in the

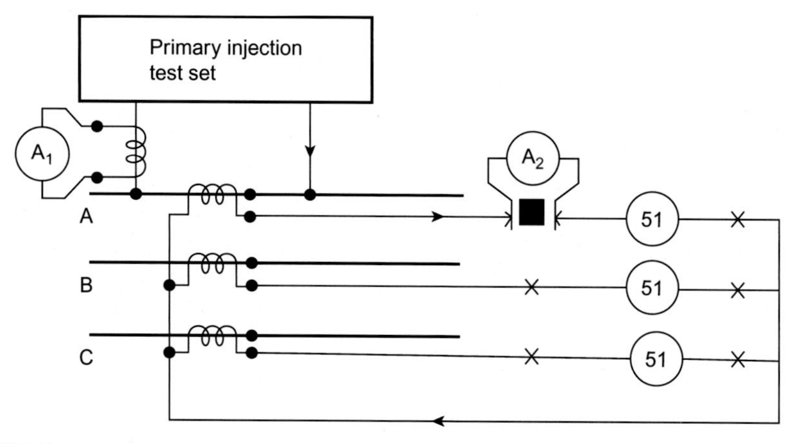

FIGURE 7.4.8

Primary injection test circuit for overcurrent relays.

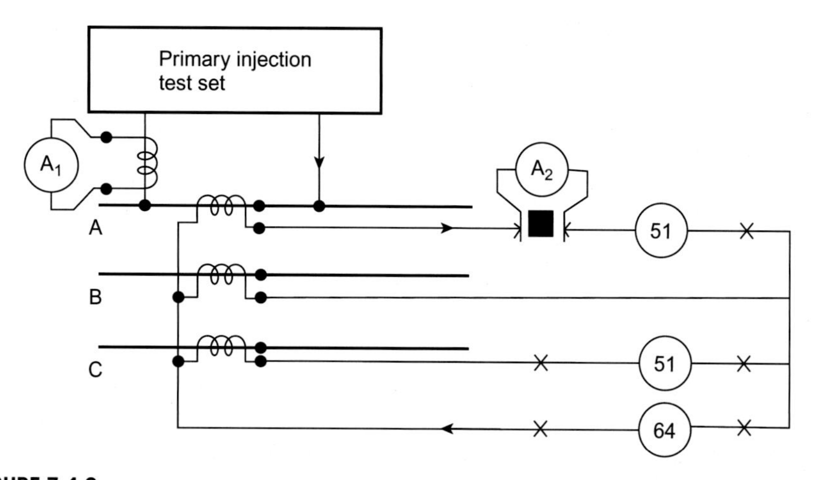

FIGURE 7.4.9

Primary injection test circuit for residually connected earth fault relays.

electrical system outside their zone of operation. Their most common application offshore is on transformer windings or as part of a transformer differential protection scheme.

The circuit for sensitivity testing is given in Fig. 7.4.10A. Each main CT is injected, and the voltage across the earth fault relay and the stabilising resistor is measured for a level of injection current which just gives relay operation.

The stability of the scheme when subject to external fault currents is tested by the application of the test circuit shown in Fig. 7.4.10B. Current is passed through the neutral CT and each phase in turn. With full primary load current flowing, the relay should remain stable and the ammeter should read only a few milliamperes.

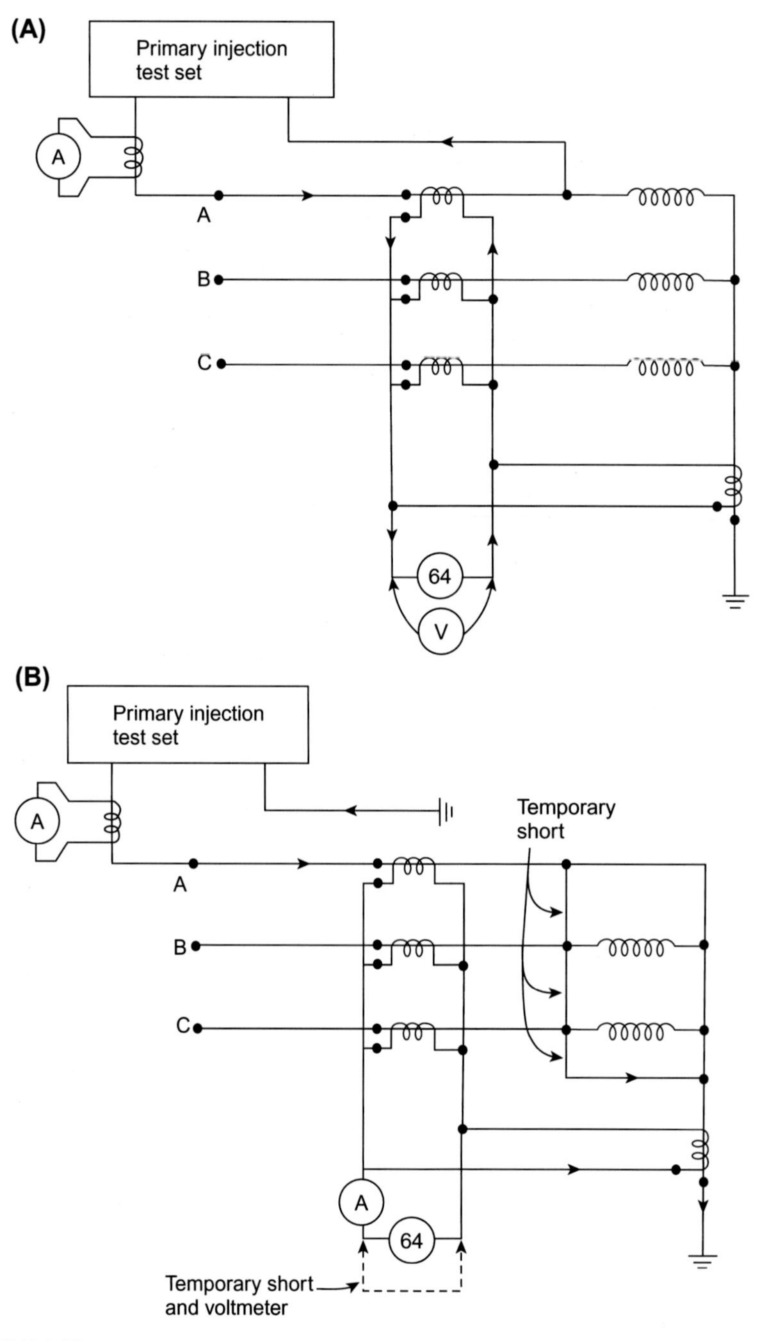

FIGURE 7.4.10

(A) Primary injection test circuit for sensitivity check on a restricted earth fault relay.
(B) Primary injection test circuit for stability check on a restricted earth fault relay.

CLOSING AND TRIPPING

The protection relays are fault sensing devices and cannot interrupt fault currents themselves. Therefore it is vital that the operation of the devices which perform this duty, i.e., the circuit breakers and their tripping and closing control circuits, are checked with their associated protection relay schemes. For the circuit breakers to be operated, it is necessary to commission the tripping and closing batteries and chargers before the primary injection test. An auxiliary DC supply is often required by the protection relays, in any case, for the operation of indication devices and as power supplies for electronic relays.

All auxiliary relays associated with each circuit breaker control system, including lockout relays and intertripping and interlocking schemes, should be functionally tested before the circuit breaker is put into service.

CHAPTER

Large Motors

5

PRECOMMISSIONING

Before commissioning, it is essential that the installation work has been properly checked before any equipment or cables are energised, or anything is allowed to rotate. Cable connections are 'rang through' and cable insulation is checked. All such work is termed precommissioning.

MOTOR COMMISSIONING TESTS

Three types of testing are recommended for the offshore commissioning of motors.

MEGOHMMETER (MEGGER) TEST

This involves the direct current (DC) measurement of the stator winding resistance to earth and between phases at the appropriate test voltage as shown in Table 7.5.1.

This test is carried out on all motor windings and is of 1 min duration.

The procedure is as follows:

1. Disconnect the neutral point and adequately insulate each phase lead from earth. On small machines, it will not be possible to disconnect the neutral point.
2. Disconnect and earth all resistance temperature devices (RTDs).
3. Ensure that the motor frame is connected to a known earthing point. Phases not under test must be earthed.
4. Measure the ambient air temperature in degree Celsius local to the motor before starting the test. If the temperature of the motor is thought to be significantly higher than the motor air temperature, then the winding temperature should be measured, if possible, by using any RTDs fitted to the motor before they are disconnected for the test.
5. If the temperature of the insulation is cold to the extent that it is below the dew point, a moisture film will form on its surface and may lower the insulation resistance, especially if the winding has developed a coating of dirt and/or salt. Therefore, if this is likely to occur, the motor anticondensation heaters should be left on for some time before the motor is tested.

Offshore Electrical Engineering Manual. https://doi.org/10.1016/B978-0-12-385499-5.00044-3

Table 7.5.1 DC Megohmmeter and Polarisation Index Tests for AC Motors

Motor Rated AC Voltage DC Test Voltage (V)	Motor Rated AC Voltage DC Test Voltage (V)
Below 100	500
Above 100 and up to 600	1000
Above 600 and up to 5000	2500
Above 5000	5000
IEEE 43-2000	
Winding Rated Voltage (V)	**Insulation Resistance Test Direct Voltage (V)**
<1000	500
1000–2500	500–1000
2501–5000	1000–2500
5001–12,000	2500–5000
>12,000	5000–10,000

AC, *alternating current;* DC, *direct current;* IEEE, *Institute of Electrical and Electronics Engineers.*

6. With a Megger or similar instrument perform the following test using the voltages given in Table 7.5.1. With the neutral point disconnected test the phase-to-phase insulation by applying the test voltage to each phase in turn with all other phases earthed. This test measures the combined insulation resistance of the tested phase, both to earth and to the other (earthed) phases.
7. Again using Table 7.5.1, with the neutral point connected, test the winding insulation to earth by applying the test voltage to each phase winding connection in turn, with all other phase connections insulated from earth. The winding insulation resistance should be corrected to 40°C by using the equation $R_c = K_t R_t$, where R_c is the insulation resistance corrected to 40°C, R_t is the insulation resistance at temperature t in °C and K_t is the insulation resistance temperature coefficient at temperature t obtained from Table 7.5.2.

POLARISATION INDEX

This test is a continuation of the above-mentioned Megger test and requires the application of the appropriate test voltage for 10 min. The ratio of the 10-min to the 1-min reading is termed the polarisation index (PI) and gives some indication of the dryness of the windings and their ability to withstand overvoltage tests.

The procedure is as follows. Using a motor-driven Megger or similar instrument measure the resistance to earth and between the phases of the motor windings. The test voltage (obtained from Table 7.5.1) should be applied for 10-min readings and those noted after 1 and 10 min.

Table 7.5.2 Winding Temperature Coefficients for Insulation Resistance Measurements at Various Temperatures

Winding Temperature (°C)	Multiplier Coefficient k_t
0	0.06
10	0.12
20	0.25
30	0.5
40	1.0
50	2.0
60	4.0
70	8.0
80	16.0
90	32.0
100	64.0

The PI can be calculated as follows:

$$PI = \frac{10\text{-min reading}}{1\text{-min reading}}$$

In the case of low-voltage machines, i.e., machines whose rated voltage is 600 V or less, the PI is calculated using the 30-s and 1-min readings and the application time is limited to 1 min.

OVERPOTENTIAL TESTS

Overpotential tests should not be carried out on motors unless the insulation resistance is greater than 100 MΩ and the PI values are greater than 2.0 for class B and class F insulation, or 1.5 for class A insulation.

High DC voltages are both lethal and, if the motor under test is in an area where flammable gases may be present, an ignition hazard. Carefully follow all operational procedures and safety precautions.

High-voltage DC tests should be applied as specified in Table 7.5.3, and the test should be made between the windings under test and the frame of the machine (which must be adequately earthed).

The test procedure is as follows. To start with, a test voltage of preferably about one-third but not more than one-half of the full test voltage quoted in Table 7.5.3 should be applied. This should be increased to the full test voltage as rapidly as the indicating instrument will allow without overshooting beyond the quoted value. The full test voltage must be maintained for 1 min, and the leakage current recorded. At the end of 1 min, the test voltage must be reduced rapidly to not more than one-third of its full value before switching off. This method of test voltage application is important to avoid overstressing the insulation with high transient voltages produced by switching, etc.

Table 7.5.3 High-Voltage DC Tests

Machine Rating	Maximum Recommended DC High-Voltage Tests (V)
Less than 1 kVA (or kW) with rated voltage below 100 V	500 + (2 × rated voltage)
Less than 1 kVA (or kW) with rated voltage 100 V and above	1000 + (2 × rated voltage)
1 kVA (or kW) and above, but less than 10 MVA (or MW)	1000 + (2 × rated voltage)
10 MVA (or MW) and above, at rated voltages	
Up to 2 kV	1000 + (2 × rated voltage)
Above 2 kV and up to 6 kV	2.5 × rated voltage
Above 6 kV and up to 17 kV	3000 + (2 × rated voltage)
Above 17 kV	Subject to special agreement

Earths should be applied to each test winding for at least 10 min after completion of the test and the complete windings for 1 h after final completion of all tests.

BS EN 60034:18-31:2012 permits DC onsite high-voltage tests to be carried out by agreement between the manufacturer and the purchaser. The test voltages must not be greater than 1.6×0.8 (i.e., 1.28) times the root mean square (RMS) value of the alternating current voltage specified in BS EN 60034. The values recommended in Table 7.5.3 are the RMS values quoted and are therefore 80% of the maxima allowed. These values must still be agreed with the machine manufacturer, however, before testing is undertaken.

EVALUATION OF MOTOR TEST RESULTS

When the tests described earlier have been carried out, the results must be interpreted to make a decision on whether or not to accept the machine for operation. An estimation of the machine's acceptability may be based on a comparison of present and previous values of insulation resistance and PI, corrected to 40°C in each case.

When the insulation history is not available, recommended minimum values of PI or of the 1-min insulation resistance value may be used. These values are given in Table 7.5.4.

The insulation resistance value of one phase of a three-phase winding, with the other two phases earthed, is approximately twice of that of the entire winding. Therefore, when the three phases are tested separately, the observed resistance of each phase should be halved to obtain a value for comparison with the appropriate value in Table 7.5.4(a).

The acceptance criteria in Table 7.5.4 must be modified, however, owing to the following limitations. First, insulation resistance is not directly related to dielectric strength, and it is therefore impossible to specify the value at which a winding will fail electrically. Second, the windings of large or slow-speed machines have extremely large surface areas, with healthy values of insulation resistance less than the recommended minimum.

Table 7.5.4 Recommended Minimum Values of PI

(a) Minimum Insulation Resistance for the Entire Winding at 40°C				
Motor rated voltage (V)	11,000	6600	3300	440
Minimum insulation resistance (MΩ)	12	7.6	4.5	1.5
(b) Minimum PI Values				
Motor insulation class	A	B	F	
PI	1.5	2.0	2.0	

PI, polarisation index.
For details of guard circuits see Thorn EMI's: A Simple Guide to Insulation and Continuity Testing, 1984.

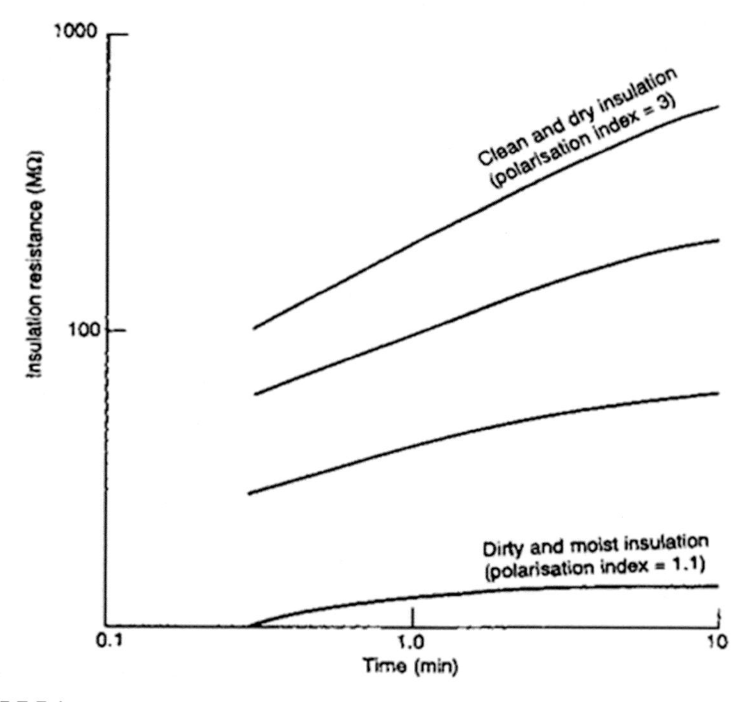

FIGURE 7.5.1

Polarisation index curve for motor windings.

Therefore, to some extent, acceptance will depend on the individual circumstances, such as motor size, number of poles, motor operating environment and frequency of operation.

Typical resistance/time characteristics are shown in Figs 7.5.1 and 7.5.2, which illustrate the behaviour of insulation under different conditions and show the significance of PIs. It must be stressed that the PI depends on a number of contributory factors, such as the winding condition, the insulation class and the type of machine.

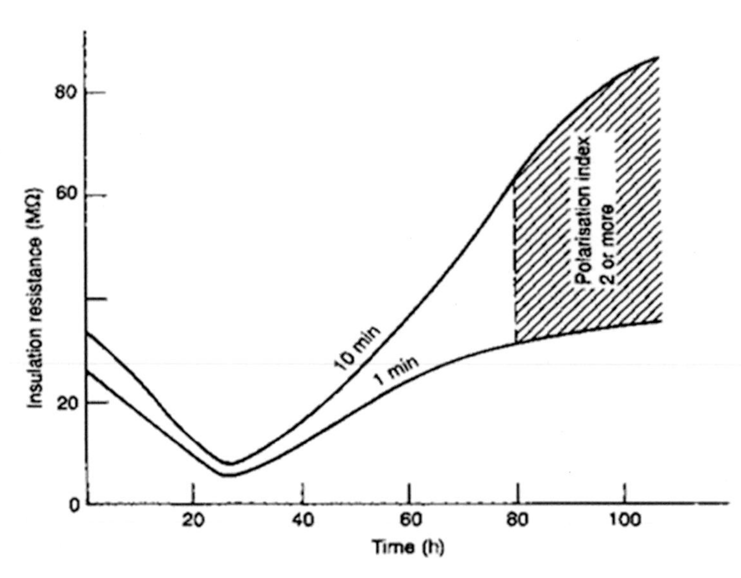

FIGURE 7.5.2

Drying curve for motor windings.

Figs. 7.5.1 and 7.5.2 are typical for class B insulation. Where the PI is low because of dirt or moisture contamination, it can be significantly improved by cleaning and drying. Drying curves are shown in Fig. 7.5.2, and such curves can be used to indicate the drying time required.

CHAPTER

Protection, Monitoring and Control

6

PRECOMMISSIONING

Before commissioning, it is essential that the installation work has been properly checked before any equipment or cables are energised. Cable connections are 'rang through' and cable insulation is checked where possible. Hard-wired logic and control and monitoring software must be checked, although this may be limited to critical circuits if the factory acceptance tests are comprehensive and well documented. The 'sense' of any current transformers must be checked to ensure terminals are not reversed.

All such work is termed precommissioning.

PERMIT-TO-WORK SYSTEMS

Monitoring of all work being done (and being aware of all possible clashes) during the major shutdown of an installation is vital to the safety of all personnel, and a failure in the control of work was a major contributor to the Piper Alpha disaster. The three-dimensional aspect of working offshore can make the control of work even more difficult, and the permit controller is second only to the offshore installation manager in having the most responsible job during a shutdown.

Let us consider two examples:

- Radiography with radioactive sources – what is the extent of the area that must be barriered off, and how many platform decks are affected at any one time?
- What isolations are required to replace a valve on a fuel-gas system – how many other systems are affected? Will main generation be affected? How many disciplines are involved?

These are only a few of the questions which require consideration if such tasks are to be completed safely. Even forgetting one could lead to injury or death.

For more information refer to the Health and Safety Executive's 'The safe isolation of plant and equipment'.

Offshore Electrical Engineering Manual. https://doi.org/10.1016/B978-0-12-385499-5.00045-5

CONTROL AND MONITORING SYSTEM COMMISSIONING

It is essential that initially all outputs are inhibited before the start of commissioning. The hierarchy of the installation's systems must be borne in mind, so that systems are brought online progressively. For example, generator prime movers will require fuel (gas or liquid), cooling water, lube oil, inlet and outlet air ducting and filtration. If the prime mover is to be load-tested with the generator, at least part of the electrical system will need to have been commissioned and a suitable electrical load provided.

CAUSE-AND-EFFECT MATRICES

The control and monitoring should be checked by progressing through the cause-and-effect charts line by line. Fig. 7.6.1 is a simple example of such a matrix.

It should be noted that complex systems may have cause-and-effect matrices extending to many pages and may take a week or more to verify.

All input causes and output effects in each line need to be confirmed by checking, for example, if valves have operated as expected. Once all causes and effects on the line have been confirmed, a highlighter can be used to indicate that the line has been checked. If an effect is found not to be present, then the actual effect and/or missing

TYPICAL ACCOMMODATION FIRE ALARM SYSTEM CAUSE & EFFECT MATRIX																						
			ANNUNCIATION		NOTIFICATION					REQUIRED FIRE SAFETY CONTROL								SUPPLEMENTARY				
CAUSE← EFFECT↓	LOCATION	SIGNAL TYPE	ALARM INDICATION & BUZZER ON CONTROL PANEL	ACTIVATION OF VOICE EVACUATION	ACTIVATE EVACUATION SIGNAL	DISPLAY/ PRINTER LOG	TRANSMIT F&G ALARM SIGNAL TO PLATFORM CONTROL ROOM	TRANSMIT TELEMETRY SIGNAL TO PLATFORM F&G SYSTEM	INDICATE ON PLATFORM CONTROL ROOM PANEL	RELEASE ALL FIRE DOORS	RELEASE MAGNETICALLY HELD SMOKE DOORS	START INTERNAL CIRCULATION FANS	SHUTDOWN LIFTS	CLOSE SMOKE/FIRE DAMPERS HVAC INTAKES	ACTUATE SMOKE EXHAUST	UNCONFIRMED FIRE OR GAS DETECTED	INITIATE SUPPRESSING SYSTEM RELEASING SEQUENCE	ACTUATE GRAPHIC SYSTEM DISPLAY FLOOR MAP	PRESSURIZE STAIRWELLS	OTHER SYSTEM SHUTDOWN PROCESS	ACTIVATE PLATFORM ALARM	SHUTDOWN EXTERNAL FIRE DAMPERS
CONFIRMED GAS DETECTED (2oo3 VOTING)	HVAC RM	GAS	X	X	X	X	X					X		X								
CONFIRMED GAS DETECTED (2oo3 VOTING)	GND. FLR.	GAS	X	X	X	X	X					X		X								
CONFIRMED GAS DETECTED (2oo3 VOTING)	1ST FLR	GAS	X	X	X	X	X					X		X								
CONFIRMED SMOKE DETECTED (2oo3 VOTING)	1ST FLR	FIRE	X	X	X	X	X			X	X	X	X		X			X			X	X
CONFIRMED SMOKE DETECTED (2oo3 VOTING)	GND. FLR.	FIRE	X	X	X	X	X			X	X	X	X		X			X			X	X
CONFIRMED SMOKE DETECTED (2oo3 VOTING)	1ST FLR	FIRE	X	X	X	X	X			X	X	X	X					X			X	X
CONFIRMED SMOKE DETECTED (2oo3 VOTING)	HVAC RM	FIRE	X	X	X	X	X			X	X	X	X					X			X	X
CONFIRMED HEAT DETECTED (2oo3 VOTING)	1ST FLR	FIRE	X	X	X	X	X			X	X	X	X					X			X	X
CONFIRMED HEAT DETECTED (2oo3 VOTING)	GND. FLR.	FIRE	X	X	X	X	X			X	X	X	X					X			X	X
ACTIVATION OF MANUAL CALL POINT	1ST FLR	FIRE	X	X	X	X	X			X	X	X	X					X			X	X
ACTIVATION OF MANUAL CALL POINT	GND. FLR.	FIRE	X	X	X	X	X			X	X	X	X					X			X	X
OTHER STATUS																						
ELECTRIC FIRE PUMP - RUN		FAULT	X			X		X	X		X	X						X				
DETECTION LOOP OPEN CIRCUIT		FAULT	X						X													
DETECTION LOOP OPEN CIRCUIT		FAULT	X						X													
MASTER ALARM OPEN CIRCUIT		FAULT	X						X													
MASTER ALARM SHORT CIRCUIT		FAULT	X						X													
EARTH FAULT DETECTED		FAULT	X						X													
FIRE PUMP SYSTEM LOW BATTERY		FAULT	X						X													
FIRE PUMP POWER FAILURE		FAULT	X					X														
FIRE PUMP PHASE REVERSAL		FAULT	X			X		X										X				
SPRINKLER CONTROL VALVE- LVL. ROOF		FAULT	X			X		X										X				
SPRINKLER CONTROL VALVE- LVL.1		FAULT	X			X		X										X				
SPRINKLER CONTROL VALVE-LVL. GRND		FAULT	X			X		X										X				
WATER FLOW- LVL. ROOF		FAULT	X			X	X				X	X		X	X				X		X	X
WATER FLOW - LVL.1		FAULT	X			X	X				X	X		X	X				X		X	X
WATER FLOW -LVL. GRND		FAULT	X			X	X				X	X		X	X				X		X	X

<u>Fire Alarm Sequence</u>

1.The fire alarm system operates as follows:

A. Manual Operation - operation of manual fire alarm (break glass p/b) shall:

1) Identify the individual station on the system control panel and sound the alarm bell at the control panel and remote alarm panel.

2) Sound audible alarms and actuate strobes

3) Transmit a signal to the platform control room

B. Confirmed Activation of area smoke or heat detection by 2oo3 voting of detectors, causes the following:

1) Provide operation as described for manual call point.

C. Activation of any supervisory circuit shall cause the following actions and indications:

1) Logging by printer of the event occurrence, including time, date and the device initiating the event.

D. Receipt of a fault signal at the fire alarm control panel causes the following:

1)Display at the control panel and platform annunciator panels alphanumeric LCD display, the origin or the fault condition report.

2) Activate trouble, audible and visual signals at the fire control panel and remote annunciator.

3) Record the occurrence of the event, the time of occurrence and the device initiating the event.

Note: Unconfirmed detector activation is not shown.

FIGURE 7.6.1

Example of a cause-and-effect matrix.

effects must be recorded. It is usually possible to continue through the lines of logic on the understanding that the exercise will need to be repeated once the issue or issues have been resolved.

Software aids are available to assist the commissioning engineer, such as logic simulators, which by allowing a virtual input of the particular cause indicate the resultant effects, intended or otherwise.

SAFETY, SHUTDOWN AND PROTECTIVE CIRCUITS AND DEVICES

It is important to prove these circuits and devices before starting up the system. This should include relevant fire and gas monitoring, active firefighting and pressure relief and flaring systems, where the system may contain hydrocarbons or pressure vessels, or both.

PART

8

Reliability, Maintenance and Logistics

CHAPTER

Reliability

1

INTRODUCTION

This chapter has been included as an introduction to the use of reliability analysis in aiding the offshore electrical system designer in his or her quest to produce a system that not only is fit for its purpose when all its component parts are working but also will continue to function in some fashion dependent on the severity of the problem when certain of its components fail or incidents occur on the installation which affect the system's integrity.

Reliability analysis encompasses a number of graphical, mathematical and textual operations which present the known facts, statistical data and/or experience about the proposed system or similar systems in a way which highlights its weaknesses or ranks the effectiveness of the options available to the designer.

It is very important when carrying out any form of reliability analysis that specific goals are selected. The availability of an electrical supply at a particular point in a system will be very much dependent on the configuration of the system and where that point is located in it. Therefore a vague instruction such as 'quantify the reliability of the supply system' is very unhelpful and it would be necessary to discuss with the client concerned about what particular supplies were of particular concern and then consider each in turn. For example, if the object was to maintain full production then only the equipment required to maintain supplies to the production switchboards would need to be considered. However, if supplies to essential equipment only were being analysed, the availability of all generation and all the routes which could be used to connect it to the essential equipment would need to be taken into account and, hopefully, in comparison, a much higher figure for the availability of supply would be obtained.

DUPLICATION AND REDUNDANCY

Provided that good quality assurance procedures are adhered to throughout a design and construction project, to the extent that all equipment purchased for installation on the platform is fit for its purpose, theoretically there should be no benefit to be gained by the replacement of one item of equipment with another one of similar function from another manufacturer. Practically, this is not the case, as we know by experience that certain devices are better obtained from a particular company. For obvious reasons, information of this kind in the form of usable data is difficult to

Offshore Electrical Engineering Manual. https://doi.org/10.1016/B978-0-12-385499-5.00046-7

obtain, and therefore, for the purposes of this introduction, it will be assumed that every component with the same function has the same reliability properties, i.e., its failure rate and its mean time to repair (MTTR) is the same as any other with that particular function. Nevertheless, it is advisable to keep records of failure and times to repair of equipment installed, as analysis of this data in itself may indicate some problem when the records are compared with available generic data.

Very significant reliability improvements can be made in systems, usually much more significant than from simply replacing components, by reconfiguring them in some way. Some examples of this are given in PART 1 Chapter 2. The basic principle for any reconfiguration is to provide redundancy in such a way that during a component's outage, its function will be maintained by one or more other components of similar function, whether this outage is planned or unplanned. In most cases the 'component' is a system in its own right, such as a generator package with its associated auxiliaries or a subsea cable supply from another platform.

FAILURE MODE, EFFECTS AND CRITICALITY ANALYSIS

This procedure, like the hazard and operability analysis (Hazops) which although is really beyond the scope of this book, is worth applying to the more complex systems as a method of identifying all modes of failure, analysing their effects and where possible, evaluating the frequency of their occurrence.

The basic failure mode, effects and criticality analysis (FMECA) process is the completion of a table with column headings similar to Fig. 8.1.1.

However, before the columns can be completed, it is necessary to draw a block schematic diagram of the system to be analysed at the level of detail required. This level should be low at the initial stage and as it becomes more obvious which components or subsystems are critical, these particular areas can be broken down into more detail by producing a more detailed block diagram with more components identified and analysed.

CIRCUIT BREAKER ILLUSTRATION

If we take a circuit breaker as an example item on an FMECA, this would appear as a block in the diagram as shown in Fig. 8.1.2. A circuit breaker is a particularly difficult component to describe in an FMECA, as it has several functions and several failure modes.

CIRCUIT BREAKER FUNCTIONS

(i) Connects and isolates an electrical circuit as required.
(ii) In the event of a fault, automatically in conjunction with a sensing device, interrupts the flow of fault current.

System: Generation
Subsystem: Main HV generation
Drawing no:

Failure mode and effect analysis worksheet
Job no:
Job title: Reliability study
Client

Page 1 of 8
Issue A
Prepared by: Date:
Checked by: Date:

(1) Item code	(2) Component	(3) Function	(4) Failure mode	Failure effect (5) Local level	(6) Plant level	(7) Means of detection	(8) Severity of effect	(9) Compensating provisions or measures in operation	(10) Remedial measures in design
1	Generator UCP	a) Trip/ Inhibit Start under fault or pre-set conditions	Unit starts under fault conditions	Damage to unit	Down time due to original fault extended Reduced production	Operator	II	ST/BY GT unit	Instrumentation and controls, plus maintenance procedures and periods seem adequate from information available
			Unit fails to trip under fault conditions	Damage to unit	Initially no effect. Severe unit damage results in loss of electrical output to distribution reduced production	Operator	II	ST/BY GT Unit if available Load shed	— " —
			Unit trips under healthy conditions	Loss of output	Loss of output to distribution load shed Initiate	Operator	I	— " —	— " —
		b) Auto sequence control brings set on load through pre-set conditions	Wrong Sequence	Damage to unit	— " —	Operator	II	Second GT unit	— " —
				Fire possible	Fire hazard	Operator/ Control system	III	— " —	— " —

FIGURE 8.1.1

Typical sheet for use with failure mode, effects and criticality analysis.

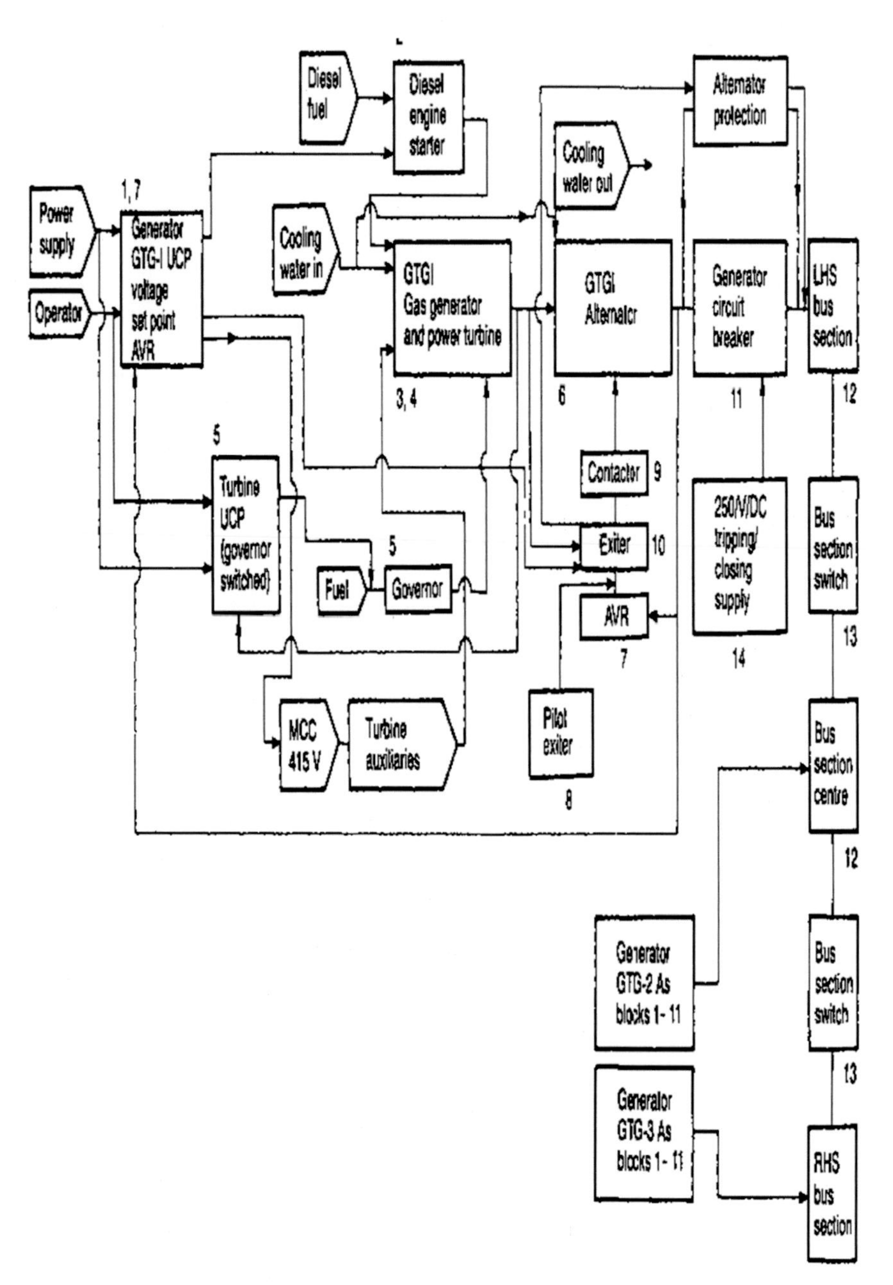

FIGURE 8.1.2

Typical functional block diagram for use with failure mode, effects and criticality analyses. *AVR*, automatic voltage regulator; *DC*, direct current; *GTG*, gas turbine generator; *LHS*, left-hand side; *MCC*, motor control centre; *RHS*, right-hand side; *UCP*, unit control panel.

Note that the adequacy of load and fault rating is assumed, as the purpose of this analysis is not to question the system design calculations but is concerned with the effects on the system of a particular mode of failure.

CIRCUIT BREAKER MODES OF FAILURE

There are two modes of failure associated with the above-mentioned first function, namely,

(a) circuit breaker fails to open when isolation/disconnection is required,
(b) circuit breaker fails to close when power supply is required.
For the fault clearing function, there is one failure mode:
(c) circuit breaker fails to interrupt fault current.
If required this could be split into
(a) fails to interrupt make fault,
(b) fails to interrupt break fault.
Lastly, there are faults associated with the breakdown of insulation within the circuit breaker cubicle:
(d) interphase fault in circuit breaker cubicle,
(e) earth fault in circuit breaker cubicle.

Some of these faults may be due to the failure of the trip/close battery supply or a protection relay. Depending on the level of detail required, however, these components can be included as part of the circuit breaker 'block' or if necessary a new block diagram may be drawn showing individual blocks for each component in the 'circuit breaking system'.

Having identified the failure modes, the next stage is to describe the failure effects. This is usually done at three arbitrary levels:

(a) effect on the system
(b) effect on adjacent equipment
(c) effect on the offshore installation

If the system is well designed, there should be no effects on the installation and only one or two effects on adjacent equipment, usually associated with fire. All these effects should have some compensating provision or remedial measure available and this should be listed as shown in Fig. 8.1.1, column 9. Where no such provision or remedy exists, column 10 should be used to propose a solution which prevents such an effect from occurring or alleviates the problem. At the end of the study, all such proposals should be listed.

If a significant number of proposals appear in column 10, it will probably be necessary to rank the associated effects in the order of criticality. This is usually done by drawing a criticality matrix as shown in Fig. 8.1.3 (extracted from MIL-STD-1629A: *ISO 9000, and IEC 61508 methodologies are other and more modern approaches*

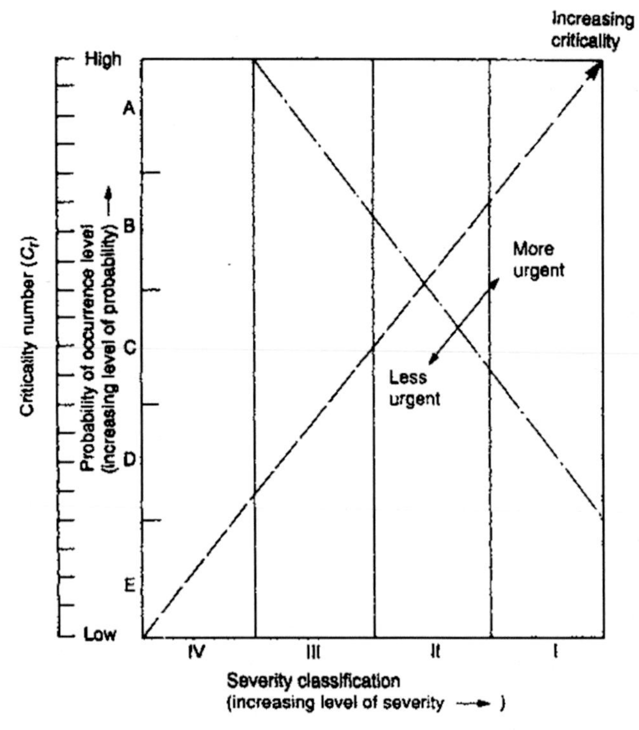

FIGURE 8.1.3

Typical criticality matrix.

to performing this kind of analysis). This is a graph with axes of severity of effect against the probability of its occurrence.

The severity is given four arbitrary categories as follows:

(a) *Category I*: *Catastrophic* – A failure which may endanger the whole installation and involve loss of lives.
(b) *Category II*: *Critical* – A failure which leads to prolonged loss of production and may cause severe injury, major equipment and/or system damage.
(c) *Category III*: *Marginal* – A failure which may cause minor injury and minor system/equipment damage and will result in delays to or partial loss of production.
(d) *Category IV*: *Minor* – A failure not serious enough to cause injury or any equipment/system damage but which will result in an unscheduled outage for maintenance or repair.

The frequency of occurrence may either be a numerical value obtained from a reliability data bank or a qualitative level assigned as follows:

Level A	Frequent	A high probability of occurrence during the time the system is running
Level B	Reasonably probable	A moderate probability of occurrence during the time the system is running
Level C	Occasional	An occasional probability of occurrence during the time the system is running
Level D	Remote	An unlikely probability of occurrence during the time the system is running
Level E	Extremely unlikely	A failure whose probability of occurrence is essentially zero during the time the system is running

A diagonal line is drawn on the matrix and each failure mode effect which has no existing compensating provision should be represented by a cross drawn on the matrix. Failure effects appearing in the upper right-hand area of the graph have the greatest criticality and hence the most urgent need for corrective action.

FAULT TREES

A fault tree is a graphical method of describing how faults in system components relate to overall system failures. The rules for constructing a fault tree will, if rigidly adhered to, lead to the creation of a graphical model of the system which will greatly assist the system designer in ensuring that as many failure modes as possible are remedied at the design stage.

A typical fault tree model is shown in Fig. 8.1.4. The diamond or circle shapes represent basic events which are component failure modes.

Combination events, which are the logical result of a combination of basic events, are represented by rectangles. The logic of the tree is shown by using mainly 'AND' and 'OR' gates.

The AND gate represents a situation where a combination event can only exist if the basic events connected to it by the AND gate all exist simultaneously.

The OR gate represents the situation where a combination event can exist if one or more of the basic events connected to it by the OR gate exist. Other logic devices can be used to represent standby equipment, voting systems, etc., and in most cases, normal Boolean algebra rules apply.

FAULT TREE CONSTRUCTION

As with all methods of reliability analysis, the system boundaries must be carefully defined. A block diagram of the system should be drawn which shows all external inputs, as these should be identified in the fault tree construction. This process assists the analyst to understand the function of the system. An FMECA should be

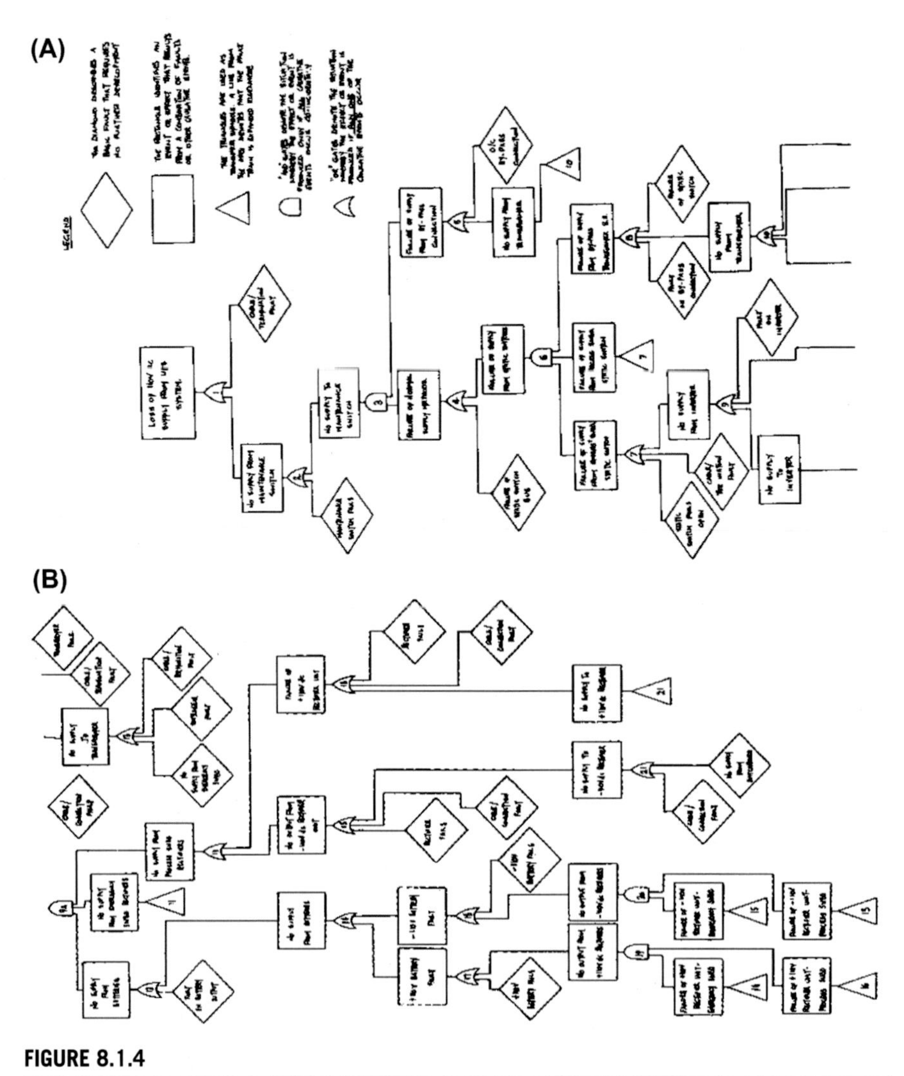

FIGURE 8.1.4

(A and B) Typical fault model.

undertaken at this stage if the failure modes of the various components are not clear. If complex logic is involved, truth tables should also be constructed.

The top event of the tree must be carefully defined, as an ambiguous or vague title will confuse and lead to much wasted effort. A method that the author has used to check the logic of the fault tree is to convert the fault tree into a *reliability block diagram* (see later discussion) by changing parallel connections to series and vice versa. This can show up any defective logic, as the function block created may not make sense.

Having defined the top event, each combination event is defined by asking the following questions:

1. '*What events taken singly will directly result in the event under consideration?*' These events are connected to the event under consideration by an OR gate.
2. '*What combination of events will directly result in the event under consideration?*' These events are connected to the event under consideration by an AND gate.

By definition, all combination events will be the consequence of events drawn below them and connected in some way via logic gates. The fault tree is complete when all combination events are shown to be the cause of two or more basic events. When the fault tree has been drawn, various methods can be applied to it to identify problems in the system that it represents, as described in the following section.

COMMON MODE FAILURE IDENTIFICATION

If the same event appears in several places on the fault tree, common mode failures may be identified. Some examples of common mode failures are as follows:

(a) Fire in a switchroom leading to the loss of two or more distribution routes.
(b) Fuel gas system fault leading to all main generators failing.
(c) Loss of hazardous area ventilation leading to shutdown of all main oil line pumps and/or all gas export compressors.

QUALITATIVE ANALYSIS

This can be carried out by inspecting a simple fault tree to identify any *minimal cut sets*. These sets consist of a group of basic events which will cause the top event to occur if and only if they exist simultaneously. Boolean reduction may also be used, provided there are no 'exotic' logic gates such as those with a sequential or timed inhibit function.

QUANTITATIVE ANALYSIS

There are two main methods of analysing a fault tree quantitatively.

Evaluation Using Event Probabilities

This method is used when the event failure data is available in terms of probabilities.

Where two events are output through an AND gate (see Fig. 8.1.5), the failure probabilities may be added as follows:

$$P(a \cdot b) = P(a) \cdot P(b) \text{ where } (a \cdot b) \text{ represents } a \text{ AND } b$$

Where two events are output through an OR gate, the resultant probability of failure is as follows:

$$P(a+b) = P(a) + P(b) - P(a)P(b) \text{ where } (a+b) \text{ represents } a \text{ OR } b$$

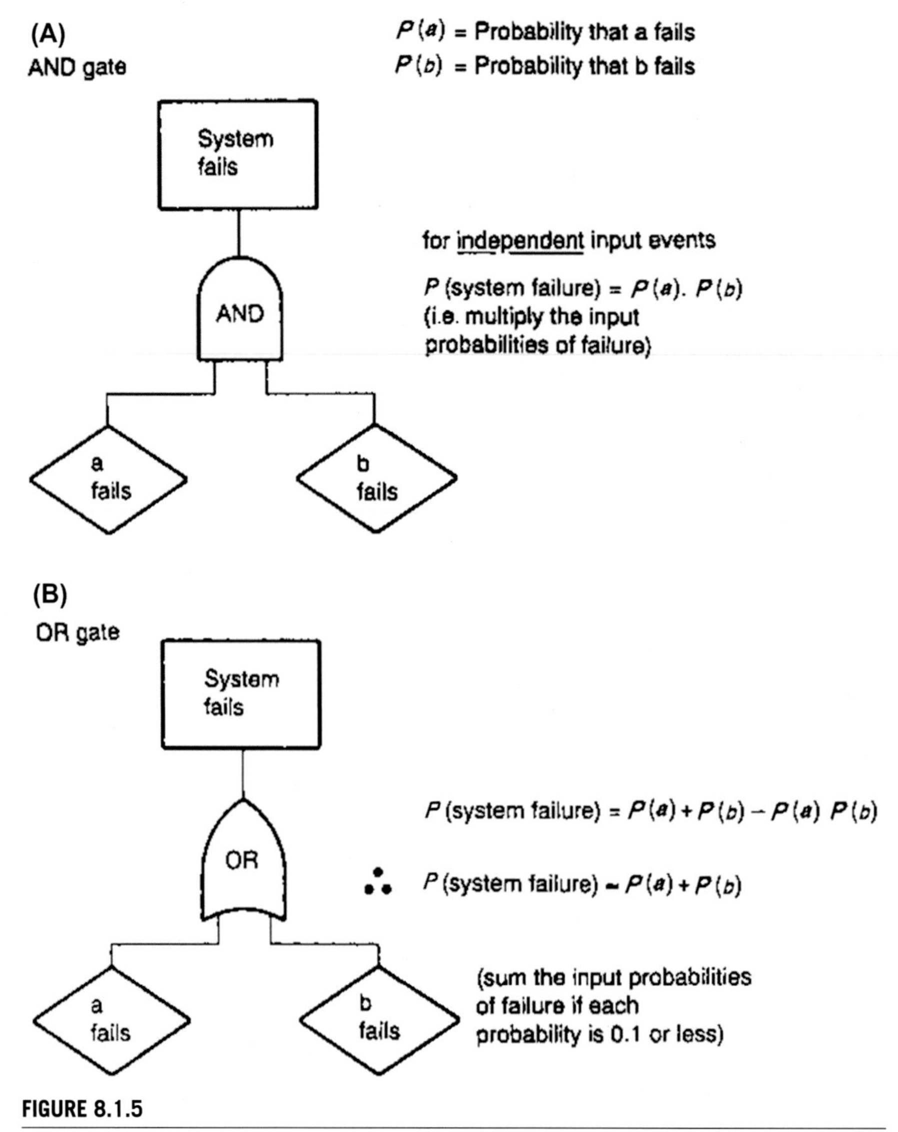

FIGURE 8.1.5

(A) AND and (B) OR logic for fault tree evaluation.

If the probabilities $P(a)$ and $P(b)$ are each 0.1 or less, then the product $P(a) \cdot P(b)$ is very small compared with the sum $P(a) + P(b)$ and a good approximation may be obtained from

$$P(a+b) = P(a) + P(b)$$

This may be proved for n events (see Reference 1) as follows:

$$P(a \cdot b \ldots n) = P(a)\,P(b)\ \ldots P(n) \text{ for AND gates}$$

$$P(a + b \ldots n) = P(a) + P(b) + \ldots P(n) \text{ for OR gates}$$

Note that this is only true for *independent* events

Evaluation Using Event Failure Rates

For any number of events with constant failure rates input to an OR gate, it can be proved (see Reference 1) that the output has a constant failure rate which is the sum of the failure rates of the inputs. For any number of events with constant failure rates input to an AND gate, it can be proved (see Reference 1) that the output failure rate after a given time t will be a function of t. If each of the events is identical, as would be the case with the failure rates for a number of generators in a system where each is capable of maintaining the full system load, then without maintenance the output failure rate would tend to approach the single unit failure rate after a certain number of hours (see Fig. 8.1.6). In a real situation

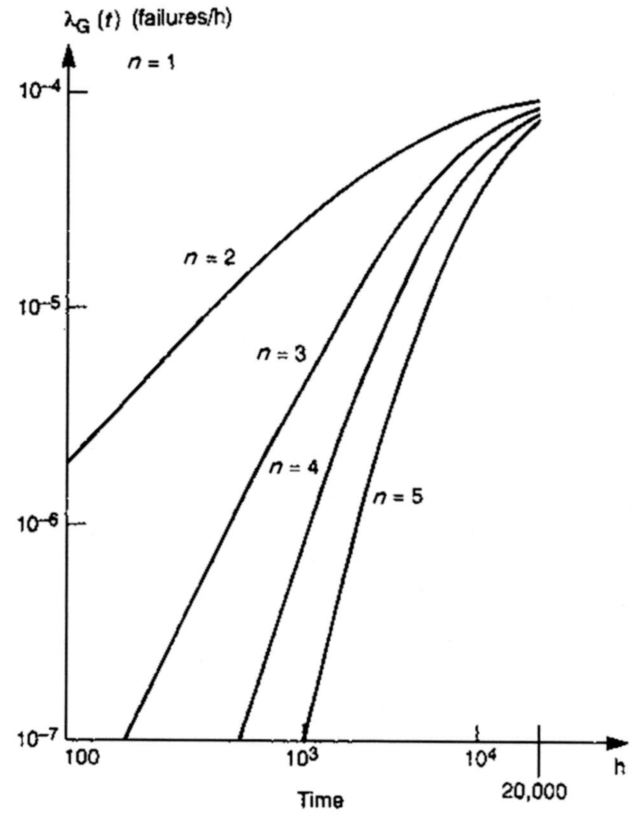

FIGURE 8.1.6

Graph of system failure rate against unit numbers, without maintenance.

where regular maintenance is carried out, as a good approximation, it is acceptable to take the output of an AND gate as the product of the input event failure probability, provided the MTTRs are very much shorter than the mean time between failures (MTBFs).

RELIABILITY BLOCK DIAGRAMS

If a system is broken down into key components or elements, the failure of which would have some effect on the system's availability, a block diagram may be drawn such that for the continuing operation of the system, a continuous string of elements must exist from one side of the diagram to the other. An example of such a block diagram is shown in Fig. 8.1.7. A reliability block diagram produced in this way provides a clear indication of any critical system components and often can also be numerically evaluated. In order to carry out this numerical evaluation, it is necessary to assign a value for failure rate and MTTR for each system element. An availability may then be calculated for each element as follows:

$$\text{Availability } (A) = \frac{\mu}{\lambda + \mu}$$

where μ is the reciprocal of the MTTR (in hours) and λ is the failure rate (failures per 10^6 h). The unavailability is given by

$$\overline{A} = 1 - A$$

and the unavailability of a system of n components in series is then

$$\overline{A}_s = \overline{A}_1 + \overline{A}_2 + \cdots + \overline{A}_n$$

If the unavailabilities are calculated for each component, they may be added together where the configuration consists of series components and where the component MTBFs are very much larger than their MTTRs.

The MTBF can be obtained by inverting the failure rate and multiplying by 10^6 as follows:

$$\frac{1}{\left(\lambda \frac{\text{failures}}{10^6} \text{ h}\right)} = 10^6/\lambda \text{ h/failure}$$

The unavailability of a system of n identical components in parallel, assuming only one component is required to maintain the system, is

$$\overline{A}_s = \overline{A}_1 \overline{A}_2 \cdots \overline{A}_n$$

Where more than one parallel component is required to maintain the system, the following can be shown to be the combined availability:

$$A_s = \sum_{j=m}^{n} \begin{matrix} n \\ j \end{matrix} \; A^j (1 - A)^{n-1}$$

where n is the total number of components and m is the number of components required to maintain the system.

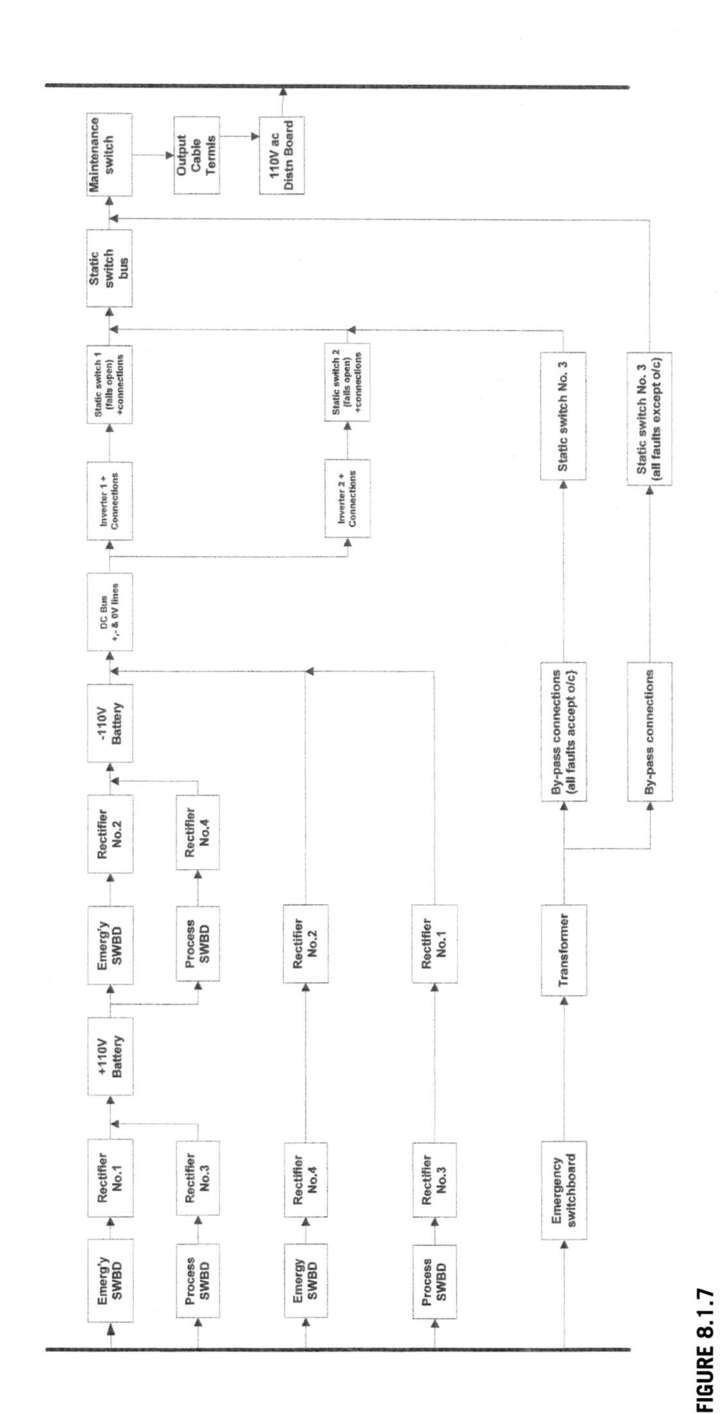

FIGURE 8.1.7
Typical reliability block diagram. *SWBD*, switchboard.

With parallel components, calculation becomes more complicated. In a system X where two parallel components are redundant, i.e., both are designed to be functioning but only one is required for system operation (Fig. 8.2.10), the unavailability of the system is

$$\overline{A}_x = \overline{A}_A \cdot \overline{A}_B$$

If there are three components in parallel and two out of the three are required for successful system operation, then

$$\overline{A}_x = 1 - \overline{A}_A \cdot \overline{A}_B - \overline{A}_B \cdot \overline{A}_C - \overline{A}_C \cdot \overline{A}_A + 2\overline{A}_A \cdot \overline{A}_B \cdot \overline{A}_C$$

With complicated systems, as in Fig. 8.1.11, a truth table can be drawn.

For the truth table in Fig. 8.1.11:

$$\overline{A}_x = \overline{A}_A \cdot \overline{A}_B + \overline{A}_A \cdot \overline{A}_B \cdot \overline{A}_C + \overline{A}_A \cdot \overline{A}_B \cdot \overline{A}_C \cdot \overline{A}_D + \overline{A}_B \cdot \overline{A}_C \cdot \overline{A}_D + \overline{A}_A \cdot \overline{A}_C + \overline{A}_B \cdot \overline{A}_D + \overline{A}_A \cdot \overline{A}_D$$

The Boolean equation may be written out directly from inspection of the diagram as follows:

$$\overline{A}_x = \left\{\left(\overline{A}_A + \overline{A}_B\right) \cdot \left(\overline{A}_C + \overline{A}_D\right)\right\} + \overline{A}_A \cdot \overline{A}_B$$

For more complex situations consult Reference 1.

As each component availability is usually a decimal fraction which approaches a value of unity with improving reliability and reducing repair time, the value is often a figure such as 0.999999945, where only the last two figures are significant. Therefore, whether calculators or computers are used, care must be exercised to ensure either that these figures are not lost because the particular hardware cannot handle the required number of decimal places or that the significant figures are not lost by rounding.

Spreadsheet programs such as Lotus 1-2-3 or Excel are very useful in producing a tabular output of failure rates, repair times, unavailabilities, etc., and some examples of evaluations are given in the following sections.

If the system unavailability is multiplied by the number of hours in a year (8736), the result is the annual system downtime in hours.

CONFIDENCE LIMITS

Failure rates are statistical values based on samples of known population. However, neither the total population, the mean value of failure rate for all components of a particular type, nor the way the values vary over the range from the worst to the least is known.

The electrical engineer needs to know how closely the sample mean ($\overline{x}$) agrees with the total population mean value of failure rate (μ).

Failures in similar components will tend to reach a peak after a certain time and then tail off, again producing a bell-shaped characteristic called a normal distribution (see Fig. 8.1.7).

Various statistics may be calculated from the data available. Those of particular interest here are as follows. First, the sample variance is given by

$$s^2 = (x_1 - \bar{x})^2 + (x_2 - \bar{x})^2 + \cdots + (x_n - \bar{x})^2 / (n-1)$$

The sample standard deviation s is obtained by taking the square root of the variance. The true population variance is usually denoted by σ.

Now 95% of the standard normal distribution lies between −1.96 and +1.96 so the interval between $x - 1.96\sigma/\sqrt{n}$ and $x + 1.96\sigma/\sqrt{n}$ is called the *95% confidence interval* for μ (Fig. 8.1.8). Given the sample mean x, this means that we are 95% confident that this interval will contain μ. The two end points of the confidence interval are called the *confidence limits*. In practice, failure rates are only known for samples, so the standard deviation is unknown and the sample standard deviation is used to estimate σ. Although the 95% confidence interval briefly discussed earlier is for cases where σ is known, the interval when σ is unknown approaches the same value as the sample size increases, as shown in Table 8.1.1.

There is, however, one more item to take care of before the confidence limits can be established. Each new sample taken will yield a new set of results with different values for the sample variance each time. It can be shown that this statistic follows a distribution called the chi-squared distribution (see Fig. 8.1.9). This distribution is related to the normal distribution and depends on a parameter known as the number of *degrees of freedom* (DF). We would say that the estimate of s^2 of σ^2 has $(n-1)$ DF.

From statistical tables, it is possible to attach confidence limits to failure rates, as in Example 8.1.2.

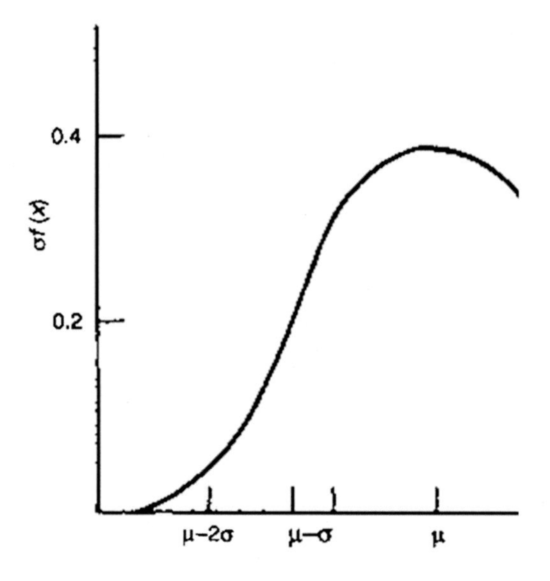

FIGURE 8.1.8

Normal distribution characteristic.

Table 8.1.1 Values of the Percentage Point of Distribution t_c for 95% Confidence Interval

n	t_c
4	3.18
8	2.36
12	2.2
20	2.09
Infinity	1.96

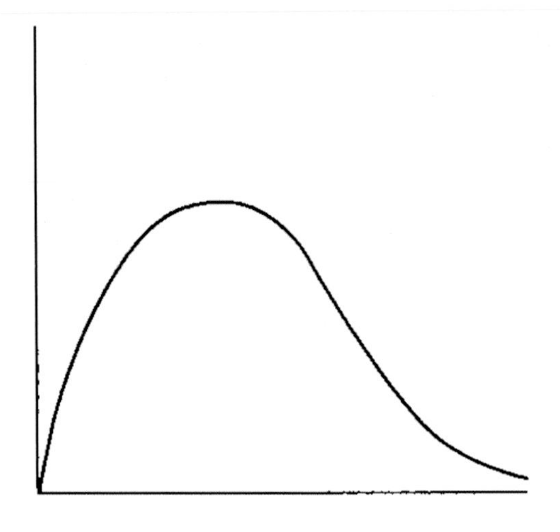

FIGURE 8.1.9

Chi-squared distribution characteristic.

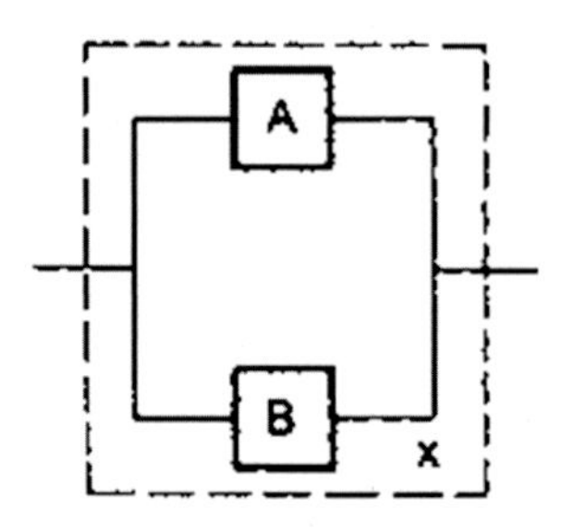

FIGURE 8.1.10

Reliability block diagram for two components in parallel.

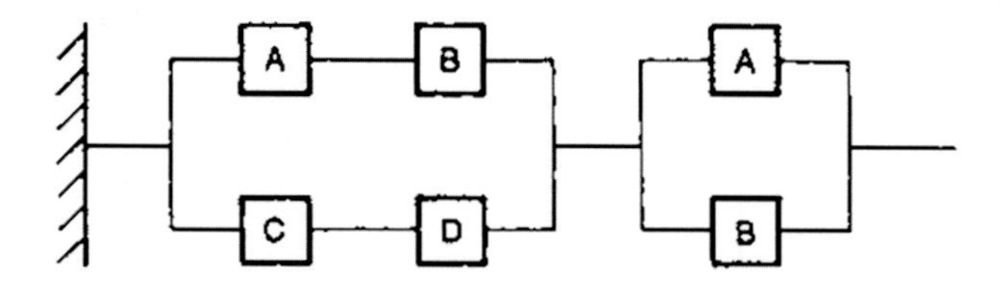

A	*B*	*C*	*D*	*System success*	
1	1	1	1	1	
×	1	1	1	1	
×	×	1	1	×	$\overline{A}_A \cdot \overline{A}_B$
×	×	×	1	×	$\overline{A}_A \cdot \overline{A}_B \cdot \overline{A}_C$
×	×	×	×	×	$\overline{A}_A \cdot \overline{A}_B \cdot \overline{A}_C \cdot \overline{A}_D$
1	×	×	×	×	$\overline{A}_B \cdot \overline{A}_C \cdot \overline{A}_D$
1	1	×	×	1	
1	1	1	×	1	
×	1	×	1	×	$\overline{A}_A \cdot \overline{A}_C$
1	×	1	×	×	$\overline{A}_B \cdot \overline{A}_D$
1	×	×	1	×	$\overline{A}_B \cdot \overline{A}_C$
×	1	1	×	×	$\overline{A}_A \cdot \overline{A}_D$

FIGURE 8.1.11

For more complex arrangements a truth table may be used.

EXAMPLE 8.1.1

A sample of 12 equipment failure rates is proved in a data bank. The mean failure rate $\bar{x}$ value given is 70 per 10^6 h. The sample standard deviation s is 0.56 per 10^6 h. Calculate the 95% confidence limits.

SOLUTION

From Table 8.1.1 a sample of size 12, the 95% confidence interval for the true population mean μ is given by

$$\bar{x} \pm 2.2\left(s\sqrt{n}\right) \quad \text{or} \quad 70 \pm 2.2\,(0.56/3.464)$$

which gives (70 ± 0.356) failures per million hours.

EXAMPLE 8.1.2

Evaluation of alternative configurations

The following numerical values are taken for this example:

Component	μ=1/MTTR	$\lambda(\times 10^{-6})$	*A*	$\overline{A}$
Generator	0.04166	50	0.998801	0.001199
Cable	0.1666	10	0.999940	0.00006
Busbar	0.1666	0.175	0.9996389	0.000361
Transformer	0.01	0.13	0.999998	0.000002

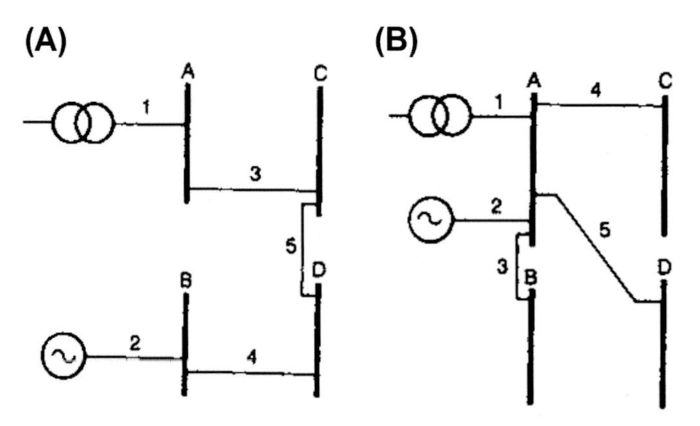

FIGURE 8.1.12

(A and B) Single-line diagrams for Example 8.1.2.

A large chemical plant has four main areas of activity which are grouped geographically. It is important that interruptions in supply to any of the four areas are kept to a minimum and that those that do occur are short-lived. Assuming standby generation is required and all other considerations have equal effect, which of the two configurations in Fig. 8.1.12 gives the better reliability?

Note that each arrangement has the same number of cables, generators and transformers.

SOLUTION

With the simple example, where component availabilities are identified and the availability effects due to configuration can easily be seen, Fig. 8.1.12A may be identified by inspection as the more reliable, as at least two components must fail before any part of the system is blacked out.

Nevertheless, to illustrate the methods of calculation, the two arrangements will be analysed.

The configurations in Fig. 8.1.12A and B may be converted into reliability block diagrams as in Fig. 8.1.13A and B, respectively. As indicated earlier, a reliability block diagram represents the configuration in terms of Boolean logic. Components may be repeated so as to represent a system success (i.e., 'supplies available at all busbars') path through the diagram from left to right. As with all Boolean diagrams, the integrity of the system may be 'tested' by deleting components to see if a path or paths remain through the network.

The level of detail used in the analysis should be related to the reliability data available. For example, circuit breakers may be introduced as components which may appear on the reliability diagram in several different locations depending on the mode of failure. If the equipment is located in the same substation or cables in the

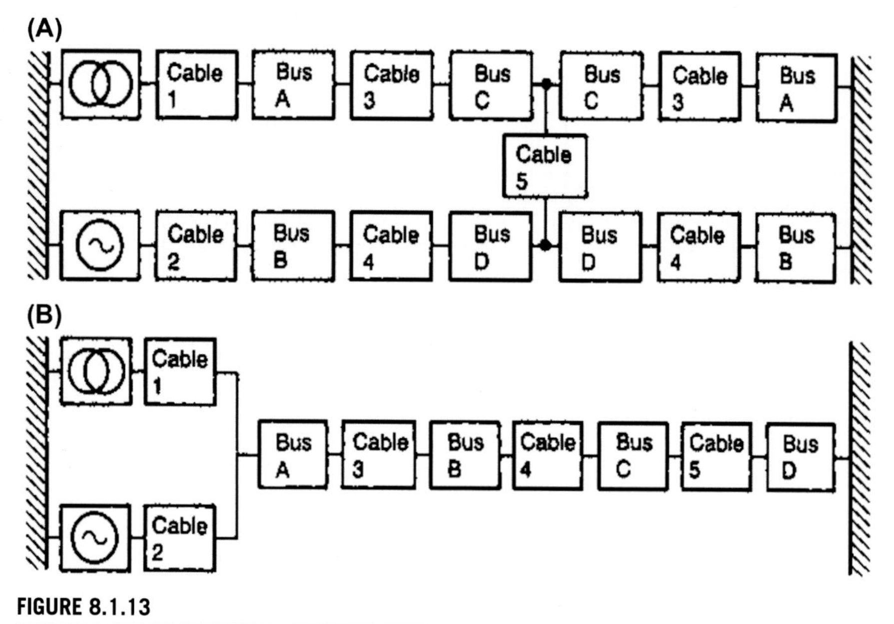

FIGURE 8.1.13

(A and B) Reliability block diagrams for Example 8.1.2.

Table 8.1.2 Evaluation of Fig. 8.1.13A

Component	μ=1/MTTR	$\lambda(\times 10^{-6})$	A	$\overline{A}$
Transformer	0.01	0.13	0.999998	0.000002
Cable 1	0.1666	10	0.999940	0.00006
Bus A	0.1666	0.175	0.9996389	0.000361
Cable 3	0.1666	10	0.999940	0.00006
Bus C	0.1666	0.175	0.9996389	0.000361
Total (LHS top)	–	–	–	0.000844
Generator	0.04166	50	0.998801	0.001199
Cable 2	0.1666	10	0.999940	0.00006
Bus B	0.1666	0.175	0.9996389	0.000361
Cable 4	0.1666	10	0.999940	0.00006
Bus D	0.1666	0.175	0.9996389	0.000361
Total (LHS bottom)	–	–	–	0.002041
Bus C	0.1666	0.175	0.9996389	0.000361
Cable 3	0.1666	10	0.999940	0.00006
Bus A	0.1666	0.175	0.9996389	0.000361
Total (RHS top)	–	–	–	0.000782
Bus D	0.1666	0.175	0.9996389	0.000361
Cable 4	0.1666	10	0.999940	0.00006
Bus B	0.1666	0.175	0.9996389	0.000361
Total (RHS bottom)	–	–	–	0.000782

LHS, *left-hand side;* MTTR, *mean time to repair;* RHS, *right-hand side.*

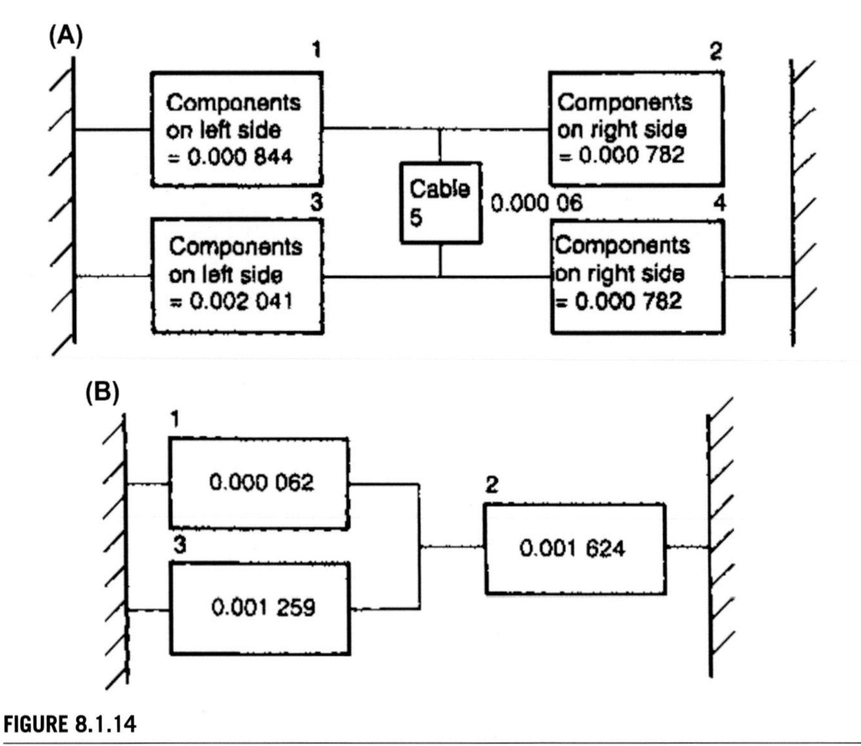

FIGURE 8.1.14

(A and B) Numerical evaluation of Example 8.1.2.

same trench, etc., blocks may be introduced to represent common mode failures such as substation fires or cable excavation accidents.

Using these illustrative values, we can set out the data for Fig. 8.1.11A as in Table 8.1.2. The total unavailabilities are summarised in Fig. 8.1.14A. The final equation is as follows, where the numbers 1–5 now refer to those in Fig. 8.1.14A:

$$\begin{aligned}\overline{A}_{(a)} &= \left\{\left(\overline{A}_1+\overline{A}_2\right) + \left(\overline{A}_1+\overline{A}_5+\overline{A}_4\right)\right\} \cdot \left\{\left(\overline{A}_3+\overline{A}_4\right) + \left(\overline{A}_3+\overline{A}_5+\overline{A}_2\right)\right\} \\ &= \left\{2\overline{A}_1+\overline{A}_2+\overline{A}_4+\overline{A}_5\right\} \cdot \left\{2\overline{A}_3+\overline{A}_2+\overline{A}_4+\overline{A}_5\right\} \\ &= \{0.003312\} \cdot \{0.005706\} \\ &= 1.8898272\times 10^{-5}\end{aligned}$$

Fig. 8.1.14A may be expressed as downtime in hours per year by multiplying by 8760 to give 0.1655 h or 10 min/year.

The configuration in Fig. 8.1.13B may be similarly summarised as in Fig. 8.1.14B, and in this case the final equation is

$$\overline{A}_{(b)} = \overline{A}_1\overline{A}_3 + \overline{A}_2 = 1.62407\times 10^{-3}$$

The downtime is therefore 14.18 h/year.

Hence the configuration of Fig. 8.1.13A is 14.18/0.1445 = 98 times more reliable than that of Fig. 8.1.13B.

CHAPTER

Maintenance and Logistics

2

RATIONALISATION OF SPARES

The weight and space limitations of an offshore installation impose restraints on electrical equipment installed. The restraints on the stocking of consumable spares and spare parts are even more severe, however, and this must be always kept in mind at the detail design stage, particularly during the procurement of the smaller drive packages such as small pumps and compressors, distribution boards and cable accessories. Sometimes a spares rationalisation can be made by using the same model of prime mover for main generators and gas lift or export compressors, or for standby diesel generation and diesel fire pumps.

The benefits may be listed as follows:

1. Common spare parts for different equipment, hence fewer need be stocked and weight of spares stored is reduced.
2. The reduced number of different types of equipment that operators have to deal with should reduce familiarisation time and hence improve safety.
3. The total weight of equipment operation and maintenance manuals and design drawings on the average offshore installation is in excess of 10 tons, even when the majority is on microfiche; therefore, some weight and space saving can also be accrued by reducing their total number.
4. The cost saving associated with items (1)–(3). The offshore spares inventory must be optimised for items most critical to
   a. safety
   b. the comfort of personnel (which may also affect safety)
   c. oil and gas production

Small light items may be flown out quickly when required, but it should be remembered that if an unscheduled helicopter call out is required, it could cost £2000/h or more depending on the size of the helicopter (in 2015).

ACCESSIBILITY AND COMMUNICATIONS

Those carrying out survey or commissioning work offshore should be warned that return travel from most, if not all, offshore installations requires customs clearance. This means that any test and measuring equipment needs proof-of-ownership documentation or a manifest for customs clearance.

Offshore Electrical Engineering Manual. https://doi.org/10.1016/B978-0-12-385499-5.00047-9

When estimating travel time for such an offshore visit, some time should be allowed for weather delays. This is often worse in the summer months when fog is more likely.

Telephone communications are by line-of-sight and/or satellite links, and access is usually available when major platform construction work is not in progress. On some of the nearer platforms, cell net telephones have been put to good use to supplement normal communications.

MAINTENANCE INTERVALS AND EQUIPMENT SPECIFICATION

When specifying and procuring offshore electrical equipment and, in particular, generator prime movers, manufacturer's recommended maintenance intervals should be carefully studied. Excessively short maintenance intervals will very quickly wipe out any savings in first cost because of the high costs of offshore servicing by manufacturer's or agent's servicing departments. Short intervals may also indicate poor reliability or unsuitability for the particular package, orientation or offshore environment in general.

SCAFFOLDING AND ABSEILING (RAPPELLING)

It is a common misconception that equipment or cabling may be accessed by scaffolding, almost invariably without a major impact on cost or completion date. In some areas such as a platform cellar deck or flare stack, the cost of erecting, maintaining and dismantling scaffolding may be the most significant cost in a small project.

Locating equipment and cables in inaccessible places, necessitating heavy use of scaffolding, has the following penalties:

1. Heavy scaffolding requirements will discourage frequent maintenance, particularly if it is going to obstruct an accessway where traffic is heavy or an area which is already congested.
2. The handling of scaffolding poles is well known in both on- and offshore petrochemical installations for causing accidents or damage to process equipment.
3. Scaffolding will also need to be erected each time it is necessary to inspect the equipment.
4. Scaffolding offshore can be very costly. Costs in excess of £300,000 for a single access structure are not unknown. Because of the need to access high-routed cable installations along their whole length in order to complete the installation, they are particularly expensive to install and should be avoided where possible. If an elevated or inaccessible location is unavoidable, consideration should be given to installing some form of permanent access structure, bearing in mind, of course, that this will itself need routine inspection, maintenance, painting, etc.

5. Abseiling may be used instead of scaffolding where maintenance intervals are infrequent and the maintenance required is within the capabilities of abseiling personnel. Poor access location jobs such as painting, testing or replacing small devices (e.g., gas detectors) are often carried out by abseiling, and the cost-saving can be very substantial.

TRANSPORT, ACCOMMODATION AND PEOPLE ONBOARD PROBLEMS

When planning an offshore construction project, the limitations of helicopter transport and accommodation for personnel, crane lifting capacity and storage/laydown space for equipment play a large part in deciding the sequence of events.

Cranes are limited in lift capability by two main variables, namely, radius of operation and sea state. If the crane is working near its maximum lift capacity, it is unlikely that the wave height in winter will allow lifting to be accomplished safely. Wind speed is also likely to affect the proceedings, but this will also depend on the wind direction and the location of the lifting operation.

Ever since the sinking of the Titanic in 1912, it has been a statutory requirement that there are sufficient lifeboats for the number of people onboard (POB). This applies just as much to fixed offshore installations as it does to the Queen Elizabeth 2 and must be borne in mind when planning a large offshore construction project. If necessary, because of a shortfall in lifeboat capacity, a floatel type accommodation vessel will need to be provided during project peak manning levels to comply with POB restrictions. This problem is often exacerbated by the need to send lifeboats ashore for maintenance during the summer months, when peak manning for platform maintenance is likely to occur. It can be overcome, however, by keeping a spare lifeboat of the same type onshore, which can be shipped out in advance, ready to replace the one being serviced.

In general, the more detailed the installation procedures are documented during the design phase, the less likely it is that logistical problems similar to those outlined earlier will occur during construction.

MAINTENANCE SCHEDULING SOFTWARE

A good reliability-centred maintenance scheduling system is essential offshore. However, the system has to take many other issues into account, for example,

1. availability of spares for the targeted equipment
2. safety criticality of the equipment
3. when the equipment will be available for maintenance
4. consequences of downtime
   a. other plant affected
   b. emergency shutdown logic affected
   c. fire and gas logic affected

5. maintenance team required
6. access requirements (e.g., scaffolding)
7. safety requirements
   a. hot-work involved
   b. other permits required
8. special tools/equipment required
9. special personal protective equipment required
10. specialist contractor/consultant required (e.g., divers/remotely operated vehicles)
11. rescheduling effects on other maintenance jobs
12. testing, commissioning and witnessing requirements

The list can be very long and complex and helpful, and intuitive software will lighten the maintenance engineer's burden.

PART

9

Statutory Requirements and Safety Practice

CHAPTER

Introduction to UK DCR Regulations and the Duties of an Electrical ICP

1

INTRODUCTION

To make good use of this chapter, it is recommended that the reference documents at the end of the chapter are at hand. Also look at the HSE webpage 'Offshore health and safety law'.

(Note 1 SCR 2015 has been extended to bring the management and control of environmental major accident hazards within its scope in line with the new EU Directive (Directive 2013/30/EU; see Reference 1)).

(Note 2 The Offshore Installations (Offshore Safety Directive) (Safety Case etc.) Regulations 2015(SCR 2015) came into force on 19 July 2015. They apply to oil and gas operations in external waters (the UK's territorial sea or designated areas within the continental shelf (UKCS)). But they only replace the Offshore Installations (Safety Case) Regulations 2005 in these waters, subject to certain transitional arrangements. Activities in internal waters (e.g., estuaries) will continue to be covered by the Offshore Installations (Safety Case) Regulations 2005 and its guide L30).

In the time of publishing of the first edition of this book, the Piper Alpha disaster had not long occurred and there was no time to include a section on the new UK legislation which came into force following the disaster. The latest (2017) versions of these documents are referenced at the end of this chapter.

This chapter is intended to cover all verification activities that electrical engineers may be requested to undertake on behalf of a duty holder, or their contractor(s). This includes verification required by the Safety Case Regulations and Prevention of Fire and Explosion, and Emergency Response Regulations. The electrical Independent Competent Person (ICP) may also be consulted on parts of the Well Examination Scheme (WES), where controls and instrumentation are involved. Note that the well examination requirements have changed in SCR 2015 (see regulations 11–13).

It is applicable to verification of UK offshore installations, for both existing facilities and new installations (i.e., during the project phase). The approach may, however, also be suitable, with adaptations as necessary, for 'verification' of installations outside of UK waters.

This chapter covers activities that engineers may be contracted to provide to a UK duty holder, or their contractor(s), including:

1. Review of safety and environmentally critical elements,
2. Review, or writing, of performance standards,

Offshore Electrical Engineering Manual. https://doi.org/10.1016/B978-0-12-385499-5.00048-0

3. Review, or writing, of verification schemes,
4. Implementation of verification schemes,
5. Updating verification schemes.

Requirements expressed here should be considered as a minimum guidance, i.e., to be followed in the absence of specific contractual or duty holder requirements. However, the independence and competence of the engineers concerned should be considered as mandatory. Note that independence does not necessarily mean that the engineer must work for a different company, but must:

1. Be outside the control of the line management engaged in the new project or operation of the installation. It is acceptable for the engineer to work for the same company but not the same line of management.
2. Have not been involved in the original design, installation or commissioning.

BACKGROUND

The Safety Case Regulations, together with the supporting regulations detailed below (DCR, PFEER, MAR, PSR), require that all Major Accident Hazards (MAHs) relating to an offshore installation are identified and adequately managed throughout the installation lifecycle. This is achieved by demonstrating the integrity of the installation, the safety of its operation and the capabilities of its preventative, mitigating, rescue and protecting systems, should an incident threatening the safety of the installation and its personnel occur.

The regulations are intended to create a 'goal setting' (or non-prescriptive) culture which looks at the topography and operating parameters of each installation in its own light and for efforts to be focused on those areas which provide the greatest contribution to risk control/management. The principal aims of the regulations are:

- Identify all credible hazards relating to an installation, and eliminate where possible.
- Assess and reduce risks from all MAHs to a level that is As Low As Reasonably Practicable (ALARP).
- Implement, maintain and verify an appropriate combination of inherent safe design, prevention, detection, control and mitigation systems and equipment, throughout the lifecycle of the installation for the management of these risks.
- Ensure that systems and equipment provided to protect personnel are suitable and capable of responding to all foreseeable safety-related incidents.
- Ensure that design, operation, maintenance and verification of these systems are undertaken by competent people who understand their responsibilities in the management of MAHs and possible escalating events.
- Assess changes to the installation which may affect the risks, and revise the systems where necessary to take account of the changes.

The SCR (as amended by DCR) and PFEER regulations further define the requirements for the establishment of performance criteria, against which those systems critical to safety can be measured and recorded, and for documented schemes, which describe how verification of adequate performance is achieved.

SCR Regulation 15 requires identification of Safety and Environmentally Critical Elements (SECEs) by the duty holder and examination of these, by Independent and Competent Persons (ICPs), to verify suitability and adequacy. Standards for verification are the performance requirements established by the duty holder specific to each asset/installation – ***Verification Scheme***.

PFEER Regulation 19 requires a ***Written Scheme*** for the systematic examination, by a competent and independent person, of plant identified within the PFEER assessment (Reference Regulation 5).

Industry practice and work processes have led to a ***Combined Scheme*** to satisfy both PFEER and SCR, although it is acceptable to write two separate schemes (duty holder decision).

A full lifecycle scheme which would cover all phases of an installation's lifecycle (design, construction, installation, commissioning, operation, decommissioning and abandonment) is an efficient way both in terms of cost and time, and provides a smooth/continuous handover of the scheme from one phase to the next.

Due to the non-prescriptive nature of the regulations, and the fact that the full responsibility for demonstration of legislative compliance rests with the installation duty holder, there are diverse routes by which compliance can be achieved.

The following **Abbreviations** are be used in this document:

ACOP	Approved Code of Practice
ALARP	As low as reasonably practicable
BOP	Blow-out preventer
CMAPP	Corporate major accident prevention policy
DCR	Design & construction regulations
DHSV	Downhole safety valve
EPIRB	Emergency position indicating radio beacon
ESD	Emergency shutdown
FOI	Fixed offshore installation
FPSO	Floating production, storage & offloading unit
FPU	Floating production unit
HIPS	High integrity protection system
HSE	Health and safety executive
HVAC	Heating ventilation & air conditioning
ICB	Independent competent body
ICP	Independent competent person
MAC	Manual alarm callpoint
MAH	Major accident hazard
MAPD	Major accident prevention document

Continued

ACOP	Approved Code of Practice
MAR	Management and administration regulations
MODU	Mobile offshore drilling unit
OPPS	Over pressure protection system
PA/GA	Public address/general alarm system
PFEER	Prevention of fire & explosion, and emergency response regulations
PPE	Personal protective equipment
PSD	Process shutdown
PSR	Pipeline safety regulations
PSV	Pressure safety valve
PUWER	Provision and use of work equipment regulations
RESDV	Riser emergency shutdown valve
RUK	Region UK
SCR	Safety case regulations
SECE	Safety and environmentally critical element
SEMS	Safety and environmental management system
SSIV	Subsea isolation valve
TEMPSC	Totally enclosed motor propelled survival craft
TR	Temporary refuge
UPS	Uninterruptible power supply
WES	Well examination scheme
WSET	Written scheme of examination and test

Definitions:

An ***Independent Competent Person (ICP)*** refers to the body appointed by the duty holder to undertake activities as required by the Safety Case Regulations – Note: referred to by some duty holders as ***Independent Competent Body*** (ICB). Note also that a duty holder may appoint more than one ICP for a single installation.

A ***major accident*** is defined as (Reference SCR Regulation 2):

1. an event involving a fire, explosion, loss of well control or the release of a dangerous substance causing, or with a significant potential to cause, death or serious personal injury to persons on the installation or engaged in an activity on or in connection with it;
2. an event involving major damage to the structure of the installation or plant affixed to it or any loss in the stability of the installation causing, or with a significant potential to cause, death or serious personal injury to persons on the installation or engaged in an activity on or in connection with it;
3. the failure of life support systems for diving operations in connection with the installation, the detachment of a diving bell used for such operations or the trapping of a diver in a diving bell or other subsea chamber used for such operations;

4. any other event arising from a work activity involving death or serious personal injury to five or more persons on the installation or engaged in an activity on or in connection with it; or
5. any major environmental incident resulting from any event referred to in paragraph (1), (2) or (4), and for the purposes of determining whether an event constitutes a major accident under paragraph (1), (2) or (5), an installation that is normally unattended is to be treated as if it were attended;

A ***Major Accident Hazard*** (MAH) is therefore:

Any hazard with the potential to give rise to a major accident.

A Safety and Environmentally Critical Element (SECE) is defined as:

'……such parts of an installation and of its plant (including computer programs), or any part of those – the failure of which could cause or contribute substantially to a major accident; or a purpose of which is to prevent, or limit the effect of, a major accident; (Reference SCR 2015 Clause 91)'.

Guidance to PFEER Regulation 5 defines *performance standards* as:

A performance standard is a statement, which can be expressed in qualitative or quantitative terms, of the performance required of a system, item of equipment, person or procedure, and which is used as the basis for managing the hazard – e.g., planning, measuring, control or audit – through the lifecycle of the installation. The regulation does not specify what performance standards should be – that is for the duty holder to decide, taking account of the circumstances on the particular installation.

Verification Scheme:

a suitable written scheme of examination of Safety and Environmentally Critical Elements as required by Regulation 15(1) of SCR.

Well Examination Scheme:

Arrangements in writing for examination of the pressure boundary of wells as required by Regulation 11 of DCR.

Written Scheme of Examination & Test (WSET):

a suitable written scheme for the systematic examination and testing of PPE and plant as required by Regulations 18 & 19 of PFEER.

The term *Examination* is used to describe any of the following:

- Inspection (e.g., survey, visual scrutiny),
- Witness (e.g., observing function test),
- Review (e.g., design review, review of records),
- Monitoring (a combination of the above, usually reserved or major activities such as structural fabrication),
- Assessment (e.g., review of a safety management process) – this is preferred to the term 'audit' which can be misconstrued.

The following offshore installation safety legislation requires verification to be undertaken:

- The Offshore Installations (Offshore Safety Directive) (Safety Case, etc.) Regulations 2015 (SCR 2015),
- The Offshore Installation and Wells (Design and Construction, etc.) Regulations 1996 (DCR),
- SI 1995 No. 743 the Offshore Installations (Prevention of Fire and Explosion, and Emergency Response) Regulations 1995 (PFEER).

The following offshore installation legislation is also applicable to parts of this document, supplemented by respective guidance and approved codes of practice:

- SI 1995 No. 738 The Offshore Installations and Pipeline Works (Management and Administration) Regulations (MAR),
- SI 1992 No. 2392 Provision and Use of Work Equipment Regulations (PUWER),
- SI 1996 No. 825 Pipeline Safety Regulations (PSR).

DUTY HOLDER

One of the duties of the duty holder is defined as follows:

> *'The duty holder shall ensure that an installation at all times possesses such integrity as is reasonably practicable'*.

This general duty is expanded upon throughout the text of the regulations. In summary, the duty holder is responsible for the determination of all risks associated with all phases of the lifecycle of an installation, and the demonstration that those risks are reduced to ALARP through implemented measures for safety management. An integral part of that risk management is the development and operation of the SCR Verification Scheme and the PFEER Written Scheme, including the selection of ICP. The duty holder must record the ICP selection process.

The duty holder must also periodically review the Verification Scheme and, where necessary, revise or replace it, in consultation with the ICP (SCR Regulation 15(3)).

INDEPENDENT AND COMPETENT PERSON (ICP)

As a minimum the duty holder is required to appoint an ICP to:

- Comment on the record of SECEs drawn up by the duty holder (SCR Regulation 15(1)).
- Draw up a Verification Scheme, or act in a consultative capacity in the drawing up of the scheme by the duty holder (SCR Regulation 15 (1)).

- Issue to the duty holder any comments in respect of the Verification Scheme (SCR Regulation 15A).
- Perform functions under the scheme (SCR Regulation 15(2)).
- Carry out a systematic examination of applicable PPE and plant (if not included in a ***Combined Scheme***) as defined within the duty holder WSET (PFEER Regulations 18 and 19).

The engineer's role as ICP could encompass some or all of the above – contract requirements will dictate the exact nature of the role, and must be clearly understood. Further, there may be more than one ICP appointed by the duty holder for a given installation. In this instance engineers should ensure that their own scope, interfaces (e.g., with other ICPs) and responsibilities are defined as far as possible.

In addition to the record of SECEs itself, the following inputs should be provided by the duty holder in order to undertake the review:

- drawings of installation location/orientation,
- description of facilities and mode(s) of operation,
- equipment list and layout,
- hazard identification report and associated studies,
- Safety Case (if available for new-build projects).

Engineers should use the above to review the completeness and extent of the MAH identification and determine whether, in their opinion as ICP, the systems and equipment which have the potential to cause, or contribute substantially to, a major accident have been identified as SECE. This assessment must be based upon consequence of failure only, not on the likelihood of failure.

The engineer should then consider any systems or items of equipment that have not been identified as safety-critical, for which the purpose is to limit the effect of a major accident have been identified as Safety and Environmentally Critical Elements. This does not mean that the only purpose of the system/item is to limit the effect of a major accident. To be in a position to consider this, the engineer should have sufficient information to understand the function of the system/item in terms of its ability to:

1. Prevent a major accident,
2. Detect a major accident,
3. Control a major accident,
4. Communicate during/after a major accident,
5. Mitigate the effects of a major accident.

The following methodology can be applied for confirming that prevention, detection, control or mitigation measures have been correctly identified as SECE:

- identify the major contributors to overall risk,
- identify the means to reduce risk,
- link the measures, the contributors to risk and the means to reduce risk to the installations' systems – these can be seen to equate to the SECE of the installation.

The record of SECEs typically provides only a list of systems and types of equipment or systems. In order for the engineer to interpret the legislative requirements, the scope of each SECE should be clearly specified such that there can be no reasonable doubt as to the precise content of each SECE. If necessary, this should be down to the level of individual tagged devices or equipment. For the purposes of subsequent management and reporting, it may be beneficial to consider allocating an identifier ('dummy' tag number) for equipment or systems that are not actually tagged.

The review should consider all phases of the installation lifecycle; for example, during the offshore commissioning phase of a new installation there may be temporary items of equipment provided which should be identified as SECE (e.g., temporary fire/gas detection).

To complement (not substitute) the above, Fig. 9.1.1 presents a list of systems (with possible electrical content) that may be appropriate for the installation under consideration. This list should be used with caution – it is not intended as a check list.

The duty holder's record of SECE may be subject to revision during the installation's lifecycle (e.g., due to modifications, changed use of equipment), and as such should be considered during the duty holder's periodic reviews of the

(A)

Safety and Environmentally Critical Elements (SECEs) (Note: Only Those With Electrical/Electronic Content Shown)	MODU	Flotel	FPSO	FOI
Drilling				
Mud Systems	X	–	–	X
Blowout Preventer System	X	–	–	X
Choke and Kill (including Emergency Blowdown)	X	–	–	X
Cement System	X	–	–	X
Marine Riser System (from drill floor to flex joint)	X	–	–	X
Well Control Instrumentation	X	–	–	X
Diverter system	X	–	–	X
Power				
Emergency Power	X	X	X	X
Battery-Backed Systems (incl. UPS)	X	X	X	X
Protection of electrical equipment	X	X	X	X
Ignition Prevention				
Electrical Earthing Continuity	X	X	X	X
Electrical Equipment in Hazardous Areas	X	X	X	X
Fire and Gas				
Gas (Flammable and Toxic) Detection System (including fire & gas logic)	X	X	X	X
Fire Detection System (including fire & gas logic etc.)	X	X	X	X
Deluge (including fire & gas logic etc.)	X	X	X	X
Sprinklers (including fire & gas logic etc.)	X	X	X	X
Fire Pumps (including fire & gas logic etc.)	X	X	X	X
Gaseous Systems (e.g. Halon, CO_2 – including fire & gas logic etc.)	X	X	X	X
Ventilation Systems (including HVAC controls and dampers)	X	X	X	X

FIGURE 9.1.1

Generic list of safety and environmentally critical elements with electrical content.

(B)

Safety and Environmentally Critical Elements (continued) (Note: Only Those With Electrical/Electronic Content Shown)	MODU	Flotel	FPSO	FOI
Emergency Response and Evacuation				
Temporary Refuge (including facilities for casualties and triage)	X	X	X	X
Emergency Lighting of Control and Safety Equipment	X	X	X	X
Helideck Systems (Markings, Nets, Obstacle Marking/Lighting etc.)	X	X	X	X
Escape (battery-backed) Lighting	X	X	X	X
Internal Communications (including PA, MACs, Alarms, Telephones)	X	X	X	X
External Communications (including Marine and Aviation)	X	X	X	X
TEMPSC (including EPIRB and Radio)	X	X	X	X
PPE (incl. Lifejackets, Survival/Immersion Suits, Helideck Crash Equip.)	X	X	X	X
Escape Systems				
Standby Vessel and associated Fast Rescue Craft	X	X	X	X
Hydrocarbon Containment				
Hydrocarbon Piping & Equipment (including Valves and Instrumentation)	–	–	X	X
Process Relief and Blowdown (including Valves and Instrumentation)	X	X	X	X
Emergency Shutdown System incl. Software, PSD, HIPS, OPPS etc.	X	X	X	X
High Speed Machinery Trips	X	–	X	X
Local Atmospheric Vents	X	–	X	X
Drains (Open and Closed Hazardous)	X	X	X	X
Hazardous Risers (including RESDV)	–	–	X	X
Hazardous Pipelines Subsea Isolation Valve	–	–	X	X
Hazardous Pipelines within 500m of Installation	–	–	X	X
Marine				
Mooring (including Heading Control System and Tension monitoring)	X	X	X	–
Navaids (including Lights, Foghorns, Marine/Weather Monitoring Systems)	X	X	X	X
Radar Early Warning System (DARPS)	X	X	X	X
Ballast and Bilge (Stability – including Ballast control and WT doors)	X	X	X	–
Inert Gas System	–	–	X	–
Dynamic Positioning System	X	X	X	–
Thrusters	X	X	X	–
Temporary Equipment				
Bridge Connections to Support Vessels	X	X	X	X
Power Generators	X	X	X	X
Temporary PA/GA Systems	X	X	X	X
Safety Management System	X	X	X	X
Corporate Major Accident Prevention Policy	X	X	X	X

FIGURE 9.1.1, cont'd

Verification Scheme. Each change to the record of SECEs should be reviewed and commented on by the ICP.

Prior to undertaking any verification engineers, ICP should formally provide agreement with, or comment on, the duty holder's record of SECE. This should be repeated with each subsequent revision to the record of SECEs, or periodically. Should there be more than one ICP appointed by the duty holder, clarification of the scope/limits of the respective ICPs should be sought before the review.

The following is equally applicable to preparing or reviewing performance standards.

It is a requirement that the duty holder establishes suitable acceptance criteria for measuring, demonstrating and recording the suitability of SECEs. The industry has in general adopted the term ***performance standards*** for these criteria, and unless the duty holder requires a different approach engineers should support this terminology.

A performance standard describes the essential requirements that a SECE (or PFEER PPE or Plant) must maintain, or provide on demand. It is a statement, which can be expressed in qualitative or quantitative terms, of the performance required of a system, item of equipment or procedure and which is used as the basis for managing the hazard and any events requiring emergency response, through the lifecycle of the installation.

For a performance standard to be suitable, it should satisfy all of the following conditions:

1. Requires measurement of the performance/capability of a parameter of the component/system,
2. The measured parameter provides evidence of the ability of the component/system to prevent, or limit the effect of, a major accident,
3. Acceptance criteria/range are defined for the parameter in question,
4. The parameter can be monitored/measured. Note that if the measurement requires special equipment or scaffolding, etc., it should be part of the planned maintenance system so that a recorded result is available to the ICP.

Experience indicates that performance standards should be at a level that sets an objective for the SECE; it should not describe how that objective is to be achieved, or how it is to be demonstrated (verified) as this is part of the Verification Scheme. They should be specified for all parts of an SECE that are critical for safe operation.

Appendix 3 lists some measurable parameters that may be appropriate to be considered for performance standards (not necessarily exhaustive).

The consequence of a performance standard *not* being met (demonstrated) should also be considered. If the consequences are such that a major accident cannot result, or a significant reduction in the effectiveness to detect, control or monitor a major accident cannot result, then the performance should not be considered as necessary.

As a minimum, the following characteristics should be considered in generating performance standards:

- Functionality – what the SECE must achieve,
- Reliability – how often it will be required to operate satisfactorily,
- Availability – how often it will be required to operate on demand,
- Survivability – the conditions under which it will be required to operate, e.g., if exposed to fire, blast, vibration, adverse weather, etc.,
- Environmental – does the equipment contain toxic materials, and if so what precautions are taken.
- Interactions – what other systems/equipment are required to enable the SECE to achieve its requirements.

In the same way that parts of an installation may be considered as SECE for certain periods of an installation's lifecycle, it is possible for performance standards to be applicable during specific phases of an installation's lifecycle only; for example, during commissioning or major in-service modifications. At all times that a part of an installation is safety or environmentally critical, there must be at least one performance standard in place.

Each duty holder is likely to have a preferred format of presenting performance standards. If the engineer is able to influence this, the most effective means is a simple series of statements each expressing performance standards (either for the entire SECE or parts of it), sequentially numbered to provide easy reference when reporting following verification itself. Example Electrical Performance Standards are shown in PART 9 Chapter 2. However, the format must comply with the duty holder's requirements, if expressed.

Performance standards presented in wordy, legalistic format tend to divert the ICP away from the best way of verifying the particular criterion; it is essential to establish prior to verification, exactly which parts of a performance standard document require verifying and are feasible for the ICP to verify, and which parts are a matter of recording electrical parameters and correct equipment function during maintenance.

Quantitative performance standards for reliability or availability of an SECE are essential for software-related SECEs/systems. Where such performance standards are specified, it is recommended that the precise requirements are fully defined, that there is a clear means of verifying that requirement and that verification can provide positive demonstration of the SECE's suitability.

PART 9 Chapter 2 gives typical electrical performance standards. Be warned, however, that performance standards must be derived from the installation's safety case and not produced in isolation.

The following guidance should be used when engaged in drawing up verification schemes on behalf of a duty holder for an installation. The minimum requirements listed below must be included but the optional requirements should only be included with the agreement of the duty holder. It is important that all parts of the duty holder's organisation with responsibility for, or who are affected by, verification are consulted in this process.

GUIDANCE ON DEVELOPING WRITTEN SCHEMES OF EXAMINATION AND TEST

The following guidance may be used by engineers engaged in drawing up written schemes of examination and test on behalf of a duty holder. The minimum requirements listed below must be included but the optional requirements should only be included with the agreement of the duty holder.

The minimum requirements to be provided for in a Written Scheme of Examination and Test (WSET) are detailed in PFEER Reg. 18 &19. These are described below:

- The scheme should clearly identify all PPE and 'plant' on the installation that should be examined *(Regulation 18(2) & 19(2));*

- The scheme should describe the nature and frequency of examination and testing. The preferred method is the use of an activity list (e.g., task dossier or detailed work instructions) which identifies the type of examination activity to be completed, to which SECE and performance standard it is related and how often it is repeated *(Regulation 18(2) & 19(3)(a));*
- The scheme should also describe how plant is to be examined prior to being brought into service or after repair *(Regulation 19(3)(b));*
- The scheme should detail how the results of the examinations are recorded *(Regulation 18(2) & 19(3)(a)).*

In addition to the above it is recommended that the WSET also includes the following:

- Details of the process implemented to identify PFEER PPE and plant to be included in the WSET,
- A copy of the performance standards for the PPE and plant covered by the WSET,
- Description of how the examination process is managed and how information is communicated between all the relevant parties,
- Details of the WSET review and revision process.

A checklist is included below (Fig. 9.1.2) as a guide or prompt for development of a WSET.

Where a *Combined Scheme* (e.g., a scheme covering a platform and jack-up drilling rig combined operation) is required by the duty holder, it should include all the requirements listed above for the Verification Scheme and the WSET.

During the development of any of the written schemes discussed above, it is important to consider all aspects of the installation's lifecycle – design, procurement, fabrication, transportation, installation, hook-up and commissioning, operation, decommissioning and abandonment. Examination activities in schemes developed for new projects should reflect all these lifecycle phases. For existing installations only, examination activities during the operation and decommissioning phases will be relevant (note – however, some assessment of the project phases will be required for confirmation of initial suitability of SECEs).

DEVELOPING THE SCHEME ACTIVITY LIST

The list of ICP examination activities is an integral part of any written scheme as it gives details of what needs to be done, by whom and when. For each performance statement or criteria there should be at least one examination activity. The information included in the activity list, its format and how it is recorded will vary from duty holder to duty holder. It is recommended that the activity list includes the following:

- Procedure – a description of the examination activity to be performed to verify that the criteria specified in each performance standard is met,
- Type of examination activity to be performed, e.g., audit, inspection, review or test,

Installation Name	
Document No.	
Verification Scheme, Written Scheme of Examination & Test or Combined Scheme?	

Item	Verification Scheme	PFEER WSET	Combined Scheme	Comments
Overview of the hazard management process		n/a		
Details of the PFEER assessment	n/a			
Description of responsibilities for development and implementation of the scheme				
Description of how verification/examination process is managed				
Criterion used for ICP selection		n/a		
Communication of information relevant to verification between all relevant parties				
Review and revision of the scheme				
Description of how results of examinations are recorded				
Issue of reports				
Procedure for distribution of reports				
Descrption of interface with PFEER WSET		n/a		
Description of interface with Well Examination Scheme		n/a		
Description of verification arrangements for:				
Temporary Equipment		n/a		
Combined Operations		n/a		
Major Projects/Installation Modifications				
Description of the interface with Classification				
Record of Safety and Evironmentally Critical Elements		n/a		
A description of the links between the Major Accident Hazards and Safety and Environmentally-Critical Elements		n/a		
Identification of all PPE and Plant to be examined	n/a			
Performance Standards included?				
Activity list (following included):				
Examination activity to be performed				
Extent of ICP involvement				
When the activity is to be done and how often is it repeated				
Unique ID for activity				
Reference to Performance Standard and Safety and Environmentally Critical Element				
List of documentation required for verification				
Identification of other activities for which credit is taken				

Reviewed by:	Approved by:
Date:	Date:

FIGURE 9.1.2

Written scheme checklist.

- Extent of ICP involvement in activity, where applicable (e.g., percentage of sample),
- How often it is to be repeated (if at all, e.g., for initial suitability only),
- Unique identifier for each activity, for control and reporting purposes,
- A clear reference to a specific performance standard and SECE for each examination activity,
- Note of specific documents which will be used as a basis for verification or for reference as part of the examination activity,
- Identification of processes for which credit is taken within the scheme, e.g., planned maintenance activities, rolling inspection programmes, etc.

The extent of examination and level of ICP involvement can be determined from:

- criticality of the SECE, or part of the SECE, based on risk,
- consequence of that particular performance criteria not being achieved,
- requirements of recognised codes or standards,
- details identified from maintenance routines or inspection plans,
- assurance activities completed by others that may be used as justification for adjusting level of independent examination.

For new projects the activity list should state which examinations will only be performed prior to the SECE being put into service (examination for 'initial suitability'). No further activities of this type will be performed during operation, unless modifications are made to the SECE. All other examination activities (examination for continued suitability), especially during the operational phase, should be repeated at intervals specified to ensure that the SECE to which it refers maintains its suitability and adequacy. The following may contribute to the specification of interval or frequency of examination:

- Requirements of recognised codes and standards,
- Criticality assessments,
- Assumptions and conclusions of risk and/or reliability studies on relevant systems,
- Extent of installation maintenance routines and inspection plans,
- Interval between of duty holder assurance activities,
- Manufacturer's recommendations for the equipment,
- Findings of previous examination activities (including those for related SECEs or performance standards).

Subsequent revisions to examination intervals should be substantiated.

COMBINED OPERATIONS

There are a number of interfaces during combined operations, such as flotel alongside a platform for commissioning or drilling rig cantilevered over platform. These interfaces can include hardware (e.g., means of access, firewater systems interconnections) and software (e.g., permit to work systems), which are likely to result in amendment to the record of SECE, performance standards and examination activities for the respective installations. These amendments should be documented by the duty holder, and accepted by the ICP, and either added as an addendum to the existing written scheme or as a separate document.

With the agreement and cooperation of all duty holders involved with the combined operations, a single joint interface document may be prepared, defining all verification arrangements for the combined operations (requirements in addition, to the written schemes for the respective installations).

At completion of combined operations, the combined scheme addendum or interface document should be reviewed by the duty holder to take account of any modifications to the installation during the combined operation. The addendum or interface document should be formally withdrawn but not discarded, as it constitutes part of the lifecycle documentation of the installation, and may be relevant for future combined operations.

The written scheme should also describe how temporary equipment brought on to the installation will be verified. It should detail how the safety criticality of temporary equipment is assessed and verification of these items is addressed. Temporary equipment should be verified prior to transportation offshore.

Reference to a scheme such as the 'Vendor's Certificate of Conformity for Temporary Plant & Equipment for use on Offshore Installations', implemented by the SNS operators forum, is an example of how this issue may be addressed.

As part of the verification process credit may be taken for assurance activities performed by or on behalf on the duty holder (e.g., planned maintenance, scheduled inspection activities, safety/technical audits, etc.), but these activities are not in themselves verification. The details of these assurance activities may be included in the written scheme, but should be clearly identified as such. All such assurance activities should be subject to examination by the ICP as a minimum (this activity must be listed in the written scheme).

Classification of MOUs, FPSOs and FPUs can be used as a basis for or in support of verification. Where Classification Society rules meet or exceed the requirements of the performance standard for an SECE, evidence of meeting Class (approval letters, product certificates, etc.) can be used as a basis for verification. Reporting of verification activities should still be completed in accordance with the requirements of the written scheme.

Where Classification Society rules do not meet the requirements of performance standards, classification should still be recognised as a contributor to verification, but additional tasks will need to be completed by the ICP.

EU Legislation (EC directives) – the CE marking of equipment in itself cannot be used as a direct substitute for verification. CE marking can however be used to demonstrate that equipment meets a particular performance standard (e.g., CE marking for conformance to the ATEX directive can be used as a basis for verifying that equipment meets a performance standard on ignition prevention). The declaration of conformity and any notified body certificate should be reviewed to confirm the relevant performance standard is addressed. Where there are any doubts regarding the applicability of CE marking as a basis for verification a review of the technical file should be completed. Examination of equipment as required by SCR and/or PFEER in no way confirms it is compliant with applicable EC directives.

PUWER – demonstrating that equipment meets the requirements of PUWER may be used as evidence to demonstrate compliance with a particular performance standard. Conversely, examination of equipment as required by SCR and/or PFEER in no way confirms it is compliant with PUWER.

PSR – there is an interface between PSR and SCR/DCR. It is likely that some parts of the pipeline covered by the MAPD under PSR will need to be verified (e.g., risers, pipeline to 500m limit, subsea isolation valves, etc.) as they are likely to be identified as SECEs.

Prior to undertaking any tasks contained within the Scheme – the ICP engineer involved should formally agree, or comment on, the content of the Verification Scheme. This should be issued whether or not the ICP's organisation has prepared the complete Verification Scheme on behalf of the duty holder.

This review should include performance standards, or other acceptance criteria.

When engineers are selected (or proposed in tenders) as ICP or ICP project manager for an installation, they should not have previously worked for the duty holder or any of his subcontractors on that particular installation.

However, engineers may undertake verification in instances where they *have* previously worked for the duty holder (either directly or through a contracting company) on that particular installation, provided their verification activities do not include the areas in which they were previously involved.

In general it is important that those carrying out verification work have appropriate levels of impartiality and take cognisance that:

- judgements regarding safety are not compromised,
- they do not verify their own work,
- management lines should be separate from those people whose work they are checking,
- judgements are not subject to financial or operational incentives

Should this screening process indicate uncertainties regarding independence, this should be discussed with the duty holder.

MULTIPLE ICPS/INTERFACE WITH OTHER ICPS

When a duty holder, and/or their contractors, appoints more than one ICP organisation (ICB) for a given installation, duties and scope of the respective ICPs should be clearly established and agreed before undertaking any verification. It is important to establish which ICP is responsible for making comment(s) on the duty holder's record of SECEs and verification scheme (could be different ICPs, although the preferred position is that only one ICP should make comment).

When 'inheriting' the role of ICP from another organisation(s), consideration should be given to ensuring that the following documents are made available by the duty holder prior to handover and acceptance of responsibilities:

- A copy of the Verification Scheme,
- Review comment(s) made by the outgoing ICP on the record of SECEs,
- Any comments in respect of the Verification Scheme made by the outgoing ICP including any reservations.

Examination (scope, methodology and timing of verification activity) must be undertaken strictly in accordance with the requirements of the Verification Scheme. If this contains insufficient or unclear detail, the duty holder's instruction should be obtained prior to progressing. However, the requirements of each activity may be treated with some flexibility (e.g., if the task can be undertaken in a slightly different way, provided that the intent of the task and the acceptance criteria remain unchanged), to enable an examination to progress without the need for a revision of the Verification Scheme.

REPORTING ACTIVITIES

Reports should be issued individually against each examination activity listed in the scheme. It is suggested the following information is included:

- Examination activity reference (including revision number),
- SECE,
- Performance standard,
- Description of the activity that has been completed,
- List of documentation examined/reviewed (including revision number),
- Description of activities still outstanding,
- Findings or conclusions (does the SECE/component meet the required performance criteria?)

Where SECE/component fails to meet the required performance criteria, recommendations for remedial action *may* be specified by engineers appointed as ICP, where required by the duty holder. Note that it is the duty holder's responsibility – not the ICP's – to define the criticality of any remedial actions recommended.

Whatever the format of the deliverable confirming suitability of an SECE, the basis for reaching that conclusion must be clearly stated, whether this be by review of design documentation or test reports, by visual examination, by witnessing testing, etc.

Document review status (e.g., status codes, approvals etc.) may be maintained during a project, but should be recognised only as a *secondary* indicator of verification status.

ICPs are often presented with results of 'verification' undertaken by other organisations as evidence of an SECE's, or part of an SECE's, suitability (e.g., equipment type approvals). In order to take credit of these for verification purposes (without unnecessary duplication of effort), the ICP needs to confirm:

1. the independence of those undertaking the review,
2. that the approval covers the actual item being provided for the installation under consideration,
3. that the terms of reference (performance criteria) correspond to those in the Verification Scheme

All communications between the ICP and the regulator on matters pertaining to verification should be via the duty holder for the installation, and also with ICP's client should they not be the duty holder.

Subject to the duty holder's agreement, the ICP should actively seek to assist the duty holder in their liaison with the regulator (within the bounds of the contract), and cooperate fully with all parties involved in such activities.

RETENTION OF VERIFICATION RECORDS

The legislation requires the duty holder of an installation to retain records of activities supporting verification for a period of 6 months after the Verification Scheme ceases to operate. However, the ICP should not rely on these as a record of our activities.

Records should be retained of the following for at least the same period:

- Verification Scheme
- Formal comment on record of SECEs and on Verification Scheme
- Activity Reports/Statements Issued
- Documentation referenced on activity reports/statements

CHAPTER

Preparation and Use of Performance Standards

2

This chapter is equally applicable to preparing or reviewing performance standards.

It is a requirement that the duty holder establishes suitable acceptance criteria for measuring, demonstrating and recording the suitability of Safety and Environmental Critical Elements (SECEs). The industry has in general adopted the term *performance standards* for these criteria, and unless the duty holder requires a different approach, engineers should support this terminology.

A performance standard describes the essential requirements that an SECE must maintain, or provide on demand. It is a statement which can be expressed in qualitative or quantitative terms of the performance required of a system, item of equipment or procedure and which is used as the basis for managing the hazard and any events requiring emergency response, through the lifecycle of the installation.

For a performance standard to be suitable, its monitored criteria should satisfy *all* the following conditions:

1. Each performance standard criterion requires measurement of a particular capability or parameter of the SECE component/system.
2. The measured parameter provides evidence of the ability of the component/system to prevent, or limit the effect of, a major accident, based on the findings of the installation's safety case.
3. Acceptance criteria/range must be defined clearly for the parameter in question.
4. Each parameter specified must be easily monitored/measured on the installation or by inspection of maintenance records/logs such that success or failure to achieve the particular criterion is clearly demonstrated.

Experience indicates that performance standards should be at a level that sets an objective for the SECE; it should *not* describe how that objective is to be achieved, or how it is to be demonstrated (verified) as this is part of the Verification Scheme. They should be specified for all parts of an SECE that are critical for safe operation.

The consequence of a performance standard *not* being met (demonstrated) should also be considered. If the consequences are such that a major accident *cannot* result, or a significant reduction in the effectiveness to detect, control or monitor a major accident *cannot* result, then the performance should not be considered as necessary. It is recommended that this is discussed with the originator and the safety authority before deletion.

Offshore Electrical Engineering Manual. https://doi.org/10.1016/B978-0-12-385499-5.00049-2

As a minimum the following characteristics should be considered in generating performance standards:

- functionality – what the SECE must achieve,
- reliability – how often it will be required to operate satisfactorily,
- availability – how often it will be required to operate on demand,
- survivability – the conditions under which it will be required to operate, e.g., if exposed to fire, blast, vibration and adverse weather,
- environmental – does the equipment contain toxic materials, and if so, what would be the consequences of a spill?
- interactions – what other systems/equipment are required to enable the SECE to achieve its requirements.

In the same way that parts of an installation may be considered as SECEs for certain periods of an installation's lifecycle, it is possible for performance standards to be applicable during specific phases of an installation's lifecycle only, for example, during commissioning or major in-service modifications. At all times that a part of an installation is considered an SECE, there must be at least one performance standard.

Each duty holder is likely to have a preferred format of presenting performance standards. If the engineer is able to influence this, the most effective means is a simple series of statements each expressing performance standards (for either the entire SECE or parts of it), sequentially numbered to provide easy reference when reporting following verification itself. However, the format must comply with the duty holder's requirements, if expressed.

Performance standards presented in wordy, legalistic format tend to lead away from the intended goal and provide and divert the independent competent person (ICP) away from the best way of verifying the particular criterion which the author of the performance standard intended; it is essential to establish before verification exactly which parts of a performance standard document require verifying by the ICP and which parts are a matter of recording maintenance and correct equipment function.

Quantitative performance standards for reliability or availability of an SECE are essential for software-related SECEs/systems. Where such performance standards are specified, it is recommended that the precise requirements are fully defined, that there is a clear means of verifying that requirement and that verification can provide positive demonstration of the SECE's suitability.

Figs 9.2.1A–C and 9.2.2 show the typical electrical performance standards. However, performance standards *must* be derived from the installation's safety case and not produced in isolation.

The guidance in the figures should be used when engaged in drawing up verification schemes on behalf of a duty holder for an installation. The minimum requirements listed in the figures must be included but the optional requirements should only be included with the agreement of the duty holder. It is important that all parts of the duty holder's organisation with responsibility for, or who are affected by, verification are consulted in this process.

(A)

INSTALLATION:	PERFORMANCE STANDARD: INTERNAL COMMUNICATIONS
The emergency role of the communications system installed on the Installation is to provide the necessar y facilities to control the emergency by: a) Enabling messages and alarms to be broadcast throughout the installation b) Communicating with nearby or physically linked installations c) Communicating with vessels nearby working on activities related to the installat ion, and the standby vessel.	

System description	
An emergency hot line telephone provides communications between: 1) the control rooms of the nearby installations 2) the radio rooms of the nearby or physically linked installations [if any] 3) the installation's OIM and the OIMs of nearby installations A UHF (FM) radio repeater system is usually installed to enhance full duplex radio communications between personnel using hand portables. In the event of failure, the hand portables are equipped with simplex chann els to allow communications, hand portable to hand portable, between personnel on the facility. A radio paging system with facility for sending verbal messages that may be accessed from the installation telephone systems enables key personnel to be paged. A duplicated, Public Address/General Alarm (PA/GA) system, fully certified for use in hazardous areas is provided to advise personnel of an emergency and the actions to be taken. Each system can be accessed from the control room, radio room, helideck and lifeboat stations. The PA/GA system carries routine and emergency speech and alarm on both halves of the system simultaneously. In an emergency, the PA system can broadcast a tone indicating the nature of the emergency (muster and prepare to evacuate) and to issue instructions to all areas where personnel may be present, including lifeboat and muster stations. The Emergency Intercom system is designed for use in hazardous areas. The system is installed in a ring, and provides communication between contr ol points and muster areas even if any one of the cables interconnecting two stations is severed, or the platform 240V- ac power supply is not available. The system provides; two channel communications (one channel each for Control points and Muster points) with the option of "All Call" simultaneous transmission both channels, full battery back-up for each outstation, full line monitoring featuring outstation health and audio line continuity. Telephones are located at strategic points throughout the operati ng and accommodation areas. In hazardous areas, weatherproof and intrinsically safe telephone sets are used.	

	Functions	Element	Assurance
1	To provide clear signals to identify emergencies on the installation.	a) Two emergency signals are provided through the PA/GA system. The PA system is used to broadcast a tone indicating the nature of the emergency. b) Instructions can be issued to all areas where personnel may be present. *The communications system has been designed such that the parts essential to the safety of the Installation personnel will continue to function in any emergency including shutdown situations.* *The visual and audible alarms are in accordance with the requirements of PFEER: SI 743 (1995).*	Telecommunications maintenance instructions Reference maintenance procedure XXXXX.
2	To ensure that under gas escape conditions emergency communications are maintained and no spark ignition is possible due to their use	a. Radio equipment is designed to limit radio frequency propagation to a safe level and prevent ignition. b. The public address system is fully certified for use in hazardous areas. c. Telephones located external to the control room or temporary refuge areas are certified for use in hazardous areas that remain operational in an ESD situation. d. Portable systems are all intrinsically safe. *e.* All flashing beacons and loudspeakers used as part of the PA/GA systems are certified for use in hazardous areas, even when installed in areas normally classified as non-hazardous. *f. PD CLC/TR 50427:2006 Assessment of inadvertent ignition of flammable atmospheres by radio -frequency radiation. Guide.*	See Ignition Prevention Performance Standard Also by design

FIGURE 9.2.1

(A) Typical internal communications performance standard (based on a floating production, storage and offloading (FPSO) with central turret, aft engine room and forward accommodation). (B) Typical fire and gas system performance standard for an FPSO. (C) Typical emergency shutdown (ESD) system performance standard for an FPSO. *HLG*, high level gas; *HVAC*, heating, ventilation and air conditioning; *LEL*, lower explosive limit; *LLG*, low level gas; *UPS*, uninterruptible power supply.

INSTALLATION:		PERFORMANCE STANDARD: INTERNAL COMMUNICATIONS (cont.)	
Functions		**Element**	**Assurance**
3	To provide essential control equipment for the communications systems within a designated safe area such that communications may be maintained in the event of hazard.	a) The central control systems for the communications are all located within the TR *During ESD situations, non-essential equipment and supplies are disconnected for safety reasons. Equipment that remains operational includes: PA, PABX, hand portables, INMARSAT, VHF radio, repeater and status systems, and navigation equipment.*	By Design
4	To provide a redundant link between the Installation and other installations in the field	a) A fibre optic link with redundant fibres is provided between these installations to provide as a minimum data and status of the safety systems.	Reference maintenance procedure XXXXX.
5	To provide visual announcement of emergency in high noise areas	a) Visual warning in the form of high intensity amber coloured flashing beacons are provided to supplement audible alarm signals and emergency announcements in all areas, where ambient noise is 85 dBA or greater.	Reference maintenance procedure XXXXX.
6	To provide safe mobile communications between control rooms and all areas	*An UHF (FM) radio repeater system is installed to enhance full duplex radio communications between personnel using hand portables. In the event of failure the hand portables are equipped with simplex channels to allow communications, hand portable to hand portable, between personnel on the facility.* a) Intrinsically safe multi -channel mobile FM VHF & UHF hand portable radios with mains, single and multi, chargers are provided for use by production, maintenance, crane operating and safety personnel. *Private channels are used for communications with the shuttle tanker, supply vessels, crane cabs and hand portables. Each crane is equipped for communication with supply vessels.*	Reference maintenance procedure XXXXX.
7	To provide safe communications between control rooms and muster areas.	A distributed emergency intercom system, fully certified for use in hazardous areas, provides voice communication between control centres, muster areas and lifeboat locations.	Reference maintenance procedure XXXXX.
8	To provide a call system to essential personnel in the event of an incident	a) A radio paging system with facility for sending verbal messages, which may be accessed from the installation telephone systems, enables key personnel to be paged. b) Telephones shall be clearly marked and sign-posted.	Reference maintenance procedure XXXXX.
9	To provide visual monitoring for Control Room Personnel.	*Closed circuit television (CCTV) is installed at various strategic positions on the Installation to allow control room operators to watch high-risk areas of the installation.* High risk areas of the FPSO covered by CCTV include CCTV viewing of cargo offloading areas, flare and PAUs, and also helideck.	Reference maintenance procedure XXXXX.
10	Provide fail-safe operations and shutdown of cargo in the event of a problem on the shuttle tanker.	a) A telemetry system is installed on the FPSO to facilitate supervisory and control of the off-loading system. b) The CCR also provides voice communication between the FPSO control room, Machinery, and the shuttle tanker loading control point (aft).	Reference maintenance procedure XXXXX.
11	Prevent toxic spill to Environment	Chemicals and fuel to be provided with secondary containment where necessary [No inventory related to internal coms]	-

Reliability & Availability	**Reference**
The Internal coms system is based on a duplex system, operating in parallel. The system monitors its own status and its power supplies and alarms all faults. The specified availability of the Internal system is as follows: • 99.1% The availability figure specified is the equipment availability of communication to each field instrument.	Safety Case Sect. XXXXX Reliability Report Doc No. XXXX

FIGURE 9.2.1, cont'd

INSTALLATION:	PERFORMANCE STANDARD: INTERNAL COMMUNICATIONS (cont.)	
Survivability	**Requirement**	**Assurance**
Explosion	Individual field instruments are not expected to survey a local explosion	Design drawings and inspection.
Fire	Cabling Circuits with safety functions such as PA / alarm systems have fire resistant characteristics to IEC 331. Cable routing is designed to minimise the probability of damage from fire. Duplicate redundant circuits or supplies provide compliance with the requirements of protection of safety or essential systems and follow separated as far as practicable.	Verified through inspection and testing. Design drawings and inspection.
Pool/jet fire	Damage to individual communications units or cabling will alarm through the line monitoring/self-checking system.	Verified through inspection and testing. Design drawings and inspection. See standard XXXXX
Environmental	**Requirement**	**Reference**
Prevent toxic spill to Environment	Chemicals and fuel to be provided with secondary containment where necessary [No significant inventory related to internal coms systems]	-
Interacting systems	**Dependence/Interaction**	**Reference**
UPS Systems	Dependent on UPS power supplies	Verified through inspection and testing. Design drawings and inspection. See standard XXXXX
F&G Systems	PA system broadcasts vessel-wide alarms as required	

FIGURE 9.2.1, cont'd

(B)

INSTALLATION:	PERFORMANCE STANDARD: FIRE AND GAS SYSTEM

The principal aims of the Fire and Gas (F & G) systems are the detection of hazardous conditions, primarily for the protection of personnel, and secondly for the protection of plant and equipment. The F & G systems provide the means of detecting the presence and accumulation of combustible gasses and/or the outbreak of fire, upon which it automatically executes control actions, initiates visual and audible alarms, initiates HVAC control actions, initiates extinguishant release, initiates fire pumps start and signals to

System

The F&G system has hard wired interfaces with the ESD, HVAC, PA and alarm system, Deluge Fire Fighting and CO2 protected enclosures/ rooms, Foam monitors and under pallet Spray systems.
The FPSO Primary F&G systems - is a stand-alone (operationally independent) system, but with a common operator interface to the ESD and Machinery Control Room (MCR) and SCADA sub-systems.
The F&G matrix panel provides status of the fire and gas systems on FPSO and controls operation and control of the firewater/foam system and extinguishing systems.
A Common Facilities matrix provides for system monitoring and fault/failure status.
The F&G system covering the Accommodation and Forward / Aft machinery spaces - is based on an addressable fire detection system with a dual redundant serial link interface to the vessel fire and gas system, voting logic for the living quarters and machinery spaces is executed within the vessel F&G system cabinet (bridge).
The cranes F&G system is a stand-alone system comprising of gas, smoke and heat detection with a common alarm annunciated at the F&G matrix and a serial link to the MCR. A trip signal is initiated to the ESD system if confirmed gas or fire is detected but only when the crane is unoccupied.
Reference Drawings: XXXXXXXXXXX

The F&G system detection devices and logic actions are defined within the fire and gas cause and effects document numbers:
Primary system:
Accommodation:
Forward/Aft Machinery Spaces:
Scope: This performance standard covers the following:
Fire and gas detection input devices:

- gas detectors (point)
- oil mist detectors
- flame detectors (UV)
- smoke detectors (ionisation)
- smoke detectors (optical)
- heat detectors (rate of rise)
- heat detectors (rate compensated)
- manual alarm call points
- Fire protection system pressure switches

Status signals from equipment such as fan, fire dampers, fire pumps etc.
logic to provide outputs to:

- Vessel (bridge) annunciator (visual/audible)
- fire and gas matrix
- activate public address system and facility alarms
- start fire pumps and open deluge valves, as required
- close HVAC fire dampers and stop fans
- initiate appropriate ESD actions via ESD system
- activate fire extinguishing systems, as required

Document references: XXXXXXXX
FPSO Safety Case
FPSO Operating Procedure
Fire & Gas System Planned Maintenance Routine
Fire & Gas System Functional Design Specification:
Fire and Gas Maintenance Routines

	Functions	Requirement	Assurance
1	Gas detection in open areas.	To continually monitor for flammable gas clouds within open areas (via point detectors) i.e. Separation Production Deck, Produced Water Production Deck, Export Metering Production Deck, Turret, Cargo Tanks, Fired Heaters, Cranes and initiate appropriate actions, status and alarm signals. Point detectors shall operate on LLG at 20% LEL Point detectors shall operate on HLG at 50% LEL	Reference maintenance procedure XXXXX.

FIGURE 9.2.1, cont'd

INSTALLATION:		PERFORMANCE STANDARD: FIRE AND GAS SYSTEM (cont.)	
Functions		**Requirement**	**Assurance**
2	Gas detection in enclosed areas.	To continually monitor for flammable gas within enclosed areas: i.e. vent intakes for the Accommodation HVAC room, Emergency Diesel Generator Room, Fire Pump room , Ballast Pump room, Inert Gas Generator room, Power Generator Enclosure and Coolers room, Machinery Spaces, Battery room, Cranes and initiate appropriate actions, status and alarm signals, i.e. combustion air intakes for the Power Generator Enclosures and initiate appropriate actions, status and alarms Point detectors shall operate on LLG at 10% LEL Point detectors shall operate on HLG at 20% LEL	Reference maintenance procedure XXXXX.
3	Oil Mist detection	To continually monitor for diesel fuel spillage associated with the Diesel Generators and initiate status and alarms only.	Reference maintenance procedure XXXXX.
4	Smoke detection in open areas.	To continually monitor for smoke (via ionisation/optical detectors) within defined areas: i.e. Accommodation, Forward/Aft Machinery spaces, Cargo Offloading Deck, MCR, Cranes, CO2 room, Turret Winch room, Deck Store room, Escape tunnel and air locks and initiate appropriate actions, status and alarm signals.	Reference maintenance procedure XXXXX.
5	Smoke detection in enclosed areas.	To continually monitor for smoke within enclosed areas, i.e. vent intakes for the Accommodation HVAC room, Emergency Diesel Generator room, Machinery spaces, Ballast pump room, Production Lab/Utilities, Escape tunnel and initiate appropriate actions, status and alarm signals.	Reference maintenance procedure XXXXX.
6	Fire detection (Flame/Heat)	To continually monitor for fires in all locations (via flame/heat detection) containing a fire hazard and initiate appropriate actions, status and alarm signals.	Reference maintenance procedure XXXXX.
7	Manual alarm call point (MAC)	Provision of pushbuttons to alert the CCR of a potentially hazardous situation at strategic locations on the installation and initiate appropriate status and alarm signals.	Reference maintenance procedure XXXXX.
8	Status display	Provision of status indication of fire and gas detection devices as described in requirements 1 to 6 above is provided in the Central Control Room (CCR). Audible and visual annunciation is provided to the operator to indicate zone status and internal faults on a F&G matrix located in the CCR and in the PCS/SCADA. Addressable system – gives detector locations	Reference maintenance procedure XXXXX. Refer Process Control Performance Standard
9	Initiate appropriate output actions	The F&G logic system shall perform all functions as specified on the F&G cause & effect diagrams: • to activate PA and facility alarms • to start fire pumps and open deluge valves • to close HVAC fire dampers and stop fans • to initiate appropriate ESD actions • to activate fire extinguishing systems • to initiate status signals Gas detection settings are clearly defined as a percentage within the applicable cause & effect for that zone.	Refer F&G cause & effect diagrams
10	Monitoring of field devices	All field input loops and extinguishant loops are continuously fault monitored (self-test facility). Common field faults are logged at the ICS operator station. Input loops have hardware inhibits on the control module to prevent signals acting within the fire and gas logic. All inhibits are logged at the console on a fire zone basis for operator awareness. A Control action inhibit keyswitch, per fire zone, mounted on the F&G matrix enables live testing of detector inputs without initiating logic actions. Addressable system - Refer XXXXXX General Monitors equipment - Refer XXXXXX Vessel system – Refer XXXXX The F&G system is designed to fully and automatically test all essential functions and hardware in the systems. In the event of a true fire or gas signal is received during system testing, the testing operations would be aborted and the appropriate executive actions initiated.	Reference maintenance procedure XXXXX.

FIGURE 9.2.1, cont'd

INSTALLATION:		PERFORMANCE STANDARD: FIRE AND GAS SYSTEM (cont.)	
Functions		**Requirement**	**Assurance**
11	Safety EMC requirements	Must not be affected by the operation of other equipment in close proximity.	By design, Reference specification XXXXX.
12	Have suitable IP rating	External equipment to be rated IP56 minimum.	By design, Reference specification XXXXX. By maintenance, Performance Standard F/U5
Reliability & Availability			**Reference**
The F&G system is based on a duplex processor, Input / Output (I/O) and logic, operating in parallel. A hardwired link exists between the primary fire and gas detection system and the ESD system. The system monitors the status of all power supplies and shutdown paths (hardware and software) and alarms all faults. The specified availability of the F&G system is as follows: • 99.997% The availability figure specified is the equipment (panel) specified availability and does not include the interaction with the field inputs/outputs.			Safety Case Sect. XXXXX Reliability Report Doc No. XXXX

Survivability	**Requirement**	**Assurance**
Explosion	Fixed detectors are wired back to appropriate monitoring modules in the F&G cabinets located in a safe area. Initiating alarms are latched within logic	Design drawings and inspection.
Fire	Cabling Circuits with safety functions such as fire and gas detection emergency shutdown and PA / alarm systems have fire resistant characteristics to IEC 331. Cable routing is designed to minimise the probability of damage from fire. Duplicate redundant circuits or supplies provide compliance with the requirements of protection of safety or essential systems and follow separated as far as practicable.	Verified through inspection and testing. Design drawings and inspection.
Pool/jet fire	The main F & G control equipment is protected by the H-120J blast/fire wall. Damage to individual detector heads or cabling will alarm through the line monitoring/self-checking system.	Verified through inspection and testing. Design drawings and inspection. See standard XXXXX
Environmental	**Requirement**	**Reference**
Prevent toxic spill to Environment	Chemicals and fuel to be provided with secondary containment where necessary [No significant inventory related to fire and gas systems]	-
		-
Interacting systems	**Dependence/Interaction**	**Reference**
Internal communications system	To implement installation alarms.	
ESD & blowdown system	To implement F & G system executive actions with respect to minimising the inventory of hydrocarbon released.	
Firewater deluge	To implement F & G system executive actions with respect to mitigating the effects of fire.	
HVAC	To implement executive actins with respect to removing flammable gas from the area and minimising the spread of smoke.	
UPS	In the event of normal installation power failure the F&G system is supplied from UPS system XXXXX. The minimum duration for the UPS supply to the HCS system is two hours based on the TR endurance time.	
Ex. certified Hardware	Necessary precautions taken to prevent ignition of hydrocarbons	
Excluded components	**Justification**	**Reference**
Emergency Power	Facilities operate under UPS supplies or in fail safe mode.	

FIGURE 9.2.1, cont'd

(C)

INSTALLATION:	PERFORMANCE STANDARD: EMERGENCY SHUTDOWN SYSTEM
The Emergency Shutdown (ESD) system ensures the safe isolation of the critical portions of the installation to reduce the consequences of an incident or a hazard in order to ensure the: • Protection of personnel • Integrity of plant and equipment	
System description	

The ESD system is an independent autonomous system but with common operator interfaces with the process control system. The ESD system may be linked by subsea fibre optic cable to other installations nearby. The ESD and F&G sub-systems may be combined into a "Safety and Environmental-Critical" section (solid-state/non-programmable level. 1 system) and a "safety-related" section (microprocessor based / programmable level. 2 system).
The ESD system has three functions:
• To rapidly isolate equipment in the event of a process or utility upset to avoid development of an undesirable or unsafe operating condition.
• To isolate all hydrocarbon sources including all production wells and hydrocarbon pipelines.
• To sectionalise the plant in the event of a hazardous situation occurring, in order to reduce the potential escape of inventory and to permit selective depressurisation of sections of the plant.
 	
The facilities on the installation have different levels of emergency shutdown depending on the severity of the hazard.
 	
Level 3 - **Unit Shutdown**
• Level 3 - Isolation of a utility or process function which would not immediately affect the main production system.
• Level 3B - Automatic shutdown and isolation of a process train.
 	
Level 2 - **Process Shutdown**
• Automatic shutdown of all production and enabling depressurisation of the production system.
 	
Level 1 - **Total Shutdown**
• Manual pushbutton or confirmed fire & gas input initiates closure of all ESD valves, sub-sea valves and open selected blowdown valves. Non-essential electrical supplies are isolated.
 	
Level 0 - **Catastrophic Shutdown**
• Manual initiation of total facility shutdown and isolation prior to abandonment.
• Emergency services supplied by dual redundant UPS systems would remain operational for a finite period:
    • Diesel firewater pumps
    • Battery backed emergency lighting
    • Navaids
    • PA/GA system
    • Heading Control System.
 	
Operator Interfaces
The SCADA Operator Station includes the facility to interface with the ESD system for monitoring and display purposes. The displays show ESD Inputs and Outputs and the System Alarm Status. Application or removal of overrides and to trip each output individually is only allowed by using a secure system of passwords and/or key switches which provide the level of security required under the installation Permit-To-Work System.
 	
Further details about the position of ESD valves and the levels of shutdown may be found in Section XXXX of the Installation Safety Case.	

FIGURE 9.2.1, cont'd

INSTALLATION:		PERFORMANCE STANDARD: EMERGENCY SHUTDOWN SYSTEM (cont.)	
Functions		**Element**	**Assurance**
1	To provide an interface with ESD systems on nearby installations	a) Independent autonomous operation of ESD system. b) *The ESD systems are independent autonomous systems but with common operator interfaces with the other process control systems and may have links with the ESD systems of nearby installations.*	a) Test ESD intertrip signals as per Functional Design Specification for Emergency Shutdown System(s) b) Test ESD interface against appropriate Cause & Effect Chart.
2	To provide ESD Level 3 - Unit Shutdown.	a) The ESD Level 3 (Unit shutdown) logic functions shall operate as specified on the cause and effect diagrams to shut down systems. There are two types of level 3 shutdown • 3 - isolation of utility or process function • 3B - isolation of process train and the causes and effects are described in the Safety Case. **FPSO** *The ESD signals from an offloading shuttle tanker operate shutdowns only on the export system on the FPSO and do not affect the production train or other auxiliary marine systems.*	Functional Design Specification for Emergency Shutdown System Function test logic functions against the current ESD cause and effect chart.
3	To provide ESD Level 2 - Process Shutdown.	a) The ESD Level 2 (Process Shutdown) logic functions as specified on the cause and effect diagrams to shut down systems. *The causes and effects of a level 2 shutdown shall be as described in the installation Safety Case.*	Functional Design Specification for Emergency Shutdown System Reference ESD Maintenance Routines Function test level 2 logic functions against ESD level 2 Cause and Effect Chart. Reference ESD Maintenance Routines
4	To provide ESD Level 1 - Total Shutdown.	a) The ESD Level 1 (Total Shutdown) logic functions shall operate as specified on the cause and effect diagrams to shut down systems. *The causes and effects of a total platform shutdown shall be as described in the appropriate section of the installation Safety Case.*	Functional Design Specification for Emergency Shutdown System Function test Category 1 logic functions against a sample from the cause and effect chart. Reference ESD Maintenance Routines

FIGURE 9.2.1, cont'd

INSTALLATION:		PERFORMANCE STANDARD: EMERGENCY SHUTDOWN SYSTEM (cont.)	
Functions		**Element**	**Assurance**
5	To provide ESD Level 0 - Catastrophic Shutdown.	a) The ESD Level 0 (Catastrophic Shutdown) logic functions shall operate as specified on the cause and effect diagrams to shut down systems. *The causes and effects of a catastrophic platform shutdown shall be as described in the installation safety case.*	Functional Design Specification for Emergency Shutdown System Function test Category 1 logic functions against a sample from the cause and effect chart. Reference ESD Maintenance Routines
6	To provide safety systems on activation of ESD at level 0.	a) PA/GA systems shall remain operational on a level 0 ESD. b) The diesel firewater pumps would remain operational until the diesel day tank is empty on a level 0 ESD. c) The forward and aft thrusters will run for over 3 hours under Heading Control System control to allow for safe evacuation on a level 0 ESD. d) All battery backed emergency lighting shall remain operational. e) Navigational aids shall remain operational Only emergency systems with a self-contained Ex 'd'-protected built-in power supply shall be left operational.	Functional Design Specification for Emergency Shutdown System a) See PA/GA performance standard b) See Active Fire Protection performance standard c) See Heading Control performance standard d) See Escape Lighting performance standard e) See Navaids performance standard
7	To isolate all hydrocarbon sources including the FPSO hydrocarbon pipelines.	a) A level 1 shutdown will close all ESD valves, and automatically open blowdown valves in the affected area only if initiated by fire or gas detection. *The ESDVs on the hydrocarbon lines are designed and located in accordance with Pipeline Safety Regulations SI 825 (1996)*	Functional Design Specification Section 3 Emergency Shutdown System Function test against ESD level 1 Cause & Effect Chart
8	To provide means to initiate ESD in the control room.	a) A level 2 Shutdown can be initiated by Operator intervention via Manual Pushbuttons in the Control Room and selected plant areas. b) A level 1 shutdown shall be activated by a Manual Emergency Pushbutton in the control room or automatically if a gas release or fire occurs. c) A level 0 shutdown shall be initiated manually from the control room or from manual stations strategically located around the installation.	Functional Design Specification for Emergency Shutdown System Function test logic functions against a sample from the cause and effect chart. Reference ESD Maintenance Routines.
9	To provide accurate feedback of ESD functions to the control room.	The Process Control Operator Stations include the facility to interface with the ESD system. It is possible to read the status of the ESD Inputs and Outputs and the system status from any window. a) Category 1 ESD system input status, output status and sequence of events initiating and following a trip. c) All the Operator actions, system alarms, system input and system output status within the plant.	Functional Design Specification for Emergency Shutdown System Function test logic functions against a sample from the cause and effect chart. Reference ESD Maintenance Routines.

FIGURE 9.2.1, cont'd

INSTALLATION:	PERFORMANCE STANDARD: EMERGENCY SHUTDOWN SYSTEM (cont.)	
Functions	**Element**	**Assurance**
Reliability & availability		**Reference**
The ESD system is a Category 1 safety system which provides protection of personnel and the installation, as defined in the maintenance management strategy. Test and maintenance procedures have been prioritised accordingly to provide a high integrity ESD system and are contained within the maintenance management system. The nominal ESD system availability is 99.997%. Contracts with Health Care providers specify reliability targets which are continuously under review.		Installation Safety Case SIL Assessment Report
Component	**Test & Frequency**	**Assurance**
Shutdown Field Devices	2/n voting is fed into the ESD system for initiation. The ESD system does not rely upon cascade effects to produce shutdowns of related process systems.	Verified through inspection and testing. Reference ESD Maintenance Routines
ESD	The ESD system is fail safe and line monitored. The ESD system is designed to fail to safety on fault detection in any single channel section of a shutdown path. Level 3 - six monthly; Other levels - yearly	Verified through inspection and testing. Reference ESD Maintenance Routines.
Cabling and Power	A hardwired matrix displays the status of system inputs, outputs, override and reset status and the health of the system. The system monitors the status of power supplies and shutdown paths and alarms all faults. A single channel fault in any section of the system produces a fail-safe response.	Verified through inspection and testing. Reference ESD Maintenance Routines

FIGURE 9.2.1, cont'd

INSTALLATION:	PERFORMANCE STANDARD: EMERGENCY SHUTDOWN SYSTEM (cont.)	
Component	**Test & Frequency**	**Assurance**
Inter-ESD System Communication Checking	A dedicated dual redundant data link connects the ESD systems on the nearby installations [if any] to carry intertrip signals identified on the cause and effect charts.	Verified through inspection and testing. Reference ESD Maintenance Routines.
Computers I/O devices	Regular shutdown path checking from the system inputs to the system outputs is automatically carried out by the system at a frequency sufficient to ensure system faults are detected, alarmed, and executive actions taken to achieve the availability/ reliability requirements of the ESD system.	System is self-checking in accordance with Functional Design Specification for Emergency Shutdown System Regular function testing of ESD System
Computer control	Each system is subdivided into a solid-state logic system for Category 1 signals and a programmable system for Category 2 signals.	Instrument Based Protective Systems IEC 61511:2016
Survivability	**Requirement**	**Assurance**
ESD Level 0	Emergency systems and essential services with Ex d protected built in power supplies would be left operational, including: • emergency lights • navigational aids • PA/GA systems Additionally the diesel firewater pumps would remain operational until the diesel supply was exhausted.	Functional Design Specification for Emergency Shutdown System
Fire & Explosion	Risk Evaluation and Protection Philosophy Study - ESDVs and Blowdown valves rated to API specification.	Risk Evaluation and Protection Philosophy, Design drawings and inspection.
UPS	Providing electrical power to computer systems	See UPS performance standard
Hydraulic Valves	Provide isolations	
Pressure Relief System	Reduces pressure in the process equipment and reduces loss of inventory.	See pressure relief performance standard
Fire and gas detection	Initiates ESD shutdown	See F&G performance standard
Hazardous Area Certified Equipment	Ex rated equipment in place	See Hazardous Area Equipment performance standard
HVAC	Executive actions carried out by ESD on HVAC system	See HVAC performance standard
Environmental	**Justification**	**Reference**
ESD	Sectionalises the plant in the event of a hazardous situation occurring, in order to reduce the potential escape of environmentally hazardous inventory	Cause & Effect Matrix
Excluded components	**Justification**	**Reference**
Emergency Power	All systems are fail safe or UPS maintained	See UPS performance standard

FIGURE 9.2.1, cont'd

Parameter	Example of Equipment/Systems Appropriate to Parameter
Equipment Performance	Generators
Standard of Equipment	TEMPSC communications, Standby Vessel communications
Suitability of equipment for operating environment	Electrical equipment in potentially hazardous areas, Ingress protection of equipment
Adequately/correctly Sized	Switchboards, cables, motors, UPS batteries
Initiate other actions	Executive actions initiated by fire/gas detection system or by detection of process deviation
Closure Time	Valves, fire dampers (by actuators)
Failure Mode	Valve position on loss of power or pressure in actuating medium
Response Time	Detectors, equipment trip, blowdown
Duration	For emergency lighting, survivability of RESDVs and TR
Location/orientation	Emergency control points, HVAC inlets relative to hazardous areas, emergency lighting.
Position/Status	Valves locked open/closed
Status Indication	F&G system, ESD valve status, fire damper status, alarm status
Reliability (or PFD)	ESD/PSD system, Fire & Gas system and components
Availability	ESD/PSD system, Fire & Gas system and components
Distribution/Coverage	Detectors, firewater, emergency lighting, PA system and beacons
Redundancy	Separate PA systems, UPS, voted sensors, valve actuation
Segregation	Dual UPS supply cabling, dual PA system cabling and PA amplifier racks, Emergency generator (from main generation), subsidiary control and communication room (from CCR)
Accessibility and operability	Safety equipment, escape routes, triage areas (including adequate emergency lighting)
Resistance to Fire	ESD valves, Temporary Refuge, fire dampers, cables
Resistance to Explosion	See Segregation
Resistance to Vibration	Fire & Gas Detectors, process instrumentation, navaid lighting
Resistance to Dropped Objects	See Segregation
Resistance to Impact	Ballast control system on MODUs and FPSOs (to preserve stability)
Resistance to Mechanical Damage	Protection of cables, valve hydraulic/pneumatic vent lines, cable armouring.
Protection against Upset Conditions	Provision and set-points of piping system instrumentation
Compatibility	Electromagnetic Compatibility
Corrosion Protection	Impressed current cathodic protection
Prevent potentially unsafe mode of operation	Ballast control system, power management system, process monitoring and control system(s)

FIGURE 9.2.2

Checklist of Parameters to be considered for performance standards relating to electrical equipment and instrumentation. *ESD*, emergency shutdown; *FPSO*, floating production, storage and offloading; *HVAC*, heating, ventilation and air conditioning; *MODU*, mobile offshore drilling unit; *PA*, public address; *PFD*, probability of failure on demand; *PSD*, process shutdown; *UPS*, uninterruptible power supply.

CHAPTER

Notes on Safety Integrity Level Assessment for Electrical Engineers

3

INTRODUCTION

As electrical engineers, we may be asked to attend Hazard and Operability (HAZOP) study and Electrical Safety and Operability (ESAFOP) study meetings, as well as safety integrity level (SIL) assessment meetings. All such meetings are run in a similar manner, and usually involve the breaking down of the system or installation into a number of areas and control loops and following a set script of questions which are applied to each item being studied. Usually software is used to present and organise the questions and pictorially highlight the location on the plan or the loop being studied. The intention of this chapter is to concentrate on the subject of SIL assessment and provide some practical guidance on the process and also the pitfalls to be avoided. For those who wish to take the subject further, I have included a number of web addresses at the end of the chapter. Some HAZOP and ESAFOP references are also provided.

SAFETY FUNCTION

There is a difference between good design of a device and the provision of functional safety. For example, an electrical motor may have an encapsulated rotor *designed* not to fly apart at high speeds, but it may also be fitted with an overspeed sensor which trips the motor at a maximum safe speed setting. The overspeed device is a provision of functional safety. A safety function is a function to be implemented by a safety instrumented function (SIF), other safety technology or external risk reduction facility, which is intended to achieve or maintain a safe state for the process in question, with respect to the specific hazard in question.

SAFETY INSTRUMENTED FUNCTION

An SIF is a safety function with a specified SIL, necessary to achieve functional safety. This can be achieved by either implementing a *safety instrumented protection function* or a *safety instrumented control function* (see IEC61511 Definition 3.2.71).

Offshore Electrical Engineering Manual. https://doi.org/10.1016/B978-0-12-385499-5.00050-9

Two requirements must be defined to achieve 'functional safety':

- A clear definition of what the safety function has to achieve (i.e., what does it do and how does it do it?) (derived from the *hazard analysis*)
- The safety integrity requirement (probability of success, or likelihood of function being performed satisfactorily) (derived from the *risk assessment*)

SAFETY INSTRUMENTED SYSTEM

A safety instrumented system (SIS) is an instrumented system used to implement one or more SIFs. An SIS may contain any combination of sensors, logic solvers and final elements.

THE SAFETY INTEGRITY LEVEL ASSESSMENT PROCESS

An SIL is a discrete level for specifying the safety integrity requirements of each SIF, as allocated to an SIS as follows:

Low Demand Mode of Operation

Safety Integrity Level (SIL)	Probability of Failure on Demand (PFD)
1	10^{-1}–10^{-2}
2	10^{-2}–10^{-3}
3	10^{-3}–10^{-4}
4	10^{-4}–10^{-5}

High Demand (or Continuous) Mode of Operation

Safety Integrity Level (SIL)	Probability of Failure on Demand (PFD/h)
1	10^{-5}–10^{-6}
2	10^{-6}–10^{-7}
3	10^{-7}–10^{-8}
4	10^{-8}–10^{-9}

For the electrical/instrument engineer, SIL assessment involvement is almost always associated with control loops such as pressure, temperature or level control or monitoring such as fire and gas detection, known as SISs. During the assessment, loops will be broken down into sensing elements (e.g., pressure transmitters), field wiring, logic solver and final control elements. Power supplies and communications interfaces may also play a part in the process, but devices should always fail to a safe condition.

The SIL assessment process should follow the guidelines in ANSI/ISA S84.00.01-2004/IEC 61511/61508.

The SIL assessment meeting should consist of a chairperson usually qualified and experienced in a safety-related discipline and representatives from design, maintenance and operations. To save time, the software should be preprogrammed with component failure data, but the sources of this data should be reviewed and agreed during the meeting.

DETERMINING SAFETY INTEGRITY LEVELS – PROCESS

When a process hazards analysis determines that an SIS is required, the level of risk reduction afforded by the SIS and the target SIL has to be assigned. The effectiveness of an SIS is described in terms of 'the probability it will fail to perform its required function when it is called upon to do so'. This is its probability of failure on demand (PFD). The average PFD (PFD_{avg}) is used for SIL evaluation. Table 9.3.1 shows the relationship between PFD_{avg}, availability of the safety system, risk reduction and the SIL values.

Various methodologies are used for assignment of target SILs. The determination must involve people with the relevant expertise and experience.

Methodologies used for determining SILs include

- simplified calculations,
- fault tree analysis,
- layer of protection analysis,
- Markov analysis.

Table 9.3.1 Relationship Between PFD_{avg}, Availability of the Safety System, Risk Reduction and the SIL Values

SIL	Availability	PFD_{avg}	Risk Reduction	Qualitative Consequence
4	>99.99%	10^{-5} to $<10^{-4}$	100,000–10,000	Potential for fatalities in the community
3	99.9%	10^{-4} to $<10^{-3}$	10,000–1000	Potential for multiple on-site fatalities
2	99%–99.9%	10^{-3} to $<10^{-2}$	1000–100	Potential for major on-site injuries or a fatality
1	90%–99%	10^{-2} to $<10^{-1}$	100–10	Potential for minor on-site injuries

Definitions: Availability, *the probability that equipment will perform its function;* PFD, *probability of failure on demand is the probability of a system failing to respond to a demand for action arising from a potentially hazardous condition;* PFD_{avg}, *the average PFD used in calculating safety system reliability;* SIL, *safety integrity level. Both IEC and ANSI/ISA standards utilise similar tables covering the same range of PFD values. ANSI/ISA, however, does not show an SIL 4. No standard process controls have yet been defined and tested for SIL 4.*

DETERMINING SAFETY INTEGRITY LEVELS – INSTRUMENTATION

SILs for field instruments are established by one of the following two methods:

1. *FMEDA* (failure modes, effects and diagnostic analysis) is best when reviewed or certified by a third party such as Exida or TUV, although self-declarations can be carried out by the manufacturer. A systematic analysis is necessary to determine failure rates, failure modes and the diagnostic capability as defined by IEC 61508/651511.
2. *Proven in use* (also called prior use) is typically used by a customer with a mature instrument in known processes. This approach requires sufficient product operational hours, revision history, fault reporting systems and field failure data to determine if there is evidence of systematic design faults in a product. IEC 61508 provides levels of operational history required for each SIL. It is generally considered of more value when done by users in their facility when comparing similar data. It is considered less reliable when done by a device manufacturer whose data may be less relevant to the end-user's application.

SAFEGUARDS

If at all possible, the system should be inherently safe and not require the need for an SIS with a high SIL. With offshore installations, good design practice should keep control loops of SIL 2 or above to an absolute minimum. This can be done by reducing the probability of the major accident event by

- minimising the staffing level of the area where the risk is present,
- providing passive fire and/or blast protection,
- using relief devices such as relief valves and bursting disks,
- keeping flammable inventories away from areas of expected high manning (e.g., accommodation modules).
  Should the event occur, its effect can be reduced by mitigating elements some of which will be SISs, such as
- fire and gas detection systems
- ignition prevention
- emergency shutdown systems
- blowdown and flare systems
- active fire protection/suppression
- communication and alarm systems
- temporary refuge, escape and evacuation systems
  Note that the SIL is based on the *whole* loop, i.e., all the components of the loop play a part in achieving the SIL, so purchasing a logic panel with a high SIL will not guarantee that each loop has a high SIL (Tables 9.3.2 and 9.3.3).

Table 9.3.2 Typical Summary of SIL and Test Interval for Each Tested Loop

Loop Number as Seen in SIL Verification Sheet	Loop Number	Required SIL	Proposed Test Interval (h)	Achieved SIL (With Proposed Test Interval)
3.1	1	1	35,040	1
3.2	2	1	35,040	1
3.3	3	1	35,040	1
3.4	4	1	35,040	1
3.5	5	1	8,760	1
3.6	6	1	35,040	1
3.7	7	1	35,040	1
3.8	8	1	35,040	1
3.9	9	1	35,040	1
3.10	10	1	35,040	1
3.11	11	1	35,040	1
3.12	12	1	8,760	2
3.13	13	1	35,040	1
3.14	14	1	35,040	1
3.15	15	1	35,040	1
3.16	16	1	35,040	1
3.17	17	1	35,040	1
3.18	18	1	35,040	1
3.19	19	1	17,520	1
3.2	20	1	8,760	2
3.21	21	1	8,760	2
3.22	22	1	8,760	2
3.23	23	1	8,760	2
3.24	24	1	8,760	2
3.28	27	1	8,760	1
3.32	28	1	8,760	1
3.37	29	1	8,760	1
3.39	30	1	8,760	1
3.42	31	1	8,760	1
3.43	32	1	35,040	1
3.53	33	1	35,040	1

SIL, *safety integrity level.*
Obtained from SIL Assessment Meeting.

Table 9.3.3 Typical SIL Calculation

Data Report – Instrumented Safety Function Implementation Requirements										
Required SIL: 1 **The loop consists of four smoke detectors voted as 2oo4, one fire and gas logic controller and four actuators as listed in Table below**										
Calculation Data and Architectural Constraints on Hardware Safety Integrity Data:										
Group Name	**Voting**	**Reference**	λ_S	λ_D	β	β_D	**DC_S**	**DC_D**	**MTTR**	**PFD**
SD95	1001	Smoke detector	2.40e-06	1.30e-06	–	–	0.40	0.40	8.0	1.368e-02
SD66	1001	Smoke detector	2.40e-06	1.30e-06	–	–	0.40	0.40	8.0	1.368e-02
SD64	1001	Smoke detector	2.40e-06	1.30e-06	–	–	0.40	0.40	8.0	1.368e-02
SD63	1001	Smoke detector	2.40e-06	1.30e-06	–	–	0.40	0.40	8.0	1.368e-02
Logic	1001	Programmable safety System	1.00e-05	1.00e-05	–	–	0.90	0.20	4.0	1.756e-02
Release of magnetic door holders 1	1001	Relay	3.00e-07	2.00e-07	–	–	0.30	0.00	24.0	2.458e-03
Release of magnetic door holders 2	1001	Relay	3.00e-07	2.00e-07	–	–	0.30	0.00	24.0	2.458e-03
Start of duty FWP	1001	Circuit breaker	5.00e-07	3.00e-07	–	–	0.00	0.00	24.0	5.263e-03
Shutdown of ventilation dampers (inlet damper) + solenoid valve	1001				–	–				2.630e-02
Shutdown of ventilation dampers (inlet damper) + solenoid valve	1001				–	–				2.630e-02
Total										7.91e-02

Group Name	Type A/B	Calculated SIL	SFF	HFT	SIL max.	Group SIL	Final SIL
SD95	Type B	1	0.79	0	1	1	3
SD66	Type B	1	0.79	0	1	1	
SD64	Type B	1	0.79	0	1	1	
SD63	Type B	1	0.79	0	1	1	
Logic	Type A	1	0.95	0	3	1	1
Release of magnetic door holders 1	Type A	2	0.72	0	2	2	
Release of magnetic door holders 2	Type A	2	0.72	0	2	2	
Start of duty FWP	Type A	2	0.63	0	2	2	
Shutdown of ventilation dampers (inlet damper) + solenoid valve	Type A	1	0.95	0	–	1	1
Shutdown of ventilation dampers (inlet damper) + solenoid valve	Type A	1	0.63	0	–	1	
Total							1

Calculated SIL: 1.
Voting within initiators group: 2004.
Voting within logic group: 1001.
Voting within actuators group: 5005.

Continued

Table 9.3.3 Typical SIL Calculation—cont'd

Partial Test: No			Nonperfect Test: No	
C1 = 0.00	C2 = 1.00	T1 (h) =	T2 (h) =	T3 (h) = - 35040.00

Block Diagram

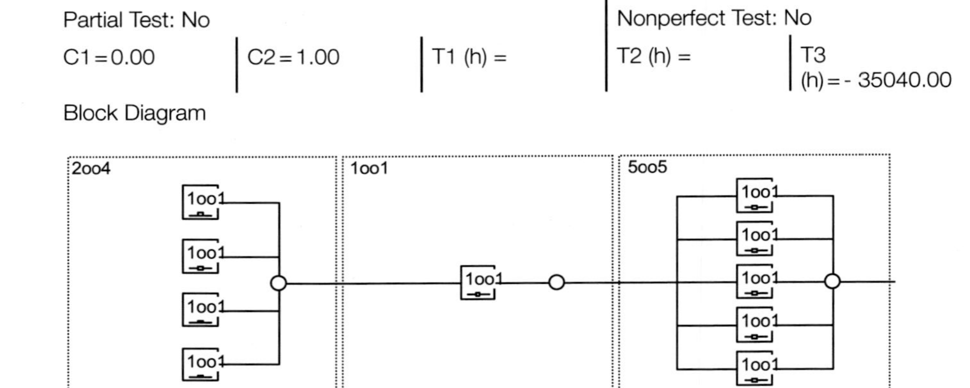

Maximum interval between tests (h): 35040

PFD according to proof test interval (h):

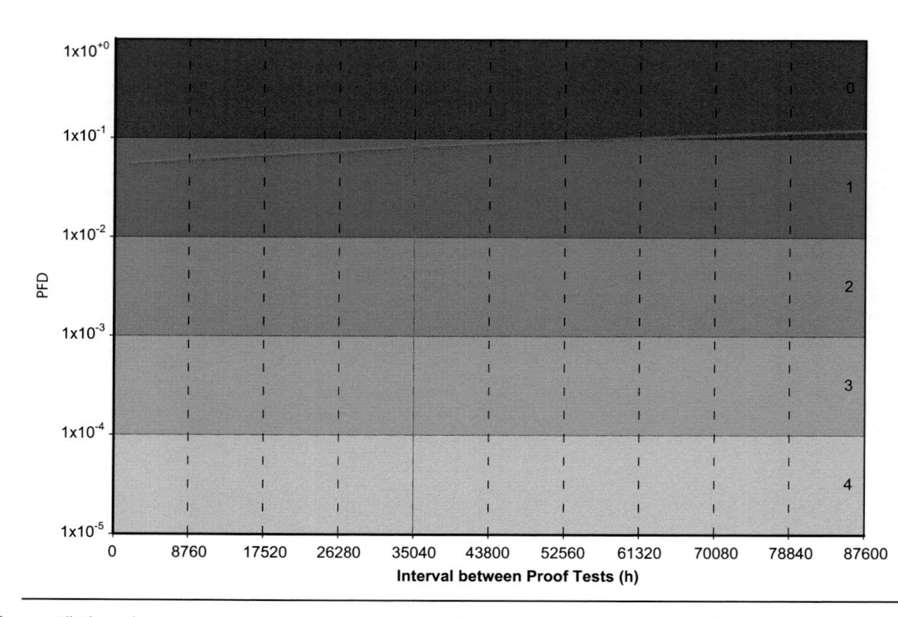

Comments: The PFD for ventilation dampers and associated solenoid valves were taken from the SIL Verification Report and accepted as adequate for the modelling. These were then added together as one block modelling the fire damper. PFD of block was then found to be PFD (FD)+PFD (SV) $=2.2 \cdot 10^{-2}+4.3 \cdot 10^{-3}=2.63 \cdot 10^{-2}$. 2oo4, *2-out-of-4;* FWP, *fire-water pump;* HFT, *hardware fault tolerance;* MTTR, *mean time to repair;* SIL, *safety integrity level;* PFD, *probability of failure on demand;* SFF, *safe failure fraction.*
ORBIT SIL Printouts courtesy of Dr Luiz Fernando Oliveira, DNV-GL Brazil.

APPENDIX

Guide to Offshore Installations

A.1 TYPES OF INSTALLATION

SEMISUBMERSIBLES

These are by far the most common type of installation and are used as mobile drilling platforms, accommodation and construction centres and fire tenders. They are sometimes modified for use as oil production platforms. With a displacement often exceeding 20,000 tons, these float on buoyancy chambers placed well below the zone of wave action, and with a draft of up to 90 ft to dampen the movement caused by the waves. The platforms are normally held in position by eight or more 15-ton anchors and like the fixed platforms are designed to withstand extreme weather conditions.

JACK-UP OR SELF-ELEVATING PLATFORMS

A jack-up rig is a barge-like vessel with steel legs each having racking gear which allows it to be racked down to the seabed until the entire vessel is lifted out of the water. These vessels are obviously limited by the length of the legs to water depths of around 300 ft. They are inherently a temporary facility, more suited to drilling and maintenance activities than production.

FIXED PLATFORMS

A fixed platform may be described as consisting of two main components:

1. The substructure
   This consists of either a steel tubular jacket or a prestressed concrete structure.
2. Steel structures
   Steel jacket structures normally consist of tubular legs held together by welded tubular bracing, the whole unit being securely piled to the seabed through tubes attached to the bottom of the legs. Also in the structure are various vertical steel tubes required for obtaining seawater for platform utilities (stilling tubes), for protecting subsea electrical cabling (J-tubes) and for guiding for well risers. The jacket when first installed without superstructure may be more than 500 ft high.

3. Concrete structures

   Two proprietary designs of concrete gravity substructure are most common in the North Sea.

   a. The 'CG Doris' type consisting of a large single watertight caisson surrounded by a circular breakwater designed with a honeycomb of holes to act as a breakwater. A central column sprouts from the caisson from which the superstructure is supported.
   b. The 'Condeep' type consisting of circular towers or legs sprouting from a cellular caisson.

      In both cases, the weight of the structure ensures stability on the seabed, avoiding the need for piles or anchors and parts of the structure may be used for the storage of oil or housing production equipment.

4. The superstructure

   The superstructure usually consists of a deck truss assembly which supports all the prefabricated production process facilities, drilling, utilities and accommodation modules, communication facilities and cranes that make up the 'topsides' of the platform.

   With concrete platforms, it is normal to float out the substructure with as much of the superstructure in place as possible. This is to avoid the higher costs associated with carrying out installation work offshore.

   Again, to minimise offshore installation work, which costs in the region of four times that of similar work onshore, all superstructure is modular, the modules being fitted out as completely as possible taking into account the loading limits of available shipping and craneage capacity.

   If it is not possible to complete the superstructure onshore then, having landed and skidded the modules into position and secured them to the supporting deck assembly, the services and process connections such as electrical power, cooling water, heating, ventilation, and air-conditioning ducting, process oil, condensate and gas may be interconnected between modules. This is known as the hook-up phase.

TENSION LEG PLATFORMS

Where oil and gas fields are found in deeper water, the costs and technical problems involved in producing ever taller structures have forced producers to consider more novel methods. The tension leg system is currently in use on one installation and consists of a hull-like structure having positive buoyancy which is held to a deeper draught by means of tension legs at each corner.

The legs consist of flexible metal tubes anchored in foundations in the seabed and tensioned to hundreds of tons in the platform. A computer-controlled ballasting system prevents load movements on the platform deck causing large differences in tension and hence avoids dangerous stressing of the hull structure or the legs themselves.

A.2 DRILLING ESSENTIALS

DRAW WORKS

A rotating column of hollow steel pipe made up of sections screwed together, called a drill 'string' fitted at the bottom with a cutting bit, is hung from a 'swivel' connection inside the derrick, a strong steel pylon, usually 45–55 m high. A wire cable runs through heavy pulleys to the 'draw works' which raise and lower the drill string. The draw works is driven by a direct current (DC) motor variable-speed drive controlled from the 'driller's console'.

ROTARY TABLE

On the floor of the derrick is a rotary table with an opening in the middle through which slides the top section of the drill string, a square or hexagonal section pipe called the 'kelly'. The rotary table is also driven by a DC motor variable-speed drive controlled from the driller's console. The rotary table turns the kelly and hence the whole drill string. As the drill bit cuts deeper, extra sections of drill pipe are added to the drill string. The whole string will require to be lifted out to replace worn, damaged or inappropriate drill bits.

TOP DRIVE

As an alternative to using a rotary table to drive the rotation of the drill string, a top drive is utilised. This is composed of one or more electrical or hydraulic motors, which is connected to the drill string via a short section of pipe known as the 'quill'. Back torque from the drill string has to be prevented from twisting the top drive, so it slides on a rack attached to the derrick. One benefit of using the top drive is that it allows the drilling rig to drill a longer section or stand of drill pipe in one operation. A rotary table rig can only drill 30-ft (9.1 m) (single drill pipe) sections of drill pipe, whereas a top drive can drill 60- to 90-ft (18–27 m) stands (double and triple drill pipe, respectively, a triple being three joints of drill pipe screwed together), depending on the drilling rig size. This saves a lot of time and money each time the whole drill-string needs to be withdrawn when it is necessary to replace the drill bit, for example.

MUD PUMPS

During drilling operations, a special fluid is continuously pumped through the top of the kelly and down the hollow drill string to cool and clean the revolving bit below. The fluid also acts as a conveyor of drill cuttings to the surface and a means of sealing and supporting the wall of the hole.

The fluid is basically composed of special clay suspended in water or oil and known as 'mud'. Its composition must be varied according to conditions and may contain various additives for better lubrication of the drill bit in abrasive rock formations or better sealing in highly permeable rock formations.

SHALE SHAKER

The mud is circulated by mud pumps using yet another DC variable-speed drive controlled from the driller's console. The drill cuttings conveyed to the surface are then sieved out by a vibrating screen called a shale shaker before returning to a pit for recirculation.

The moving column of mud in the hole is vital to the safety of the drilling operation, as its weight helps prevent a 'blowout' if oil or gas at very high pressures is encountered.

The petroleum engineer may increase the density of the mud at critical stages of drilling where high-pressure gas or oil pockets are likely. This increases the weight of the column of mud in the hole and reduces the blowout risk.

BLOWOUT PREVENTER

If a pocket of high-pressure oil or gas is unexpectedly pierced by the drill, the hydrostatic pressure and viscosity of the mud in the drillhole will not be sufficient to prevent the oil and gas from blowing the mud back and erupting, often in a fierce jet from the well. If this very hazardous condition, known as a blowout, is not quickly controlled, it may lead to serious fires and explosions, and possibly the loss of the entire installation. To guard against blowouts, up to six heavy-duty valves collectively known as the blowout preventer are fixed beneath the raised floor of the derrick or on the seabed. If there is a risk of a blowout occurring, the well can be closed by these valves, even whilst drilling is in progress.

WIRELINE LOGGING

One of the temporary facilities the electrical supply of which is often problematical for electrical engineers is the wireline cabin. This is because they are almost invariably in the hazardous area created by the wellheads. They are necessary because a continuous log often needs to be made of the electrical and other properties of the rock formations drilled through. Different rock types, their fluid content and physical properties can be distinguished, and with information obtained by analysis of drill cuttings and specially drilled core samples, a map of the oil or gas field geological structure may be built up.

WELL COMPLETION

Once oil or gas is found in sufficient quantities to be economically viable, production wells may be drilled. In this case the hole is periodically lined with steel casing of different diameters. The production casing is the longest and smallest diameter, typically 9.5/8 in., followed by intermediate casings and ending up with 30-in. surface casing.

The casings are inserted in a preplanned sequence as each section of the well is drilled and then cemented in place. If the producing layer is firm, such as limestone, the final part of the hole need not be lined and the oil can be produced directly into the bore well.

PERFORATION

If the producing layer is friable, a final string of casing sections is run right down to the bottom, cemented in place and holes are shot through it with a perforating gun to let in the oil.

This last operation usually requires radio silence to be maintained for several hours on the platform, as operation of transmitters can inadvertently trigger the perforating gun before it is in place.

If the producing formation is loose sand, a slotted pipe called a 'liner' with a 'gravel pack' or 'wire filter' round it can be hung from the casing in the producing rock strata.

CHRISTMAS TREES

At this stage, the well is still full of heavy drilling mud. Because the casing and liner must remain in the well for a long time and their repair or replacement would be expensive, another string of small-bore pipe, 'the tubing', hung from the wellhead at the top of the casing, is run down to the foot of the well where it is sealed against the production casing with a packer. This is the pipe string through which the oil and gas flow to the surface.

The drilling mud is flushed out of the tubing with water, allowing the oil or gas to emerge. Because of the high well pressures, often in hundreds of bars, the oil flow needs careful regulation at the surface, and a cluster of valves and fittings which seal off the well and control the flow, known as the 'Christmas tree', is a familiar sight in documentaries about oil.

SUBSEA COMPLETIONS

Where other oil or gas fields are discovered adjacent to existing offshore production facilities, it may not be necessary to install a new platform or floating facility.

Instead, a seabed wellhead facility is installed which, after production drilling, allows oil and gas to be piped to the adjacent established production facility, either onshore or offshore, for processing and export. The adjacent facility also acts as a control point for operation of the subsea wellhead valves and the point of supply for electrical equipment located on the subsea wellhead manifold.

DEVIATED DRILLING

With present day technology, it is possible to guide the direction of drilling so that the well bore is deliberately deflected from the vertical in order to achieve good spacing at the reservoir to allow the gas to be removed in a more uniform manner. The method may also be used where satellite wells are completed underwater.

In some cases where the field is shallow, such as the Morecambe Bay gas field, wells may be drilled at an angle to the vertical using a special drilling rig with a derrick which may be tilted to the correct angle by hydraulic jacks before drilling commences.

A.3 PRODUCTION PROCESS ESSENTIALS

The processes outlined in the following are shown in the flow diagram in Fig. A3.1 and are typical of a large North Sea fixed production platform.

OIL, WATER AND GAS SEPARATION

Petroleum mixtures are usually complex and not easy to separate efficiently. Vapour recovery and evaporation control must be engineered in a hostile environment with minimum risk of fire or explosion.

Separators on offshore installations consist of large steel pressure vessels, either horizontal or vertical, which by a combination of changing flow direction many times and the action of gravity cause the heavier liquids to fall to the bottom. Special (sometimes vertical) separator vessels are used on floating installations to allow the separation process to occur in bad weather, with the ship rolling and pitching.

After the first stage of separation, the gas will still be 'wet' and require at least one further stage of separation. Liquids which would revert to a gas at atmospheric pressure are referred to as condensate and are normally re-entrained with the crude oil before exporting from the platform by a subsea pipeline or tanker. On some installations, the gas is liquefied and also injected into the oil pipeline before exporting ashore.

A large proportion of the flow from any North Sea well is water, and this must be removed from the oil and gas simply because it has no economic value and it would also cause corrosion in the export subsea pipeline. As the petroleum liquids are less dense than water, they form a middle layer in the separators so that the 'produced water' can be drained off at the bottom. Level controllers ensure that the three different outputs of each separator are maintained. Naturally occurring radioactive materials (NORM) in the form of natural radioactive salts may collect in some separators, and this will need to be removed and taken to Sellafield or a similar plant to be disposed of safely.

HYDRATES

Mixtures of water vapour and natural gas at high pressure tend to form solid ice-like crystals on cooling called hydrates. Hydrates can cause restriction or even completely prevent flow of gas.

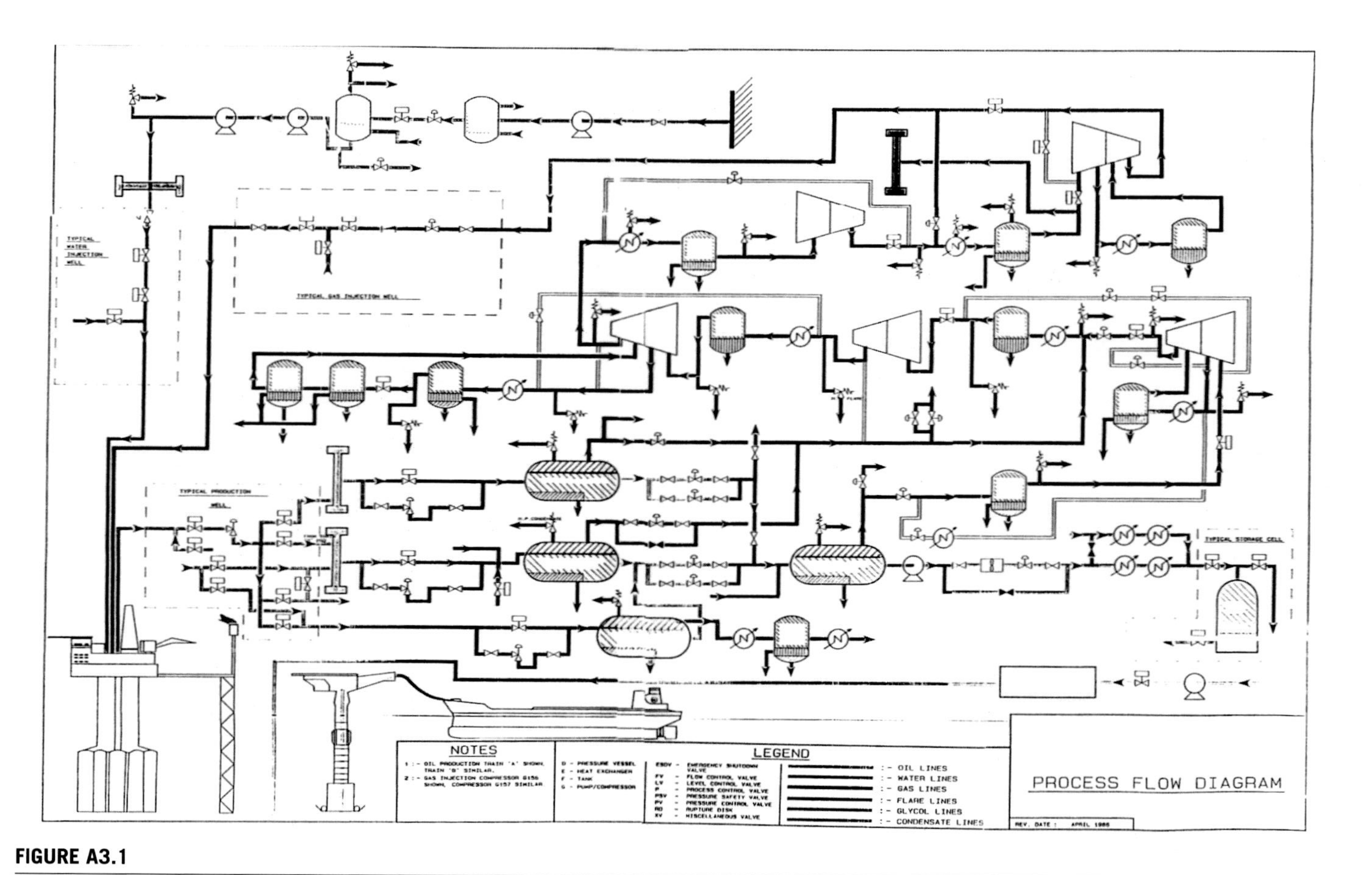

FIGURE A3.1

Typical offshore process flow diagram.

Some liquids have the property of absorbing water vapour from gases and are used to prevent or restrict formation of hydrates.

Ethylene glycol or similar agents are used for this purpose offshore. Gas is passed through a pressure vessel into which a fine spray of the agent is introduced.

WATER INJECTION

Seawater is drawn up by lift pumps and, along with treated produced water, is injected into the oil bearing formations via a separate injection well. The purpose of this is to artificially maintain well pressure by replacing the oil, gas and water removed via the producing well. The injection pumps required are usually very large, in the order of several megawatts. The reinjection of produced water has the added advantage of disposing of a potential environmental pollutant, although with high oil prices it is economically attractive to centrifuge out quite low levels of entrained oil.

GAS REINJECTION

In the early days of oil production, it was accepted that the gas removed from the separators could be flared, should it be found uneconomical to pipe the gas ashore. Nowadays, with established North Sea gas pipe networks such as 'FLAGS' it is normally possible for the producing company to tie in to an established gas line and supply the gas via its onshore terminal to the particular national gas utility. In any case, the flaring of large quantities of gas except in abnormal circumstances is strictly controlled by national regulations.

Where the gas export facilities of a platform are nonexistent or restricted to the extent that oil production would be affected, it is necessary to reinject the gas as an acceptable alternative to flaring. It also has the advantage of helping to maintain well pressure and hopefully may be largely recoverable at a later date.

A large compressor train of two or three stages is normally required to obtain the gas flow rates and pressures necessary for reinjection. A typical scheme would be of two stages utilising axial compressors driven by large induction motors, followed by a third-stage reciprocating compressor driven by a synchronous motor to ease voltage and frequency control problems for the platform generation. The problems of large process loads such as these, when powered from a small number of generators, isolated from a national grid network, are discussed in PART 1 Chapter 2.

OIL STORAGE AND EXPORT

Oil that has been through the platform processes may be exported immediately via large main oil line pumps or, where no export pipeline exists, may be stored in the subsea cells of the platform structure before offloading by a tanker. As the oil from the particular field may be above the average value of the oil from a shared export pipe network, it may be economically advantageous to export by tanker as well as by pipeline.

Where tanker offloading is carried out, a separate floating tanker loading facility is necessary, as it would be highly dangerous for the tanker to approach within normal flexible pipe loading distance of the production platform. These single buoy moorings may have their own storage and pumping facilities and are connected to the associated platform via subsea pipelines.

ARTIFICIAL LIFT FACILITIES

The output from a typical oil field will rise to a maximum after all the production wells have been completed. There will then be a plateau in the output figures for, hopefully, a number of years before outputs start to decline. The timescales involved vary greatly with different fields and even with different wells in the same field, particularly where a lot of deviation drilling has been necessary. In order to maintain the oil output, submersible downhole pumps may be inserted down wells, or gas may be injected and released low down in the well to reduce the density of the oil. Both methods tend to increase oil flow rates and are known collectively as artificial lift facilities.

A.4 OIL COMPANY OPERATIONS

FINANCING

As the cost of recovering oil from a field in the North Sea, from exploration to production can easily exceed several £1000 million, it is usually necessary for several oil companies and financial institutions to collaborate in the project. Typical costs and employee requirements are shown in Table A.4.1.

It should be noted that the decline in output may be arrested or at least slowed by the use of artificial lift facilities. As the average cost of drilling a single exploration well is £3 million, a large part of £100 million may be spent on exploring a potential field before a decision is made to go ahead with the production facility.

DIVISION OF LABOUR

When working on oil company projects, it is particularly necessary to be aware of the company management structure with respect to project funding and therefore it is mentioned briefly here.

Most oil companies have four main management streams: operations, maintenance, projects and technical facilities. Those who take responsibility for the production plant, the asset holders, are the operators. The maintenance department may be refused access to an item of plant due for maintenance if it is still required by the operations department to maintain full oil production.

Table A.4.1 Resources Required for Recovery of a Typical Oil Field

	Exploration	Construction	Production	Abandonment
Programme	Survey of 2000–4000 km^2	Planning/design	Rising output 3–5 years	Planning
	Exploration and appraisal drilling of 5–30 wells	Construction of platform structure, production and accommodation facilities	Plateau 5 years	Prepare topsides for removal Phased removal of topsides and support structure
		Drilling of production wells	Decline 8–10 years	Provision of navigation aids for remaining structure
		Construction of transport facilities		
Time	2–6 years	5–6 years	16–20 years	2 years
Direct labour	200–400 People	1000–2000 People	300–400 People	100 People
Capital outlay	£20–100 Million	£1000 Million	£250 Million (depends on the type of secondary recovery scheme, e.g., artificial lift or other operations necessary)	£30–£100 Million (assumes some outlay on heavy lift vessels and accommodation barge, but using platform accommodation as much as possible)
Operating costs			£250–500 Million	N/A (navaids for remnant structure will require servicing)

The operations department may also refuse to accept new plant from a project construction team if they feel it is not performing adequately. The technical facilities department act only in an advisory capacity on matters of design and safety, although there is usually a *separate* safety department that carries out hazard and operability (hazop) studies and liaise with national inspectorates and shipping bureaux.

Typical Commissioning Test Sheets

PLATFORM:	FIELD:	PLANT NO:	SHEET OF
TAG NO:	SERIAL NO:	LOCATION:	SYSTEM:
TYPE:	P.O. NO:	DRG NO:	MANUFR:
VOLTAGE:	FREQUENCY:	RATING	DATE:
TESTER:	TEST COMPLETED:	SIGNED:	APPROVED:

SWITCHBOARDS AND BUSBARS

BUSBAR INSULATION RESISTANCE TEST RESULTS

BETWEEN	R-Y	R-B	Y-B	R-N	Y-N	B-N	RYB-E	N-E
IR (MEGOHMS)								

BUSBAR PRESSURE TEST RESULTS

BETWEEN	RY-B-E	RB-Y-E	RYB-E	RYB-N	N-E
VOLTS AC/DC					
LEAKAGE AMPS					
DURATION MINUTES					

BUSBAR RESISTANCE MEASURED WITH A DUCTOR
SEE ATTACHED DRAWING OF BUSBARS SHOWING BUSBAR JOINTS

PHASE	MILLIOHMS	NO OF JOINTS
RED		
YELLOW		
BLUE		
NEUTRAL		
EARTH PATH RESISTANCE		

VOLTAGES

CLOSING SUPPLY VOLTAGE		VOLTS
TRIPPING SUPPLY VOLTAGE		VOLTS

TEST EQUIPMENT

DEVICE	SERIAL NO	DEVICE	SERIAL NO

BUSBAR PHASE ROTATION ....

IR TEST ON ANTI-CONDENSATION HEATERS .... MEGOHMS

Sheet 1: Busbar testing.

PLATFORM:	FIELD:	PLANT NO:	SHEET OF
TAG NO:	SERIAL NO:	LOCATION:	SYSTEM:
TYPE:	P.O. NO:	DRG NO:	MANUFR:
VOLTAGE:	FREQUENCY:	RATING	
TESTER:	TEST COMPLETED:	SIGNED:	APPROVED:

POWER TRANSFORMER

AREA CLASSIFICATION	LOCATION
PRIMARY VOLTS	MANUFACTURER
SECONDARY VOLTS	RATING
COOLING	IMPEDANCE
VECTOR GROUP	FRQUENCY

WINDING RESISTANCE TEST RESULTS

BETWEEN	HV(R-Y)	HV(Y-B)	HV(B-R)	LV(R-Y)	LV(Y-B)	LV(B-R)
RESISTANCE						

WINDING INSULATION TEST RESULTS

	BEFORE	PRESSURE	TEST	AFTER	PRESSURE	TEST
BETWEEN	HV-EARTH	LV-EARTH	HV-LV	HV-EARTH	LV-EARTH	HV-LV
TEST VOLTS						
MEGOHMS						

WINDING PRESSURE TEST RESULTS

PRIMARY TEST VOLTAGE(kV)	DURATION (MINS)	LEAKAGE(mA)
SECONDARY TEST VOLTAGE(kV)		

MANUFACTURERS TEST CERTIFICATE	NO	POLARITY AND PHASE ROTATION	
OIL TEST CERTIFICATE	NO	TEMPERATURE TRIP OPERATIONAL	SETTING
RATIO AND MAGNETISING TEST CERT.	NO	TEMPERATURE ALARM OPERATIONAL	SETTING
FEEDER CIRCUIT BREAKER TEST CERT.	NO	OVER PRESSURE DEVICE	SETTING
HV AND LV CABLE TEST CERT.	NO		

TAP CHANGE MAGNETISING CURRENT RESULTS

	HV	WINDINGS		LV	WINDINGS	
	CONSTANT MAGNETIZING	AT CURRENT	VOLTS	SECONDARY	VOLTS	
TAP POSITION	R	Y	B	R-Y	Y-B	B-R
1						
2						
3						
4						
5						

Sheet 2: Transformer testing.

PLATFORM:	FIELD:	PLANT NO:	SHEET OF
TAG NO:	SERIAL NO:	LOCATION:	SYSTEM:
TYPE:	P.O. NO:	DRG NO.	MANUFR:
VOLTAGE:	FREQUENCY:	RATING	
TESTER:	TEST COMPLETED:	SIGNED:	APPROVED:

INSPECTION TEST RECORD	CIRCUIT BREAKER SECONDARY INJECTION

CUBICLE REFERENCE			
OVERCURRENT/EF RELAY		CT RATIO	
RELAY MANUFACTURER		TYPE	
SERIAL NUMBER		COIL RATING	
CHARACTERISTIC		PROTECTION	

INJ		SETTING		RELAY		TIME		RESET	
COIL	AMPS	PSM	TM	AMPS	ERROR %	CURVE	ACTUAL	AMPS	TIME
R									
Y									
B									
N									

CIRCUIT RATING		
FULL LOAD CURRENT		
SETTING EF	PSM	TM
SETTING OC	PSM	TM

MOTOR PROTECTION RELAY			
RELAY MANUFACTURER		TYPE	
SERIAL NUMBER		COIL RATING	
CHARACTERISTIC		PROTECTION	

TEST	AMPS	PLUG SET	CURVE	ACTUAL	RESET (SECS)
START CURVE					
RUNNING CURVE					
RUNNING CURVE					
EARTH FAULT					
INSTANTANEOUS TRIP					

FULL LOAD CURRENT		% LOAD TO TRIP	
PLUG SETTING		INST TRIP	

COMMENTS/EXCEPTIONS

Sheet 3: Circuit breaker testing.

PLATFORM:	FIELD:	PLANT NO:	SHEET OF
TAG NO:	SERIAL NO:	LOCATION:	SYSTEM:
TYPE:	P.O. NO:	DRG NO:	MANUFR:
VOLTAGE:	FREQUENCY:	RATING	
TESTER:	TEST COMPLETED:	SIGNED:	APPROVED:

INSPECTION TEST RECORD	ELECTRIC MOTOR

MOTOR NUMBER			
AREA CLASSIFICATION			
FRAME SIZE		CONNECTION STAR ☐ DELTA ☐	
LUBRICATION		ROT VIEWED FROM DE CW ☐ CCW ☐	
VOLTAGE	SPEED	R.P.M.	CURRENT F.L.C. AMPS
RATING KW	FREQUENCY	H.Z.	O/L SETTING AMPS
FED FROM			

APPROPRIATE FORM COMPLETED FOR EX MOTORS

SIGNED DATE

WINDING RESISTANCE

POLARISATION INDEX MEGOHMS AFTER 1 MIN MEGOHMS AFTER 10 MINS

BEARING INSULATION RESISTANCE

ALL ALARM CIRCUITS TESTED

REMOTE CONTROL CIRCUITS TESTED

ANTI-CONDENSATION HEATER CIRCUIT TESTED

CIRCUIT BREAKER/CONTACTOR TEST CERTIFICATES AVAILABLE

CABLES TEST CERTIFICATES AVAILABLE

TIME	INSULATION RESISTANCE	3-PHASE CURRENT IN AMPS			BEARING TEMPERATURE		WINDING TEMPERATURE	MOTOR COOLING	MOTOR SPEED	VIBRATION
	MEGOHMS	B	W	R	FRONT	REAR			R.P.M.	
UNCOUPLED										
ON LOAD										

COMMENTS/EXCEPTIONS

Sheet 4: Electric motor testing.

CONTACTOR STARTER	INSPECTION TEST RECORD

PLATFORM:	FIELD:	PLANT NO:	SHEET OF
TAG NO:	SERIAL NO:	LOCATION:	SYSTEM:
TYPE:	P.O. NO:	DWG NO:	
VOLTAGE:	FREQUENCY:	RATING	
TESTER:	TEST COMPLETED:	SIGNED:	APPROVED:

MANUFACTURER		COMPARTMENT No	
MCC No.		DRIVE ALLOCATION	
OVERCURRENT TRIP SETTING	AMPS	TIMER T . . . . SETTING	SECS
DRIVE FULL LOAD CURRENT	AMPS	CONTROL FUSE SIZE	AMPS
MOTOR FUSE SIZE	AMPS	CONTROL VOLTAGE	VOLTS
SYSTEM VOLTAGE	VOLTS		

SIGNED . . . DATE . . . .

INSULATION RESISTANCE . . . . MEGOHMS TEST VOLTAGE . . . . VOLTS

CONTROL CIRCUITS CORRECT TO SCHEMATIC DIAGRAM No. . . . . REV . . . .

REMOTE CONTROL AND ANNUCIATOR TESTED AND CORRECT . . . .

ANTI-CONDENSATION HEATER CIRCUIT TESTED . . . .

THERMAL OVERLOAD TEST RESULTS					
CURRENT		TRIP TIME (SECS)			REMARKS
% FLC	AMPS	R	Y	B	
115					NO TRIPS
					MIN. TRIP AMPS
200					
600					

SINGLE PHASING TEST RESULTS (80% FLC)			
CIRCUIT	R-Y	R-B	Y-B
AMPS			
TRIP TIME (SECS)			

EARTH FAULT TEST RESULTS			
PHASE	R	Y	B
MIN. AMPS TO OPERATE			

CT RATIO PROTECTION . . . .

CT RATIO METERING . . . .

AMMETER TEST RESULTS	
LOCATION TEST AMPS	
READING	
READING	
READING	

COMMENTS/EXCEPTIONS

Sheet 5: Contactor starter testing.

APPENDIX

Comparison of Hazardous Area Equipment Certification

APPENDIX C1: COMPARISON OF EU AND US CERTIFICATION FOR HAZARDOUS AREAS

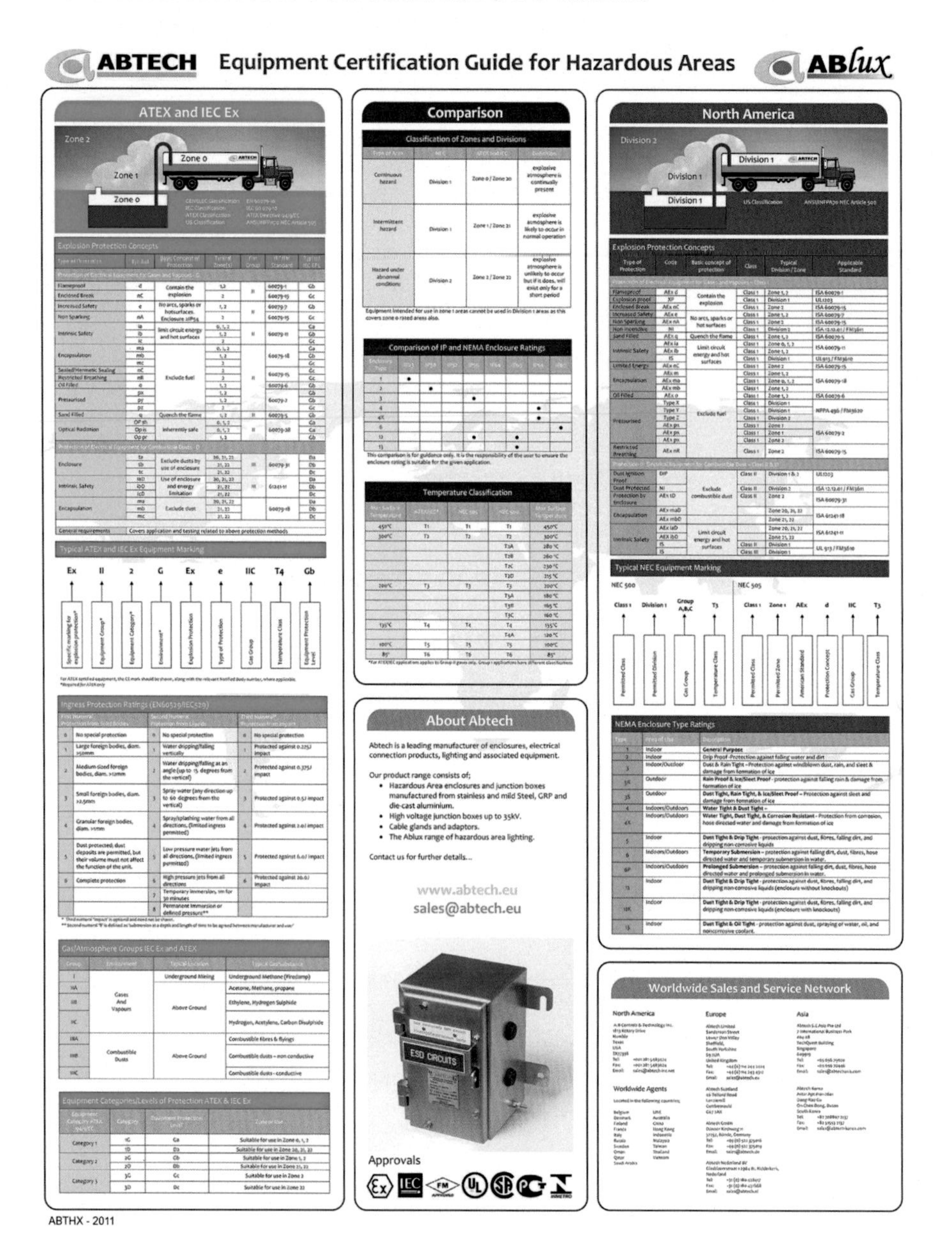

APPENDIX C2: HAZARDOUS AREA GUIDE FOR ATEX & IECEx (ZONES/GROUPS)

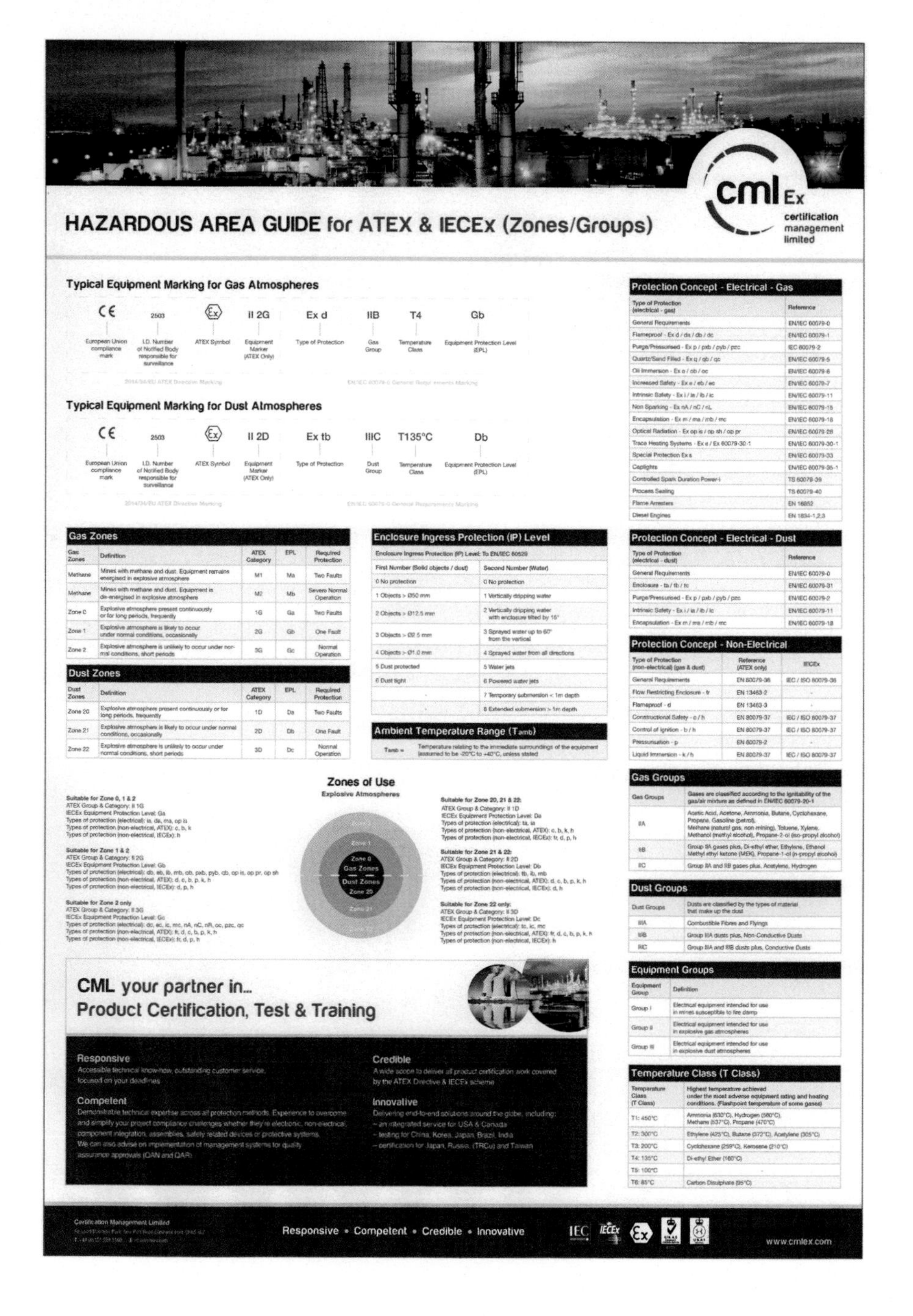

HAZARDOUS AREA GUIDE for ATEX & IECEx (Zones/Groups)

cml Ex certification management limited

Typical Equipment Marking for Gas Atmospheres

CE	2503	Ex	II 2G	Ex d	IIB	T4	Gb
European Union compliance mark	I.D. Number of Notified Body responsible for surveillance	ATEX Symbol	Equipment Marker (ATEX Only)	Type of Protection	Gas Group	Temperature Class	Equipment Protection Level (EPL)

2014/34/EU ATEX Directive Marking — EN/IEC 60079-0 General Requirements Marking

Typical Equipment Marking for Dust Atmospheres

CE	2503	Ex	II 2D	Ex tb	IIIC	T135°C	Db
European Union compliance mark	I.D. Number of Notified Body responsible for surveillance	ATEX Symbol	Equipment Marker (ATEX Only)	Type of Protection	Dust Group	Temperature Class	Equipment Protection Level (EPL)

2014/34/EU ATEX Directive Marking — EN/IEC 60079-0 General Requirements Marking

Gas Zones

Gas Zones	Definition	ATEX Category	EPL	Required Protection
Methane	Mines with methane and dust. Equipment remains energised in explosive atmosphere	M1	Ma	Two Faults
Methane	Mines with methane and dust. Equipment is de-energised in explosive atmosphere	M2	Mb	Severe Normal Operation
Zone 0	Explosive atmosphere present continuously or for long periods, frequently	1G	Ga	Two Faults
Zone 1	Explosive atmosphere is likely to occur under normal conditions, occasionally	2G	Gb	One Fault
Zone 2	Explosive atmosphere is unlikely to occur under normal conditions, short periods	3G	Gc	Normal Operation

Dust Zones

Dust Zones	Definition	ATEX Category	EPL	Required Protection
Zone 20	Explosive atmosphere present continuously or for long periods, frequently	1D	Da	Two Faults
Zone 21	Explosive atmosphere is likely to occur under normal conditions, occasionally	2D	Db	One Fault
Zone 22	Explosive atmosphere is unlikely to occur under normal conditions, short periods	3D	Dc	Normal Operation

Enclosure Ingress Protection (IP) Level

Enclosure Ingress Protection (IP) Level: To EN/IEC 60529

First Number (Solid objects / dust)	Second Number (Water)
0 No protection	0 No protection
1 Objects > Ø50 mm	1 Vertically dripping water
2 Objects > Ø12.5 mm	2 Vertically dripping water with enclosure tilted by 15°
3 Objects > Ø2.5 mm	3 Sprayed water up to 60° from the vertical
4 Objects > Ø1.0 mm	4 Sprayed water from all directions
5 Dust protected	5 Water jets
6 Dust tight	6 Powered water jets
-	7 Temporary submersion < 1m depth
-	8 Extended submersion > 1m depth

Ambient Temperature Range (Tamb)

Tamb =	Temperature relating to the immediate surroundings of the equipment (assumed to be -20°C to +40°C, unless stated

Zones of Use

Explosive Atmospheres

Suitable for Zone 0, 1 & 2
ATEX Group & Category: II 1G
IECEx Equipment Protection Level: Ga
Types of protection (electrical): ia, da, ma, op is
Types of protection (non-electrical, ATEX): c, b, k
Types of protection (non-electrical, IECEx): h

Suitable for Zone 1 & 2
ATEX Group & Category: II 2G
IECEx Equipment Protection Level: Gb
Types of protection (electrical): db, eb, ib, mb, ob, pxb, pyb, qb, op is, op pr, op sh
Types of protection (non-electrical, ATEX): d, c, b, p, k, h
Types of protection (non-electrical, IECEx): d, p, h

Suitable for Zone 2 only
ATEX Group & Category: II 3G
IECEx Equipment Protection Level: Gc
Types of protection (electrical): dc, ec, ic, mc, nA, nC, nR, oc, pzc, qc
Types of protection (non-electrical, ATEX): fr, d, c, b, p, k, h
Types of protection (non-electrical, IECEx): fr, d, p, h

Suitable for Zone 20, 21 & 22:
ATEX Group & Category: II 1D
IECEx Equipment Protection Level: Da
Types of protection (electrical): ta, ia
Types of protection (non-electrical, ATEX): c, b, k, h
Types of protection (non-electrical, IECEx): fr, d, p, h

Suitable for Zone 21 & 22:
ATEX Group & Category: II 2D
IECEx Equipment Protection Level: Db
Types of protection (electrical): tb, ib, mb
Types of protection (non-electrical, ATEX): d, c, b, p, k, h
Types of protection (non-electrical, IECEx): d, h

Suitable for Zone 22 only:
ATEX Group & Category: II 3D
IECEx Equipment Protection Level: Dc
Types of protection (electrical): tc, ic, mc
Types of protection (non-electrical, ATEX): fr, d, c, b, p, k, h
Types of protection (non-electrical, IECEx): h

Protection Concept - Electrical - Gas

Type of Protection (electrical - gas)	Reference
General Requirements	EN/IEC 60079-0
Flameproof - Ex d / da / db / dc	EN/IEC 60079-1
Purge/Pressurised - Ex p / pxb / pyb / pzc	IEC 60079-2
Quartz/Sand Filled - Ex q / qb / qc	EN/IEC 60079-5
Oil Immersion - Ex o / ob / oc	EN/IEC 60079-6
Increased Safety - Ex e / eb / ec	EN/IEC 60079-7
Intrinsic Safety - Ex i / ia / ib / ic	EN/IEC 60079-11
Non Sparking - Ex nA / nC / nL	EN/IEC 60079-15
Encapsulation - Ex m / ma / mb / mc	EN/IEC 60079-18
Optical Radiation - Ex op is / op sh / op pr	EN/IEC 60079-28
Trace Heating Systems - Ex e / Ex 60079-30-1	EN/IEC 60079-30-1
Special Protection Ex s	EN/IEC 60079-33
Caplights	EN/IEC 60079-35-1
Controlled Spark Duration Power-i	TS 60079-39
Process Sealing	TS 60079-40
Flame Arresters	EN 16852
Diesel Engines	EN 1834-1,2,3

Protection Concept - Electrical - Dust

Type of Protection (electrical - dust)	Reference
General Requirements	EN/IEC 60079-0
Enclosure - ta / tb / tc	EN/IEC 60079-31
Purge/Pressurised - Ex p / pxb / pyb / pzc	EN/IEC 60079-2
Intrinsic Safety - Ex i / ia / ib / ic	EN/IEC 60079-11
Encapsulation - Ex m / ma / mb / mc	EN/IEC 60079-18

Protection Concept - Non-Electrical

Type of Protection (non-electrical) (gas & dust)	Reference (ATEX only)	IECEx
General Requirements	EN 80079-36	IEC / ISO 80079-36
Flow Restricting Enclosure - fr	EN 13463-2	-
Flameproof - d	EN 13463-3	-
Constructional Safety - c / h	EN 80079-37	IEC / ISO 80079-37
Control of ignition - b / h	EN 80079-37	IEC / ISO 80079-37
Pressurisation - p	EN 60079-2	-
Liquid Immersion - k / h	EN 80079-37	IEC / ISO 80079-37

Gas Groups

Gas Groups	Gases are classified according to the ignitability of the gas/air mixture as defined in EN/IEC 60079-20-1
IIA	Acetic Acid, Acetone, Ammonia, Butane, Cyclohexane, Propene, Gasoline (petrol), Methane (natural gas, non-mining), Toluene, Xylene, Methanol (methyl alcohol), Propane-2-ol (iso-propyl alcohol)
IIB	Group IIA gases plus, Di-ethyl ether, Ethylene, Ethanol Methyl ethyl ketone (MEK), Propane-1-ol (n-propyl alcohol)
IIC	Group IIA and IIB gases plus, Acetylene, Hydrogen

Dust Groups

Dust Groups	Dusts are classified by the types of material that make up the dust
IIIA	Combustible Fibres and Flyings
IIIB	Group IIIA dusts plus, Non-Conductive Dusts
IIIC	Group IIIA and IIIB dusts plus, Conductive Dusts

Equipment Groups

Equipment Group	Definition
Group I	Electrical equipment intended for use in mines susceptible to fire damp
Group II	Electrical equipment intended for use in explosive gas atmospheres
Group III	Electrical equipment intended for use in explosive dust atmospheres

Temperature Class (T Class)

Temperature Class (T Class)	Highest temperature achieved under the most adverse equipment rating and heating conditions. (Flashpoint temperature of some gases)
T1: 450°C	Ammonia (630°C), Hydrogen (560°C), Methane (537°C), Propane (470°C)
T2: 300°C	Ethylene (425°C), Butane (372°C), Acetylene (305°C)
T3: 200°C	Cyclohexane (259°C), Kerosene (210°C)
T4: 135°C	Di-ethyl Ether (160°C)
T5: 100°C	-
T6: 85°C	Carbon Disulphate (95°C)

Bibliography

PART 2 CHAPTER 2

REFERENCES

- The Performance and Design of Alternating Current Machines, Transformers, Three-Phase Induction Motors and Synchronous Machines: Prof. M.G. Say PhD., MIEE
- Marine Electrical Equipment and Practice; Second edition H.D. McGeorge, CEng, FIMarE, MRINA
- Hughes Electrical and Electronic Technology 11th Edition
- Schaum's Outline of Electric Machines & Electromechanics (Schaum's Outline Series): Second Edition: Syed A. Nasar.

WEB REFERENCES

Please note that the manufacturers of smaller generators are covered in PART 2 Chapter 3 and asynchronous machines in PART 6 Chapter 6. The list below is restricted to companies manufacturing/packaging medium voltage ac generators.

#	Manufacturer	Website
1	ABB Group	http://www.abb.com/product/seitp322/9e2b171b82941136832575a5006c250b.aspx?productLanguage=us&country=GB
2	Brush Group	http://www.brush.eu/en/37/BRUSH-Group/Offerings/Generators
3	Caterpillar	http://www.cat.com/en_US/products/new/power-systems/oil-and-gas.html
4	Emerson Kato	http://www.emersonindustrial.com/en-EN/Electric-Power-Generation/Products/kato-generators/highvoltage/Pages/default.aspx
5	GE Power Generation	https://powergen.gepower.com/products/generators.html
6	Jeumont Electric	http://www.jeumontelectric.com/alternateurs/?lang=en
7	Marathon	http://www.marathonelectric.com/MGPS/standard.jsp
8	MTU	http://www.mtu-online.com/mtu/applications/oil-gas/mtu-systems-offshore/?L=%3Fr%3DPb9dqKNh
9	Parsons Peebles	http://www.parsons-peebles.com/markets/offshore-oil-gas
10	Siemens	https://www.siemens.com/global/en/home/company/about/businesses/power-gas.html
11	Solar Turbines	https://mysolar.cat.com/en_US/products/oil-and-gas/gas-turbine-packages.html
12	Wärtsilä	http://www.wartsila.com/oil-gas

PART 2 CHAPTER 3
REFERENCES

#	Title	Author	Website
1	NFPA 20: Standard for the Installation of Stationary Pumps for Fire Protection	NFPA (US)	http://www.nfpa.org/codes-and-standards/all-codes-and-standards/list-of-codes-and-standards
2	NFPA 37: Standard for the Installation and Use of Stationary Combustion Engines and Gas Turbines	NFPA (US)	http://www.nfpa.org/codes-and-standards/all-codes-and-standards/list-of-codes-and-standards
3	Offshore Installations (Prevention of Fire and Explosion, and Emergency Response) Regulations 1995	HSE (UK)	http://www.hse.gov.uk/pubns/books/l65.htm
4	NORSOK E-001:2016: Electrical Systems [plus associated Excel file EDS002]	NORSOK (Norway)	http://www.standard.no/en/sectors/energi-og-klima/petroleum/norsok-standards/
5	BS EN 1834-1: Reciprocating internal combustion engines – Safety requirements for design and construction of engines for use in potentially explosive atmospheres.	BSI	http://shop.bsigroup.com/Navigate-by/Standards/
6	IEC 61892:2015	IEC	https://webstore.iec.ch/publication/6083
7	IEC 61892: Mobile and fixed offshore units – Electrical installations	IEC	https://webstore.iec.ch/publication/6083
8	BS ISO 8528-2016: Reciprocating internal combustion engine driven alternating current generating sets.	BSI (UK)	http://shop.bsigroup.com/Navigate-by/Standards/
9	DNVGL-OS-A101: Safety principles and arrangements	DNV-GL	https://www.dnvgl.com/rules-standards/index.html
10	DNVGL-OS-D201: Electrical installations		
11	63 Facilities On Offshore Installations: 2017	ABS	http://ww2.eagle.org/en/rules-and-resources/rules-and-guides.html
12	LR Rules for Offshore Units	LR	http://www.lr.org/en/services/classification-for-oil-and-gas.aspx

USEFUL LINKS

#	Emergency Generator Manufacturer/Packager	Website
1	F.G. Wilson (Caterpillar)	https://www.fgwilson.com/en_GB/products.html
2	Finnings (Caterpillar)	http://www.finning.com/en_GB/industries/oil-gas/generators/emergency.html
3	Gen Ex	http://www.genexdesign.com/power_generators/
4	Interpower	http://www.interpower.co.uk/offshore-generator.html
5	kVA	http://www.kva.co.uk/sales/Onshore-Aux-Emerg-Generators.htm
6	Quantum Offshore	http://www.quantumoffshore.co.uk/power-generation/emergency-generator-ex-cooling
7	Stewart & Stevenson	http://www.stewartandstevenson.com/markets/oil-gas/oilfield-power-generation/oilfield-emergency-power

PART 2 CHAPTER 4

REFERENCES

#	Title	Author	Website
1	NFPA 20: Standard for the Installation of Stationary Pumps for Fire Protection	NFPA (US)	http://www.nfpa.org/codes-and-standards/all-codes-and-standards/list-of-codes-and-standards
2	NFPA 37: Standard for the Installation and Use of Stationary Combustion Engines and Gas Turbines	NFPA (US)	http://www.nfpa.org/codes-and-standards/all-codes-and-standards/list-of-codes-and-standards
3	Offshore Installations (Prevention of Fire and Explosion, and Emergency Response) Regulations 1995	HSE (UK)	http://www.hse.gov.uk/pubns/books/l65.htm
4	NORSOK E-001:2016: Electrical Systems [plus associated Excel file EDS002]	NORSOK (Norway)	http://www.standard.no/en/sectors/energi-og-klima/petroleum/norsok-standards/

Continued

—Continued

#	Title	Author	Website
5	BS ISO 19859:2016: Gas turbine applications. Requirements for power generation	BSI	http://shop.bsigroup.com/Navigate-by/Standards/
6	BS EN 60034-3:2008: Rotating electrical machines. Specific requirements for synchronous generators driven by steam turbines or combustion gas turbines		
7	IEC 61892: 2015 Mobile and fixed offshore units – Electrical installations	IEC	https://webstore.iec.ch/publication/6083
8	DNVGL-OS-A101: Safety principles and arrangements	DNV-GL	https://www.dnvgl.com/rules-standards/index.html
9	DNVGL-OS-D201: Electrical installations		
10	63 Facilities On Offshore Installations: 2017	ABS	http://ww2.eagle.org/en/rules-and-resources/rules-and-guides.html
11	LR Rules for Offshore Units	LR	http://www.lr.org/en/services/classification-for-oil-and-gas.aspx

USEFUL LINKS

#	Gas Turbine Generator Manufacturer/ Packager	Website
1	ABB Power – Generation	http://new.abb.com/power-generation/systems/turbine-control/gas-turbine-control
2	GE Power Generation	https://powergen.gepower.com/
3	Mitsubishi Hitachi Power Systems (MHPS)	http://www.mpshq.com/gas-turbines.html
4	Siemens Gas Turbine Power Plant	https://www.siemens.com/global/en/home/products/energy/power-generation/gas-turbines/offshore-floating-production.html
5	Solar Turbines	https://mysolar.cat.com/en_US/products/power-generation.html
6	Wartsila, Helsinki, Finland	https://www.wartsila.com/energy/learning-center/technical-comparisons/gas-turbine-for-power-generation-introduction

PART 2 CHAPTER 5

REFERENCES

#	Title	Author
1	The J. & P. Switchgear Book 7th Edition 1972	R.T. Lyall/C.A. Worth
2	Distribution Switchgear	Stan Stewart
3	The Fundamentals of Circuit Breaker & Protection Maintenance: Volume 1 (Switchgear Maintenance)	John Dunning
4	Introduction to Switchgear (High Voltage Engineering Series)	Henry E. Poole
5	Switching in Electrical Transmission and Distribution Systems: 2014	Rene Smeets/Lou Van Der Sluis
6	Switchgear Operation and Maintenance for Process Plants (Process Plant Maintenance Book 1) Kindle Edition	Wayne Smith

WEB REFERENCES

#	Switchgear Manufacturer	Web Page
1	ABB	http://new.abb.com/medium-voltage/switchgear
2	ABB (Current Limiters)	http://www.abb.com/product/db0003db004279/c125739900636470c1257057003259d1.aspx#!
3	Mitsubishi Electric	http://www.mitsubishielectric.com/bu/powersystems/products/switchgear/mv_switchgear/index.html
4	Powell	https://www.powellind.com/ProductsServices/Pages/Medium-Voltage-Type-298.aspx#tab-ProdInfo
5	Schneider Electric	http://www.schneider-electric.com/b2b/en/products/medium-voltage-switchgear-and-energy-automation.jsp
6	Siemens	http://w3.usa.siemens.com/powerdistribution/us/en/product-portfolio/Medium-Voltage-Switchgear/Pages/Medium-Voltage-Switchgear.aspx

PART 2 CHAPTER 6

REFERENCES

#	Title	Author
1	Marine Electrical Practice, 6th Edition (Revised by BP Shipping)	G.O. Watson CEng, FIEE, FIEEE, FIMarE
2	Marine Electrical Equipment and Practice (2nd Edition)	H.D. McGeorge CEng, FIMarE, MRINA
3	Schaum's Outline Series – Electrical Machines and Electromechanics 2nd edition	Syed Nasar PhD.

WEB REFERENCES

#	Company	Web address
1	Casey, USA	https://www.caseyusa.com/products/class_id/EDCGEN
2	Westinghouse Teco, USA	http://tecowestinghouse.com/Default.aspx
3	Parsons Peebles, Scotland, UK	http://www.parsons-peebles.com/

PART 2 CHAPTER 7

REFERENCES

#	Title	Author
1	Marine Electrical Practice Sixth edition	G.O. Watson CEng, FIEE, FIEEE, FIMarE
2	Marine Electrical Equipment and Practice Second edition	H.D. McGeorge, CEng, FIMarE, MRINA
3	Distribution Switchgear	S. Stewart

WEB REFERENCES

#	Company	Website
1	Eaton	http://www.eaton.com/Eaton/ProductsServices/Electrical/ProductsandServices/CircuitProtection/LVAirPower/Magnum/MagnumDCswitches/index.htm#tabs-2
2	GE Gerapid	http://www.geindustrial.com/products/switchgear/gerapid-dc-oem-module
3	Siemens	http://www.mobility.siemens.com/mobility/global/en/rail-solutions/rail-electrification/dc-traction-power-supply/dc-switchgears/Pages/overview.aspx
4	Powell	https://www.powellind.com/ProductsServices/Pages/DC-Switchgear.aspx
5	Mitsubishi	http://www.mitsubishielectric.com/bu/powersystems/products/switchgear/obsolete/index_09.html
6	ABB	http://www.abb.com/product/us/9AAC100800.aspx

PART 2 CHAPTER 8

REFERENCES

#	Title	Author
1	IEC 60079: Explosive atmospheres	International Electrotechnical Commission (IEC)
2	IEC 61892: Mobile and fixed offshore units – Electrical installations	
3	DNVGL-OS-D201: Electrical Installations	DNVGL
4	Facilities on Offshore Installations: 2016	ABS
5	IEE Recommendations for the Electrical and Electronic Equipment of Mobile and Fixed Offshore Installations	IET
6	UL913 Standard for Intrinsically Safe Apparatus and Associated Apparatus for Use in Class I, II, III, Division 1, Hazardous (Classified) Locations	UL
7	Rules and Regulations for the Classification of Offshore Units: July 2016	LR
8	1580-2010: IEEE Recommended Practice for Marine Cable for Use on Shipboard and Fixed or Floating Facilities	IEEE
9	NEK TS 606:2016: Cables for offshore installations – Halogen-free low smoke and flame-retardant/fire-resistant (HFFR-LS)	Norsk Standard

WEB REFERENCES

#	Company/ Organisation	Website
1	IEC	https://webstore.iec.ch/?ref=menu
2	IET	http://www.theiet.org/resources/standards/index.cfm
3	IEEE	http://standards.ieee.org/cgi-bin/lp_index?type=standard&coll_name=power_and_energy&status=active
4	DNVGL	https://rules.dnvgl.com/ServiceDocuments/dnvgl/#!/industry/1/Maritime/9/DNV%20GL%20offshore%20standards%20(OS)
5	ABS	http://ww2.eagle.org/en/rules-and-resources/rules-and-guides.html
6	UL	https://standardscatalog.ul.com/
7	LR	http://www.lr.org/en/services/rules-for-offshore-units-in-oil-and-gas.aspx
8	Norsk Standard	http://www.standard.no/en/
9	Draka Cables	http://www.draka.nl/catalogue/marine-offshore-cables/7000
10	Anixter	https://www.anixter.com/en_uk/products/wire-and-cable.html
11	Nexans	http://www.nexans.com/eservice/Corporate-en/navigate_322326_-4495_20_14078/Products_Solutions.html

PART 2 CHAPTER 9

REFERENCES

#	Title	Author
1	Prevention of fire and explosion, and emergency response on offshore installations (PFEER)	HSE
2	RP 14C4: Recommended Practicis, Design, Instale for Analyslation, and Testing of Basic Surface Safety Systems for Offshore Production Platforms	American Petroleum Institute
3	RP 14G: Recommended Practice for Fire Prevention and Control on Fixed Open-type Offshore Production Platforms	
4	NFPA 20: Standard for the Installation of Stationary Pumps for Fire Protection	US National Fire Protection Association
5	Troubleshooting Rotating Machinery: (Wiley 2016)	Robert X. Perez & Andrew P. Conkey

WEB REFERENCES

#	Organisation/Company	Website
1	UK Health & Safety Executive	http://www.hse.gov.uk/offshore/index.htm
2	API	http://www.api.org/products-and-services/standards
3	NFPA	http://www.nfpa.org/codes-and-standards
4	ABB Group	http://www.abb.com/product/seitp322/9e2b171b82941136832575a5006c250b.aspx?productLanguage=us&country=GB
5	Brush Group	http://www.brush.eu/en/37/BRUSH-Group/Offerings/Generators
6	GE Power Generation	https://powergen.gepower.com/products/generators.html
7	Jeumont Electric	http://www.jeumontelectric.com/alternateurs/?lang=en
8	Marathon	http://www.marathonelectric.com/MGPS/standard.jsp
9	MTU	http://www.mtu-online.com/mtu/applications/oil-gas/mtu-systems-offshore/?L=%3Fr%3DPb9dqKNh
10	Parsons Peebles	http://www.parsons-peebles.com/markets/offshore-oil-gas
11	Siemens	https://www.siemens.com/global/en/home/company/about/businesses/power-gas.html

—Continued

#	Organisation/Company	Website
12	Wärtsilä Hamworthy	http://www.wartsila.com/products/marine-oil-gas/pumps-valves/firefighting-pumps/wartsila-hamworthy-firewater-pump-packages
13	SPP Pumps	http://www.spppumps.com/divisions/oil-gas/electro-submersible-pumps/
14	Quantum Offshore	http://www.quantumoffshore.co.uk/

PART 2 CHAPTER 10

REFERENCES

#	Title	Author
1	Low-voltage handbook: Technical reference for switchgear, controlgear and distribution systems	Theodor Schmelcher (Published by Siemens Aktiengesellschaft (1984))
2	Electric Motor Control 10th Edition	Stephen Herman (Published by CENGAGE Delmar Learning (2014))
3	IEC 60092-60302:1997: Electrical installations in ships – Part 302: Low-voltage switchgear and controlgear assemblies	IEC
4	IEC 60947: Low-voltage switchgear and controlgear (2014)	
5	BS EN 62271-106:2011: High-voltage switchgear and controlgear. Alternating current contactors, contactor-based controllers and motor-starters	BSI
6	1683-2014: IEEE Guide for Motor Control Centers Rated up to and including 600V AC or 1000V DC with Recommendations Intended to Help Reduce Electrical Hazards	IEEE
7	NEMA ICS 18-2001 (R2007): three-phase 50 and 60Hz motor control centers rated not more than 600Vac.	NEMA

WEB REFERENCES

#	Organisation/Company	Website
1	BSI	https://www.bsigroup.com/en-GB/standards/
2	IEC	https://webstore.iec.ch/?ref=menu
3	IEEE	http://standards.ieee.org/findstds/standard/1683-2014.html

Continued

—Continued

#	Organisation/Company	Website
4	NEMA	https://www.nema.org/Standards/pages/default.aspx
5	ABB	http://new.abb.com/control-systems/system-800xa/electrical-control-system
6	Allen-Bradley/Rockwell Automation	http://ab.rockwellautomation.com/Motor-Control/Motor-Control-Centers/IEC-CENTERLINE-2500
7	Elecsis	http://www.elecsis.com/products/switchgear/motor-control-centres-(mccs)/
8	Siemens	http://manualzz.com/doc/8216176/siemens-tiastar%E2%84%A2-smart--intelligent--mcc-with-profibus
9	WEG	http://old.weg.net/uk/Products-Services/Electric-Panels/Control-Panels/Low-Voltage-Motor-Control-Center-LV-MCC

PART 2 CHAPTER 11
REFERENCES

#	Title	Author
1	Safety Related Telecommunication Systems on Fixed Offshore Installations Issue 1	Oil & Gas UK
2	CAP 437 Standards for Offshore Helicopter Landing Areas	UK CAA
3	Approved Code of Practice and guidance L65: Prevention of fire and explosion, and emergency response on offshore installations	UK Health and Safety Executive: (Third edition, 2016)
4	DNVGL-OS-A101: Safety principles and arrangements	DNV GL
5	DNVGL-OS-D201: Electrical installations	
6	DNVGL-OS-D202: Automation, safety and telecommunication systems	
7	ABS 63 Facilities On Offshore Installations: 2016	ABS
8	LR Rules and Regulations for the Classification of Offshore Units Part 6: Control and electrical engineering; Part 7: Safety systems, hazardous areas and fire;	LR
9	Variable Frequency Drives Theory, Application, And Troubleshooting	Howard W. Penrose University of Illinois at Chicago Energy Resources Center
10	Power Electronics: Converters, Applications, and Design	Ned Mohan and Tore M. Undeland: (John Wiley 2002)
11	Power Electronics in Motor Drives: Principles, Application and Design (Electrical Engineering)	Martin Brown (Elektor Electronics Publishing July 12, 2010)

WEB REFERENCES

#	Company/Organisation	Webpage
1a	ABB UPS	http://new.abb.com/ups/systems
1b	ABB Drilling Electrical Drives	http://new.abb.com/marine/systems-and-solutions/power-generation-and-distribution/drilling-drives-system
2	Cetronic	http://www.cetronicpower.com/products/acdc-secure-power-systems/
3	Dale	https://www.dalepowersolutions.com/home-page/ac-and-dc-secure-ups-solutions/
4	GE	http://www.geindustrial.com/products/uninterruptible-power-supplies-ups-three-phase
5	Gutor (Schneider Electric)	http://www.schneider-electric.com/en/product-range/61354-gutor-sdc/?parent-category-id=8000&parent-subcategory-id=88016
6	Harland Simon	http://www.harlandsimonups.com/marine-and-offshore-ups/
7	Power Systems International	http://www.powersystemsinternational.com/ups-systems-power-for-industry/

PART 2 CHAPTER 12
REFERENCES

#	Title	Author
1	Troubleshooting Rotating Machinery:	Robert Perez and Andrew Conkey (Wiley 2016)
2	Forsthoffer's Best Practice Handbook for Rotating Machinery	William Forsthoffer (Butterworth-Heinemann; 2011)
3	Electrical Submersible Pumps Manual: Design, Operations, and Maintenance	Gabor Takacs (Gulf Professional Publishing 2013)
4	NFPA 20: Standard for the Installation of Stationary Pumps for Fire Protection	US National Fire Protection Association
5	Application Guideline For Electric Motor Drive Equipment For Natural Gas Compressors	Marybeth G. Nored, Justin R. Hollingsworth, Klaus Brun (GMRC 2009)
6	Pipeline Pumping and Compression Systems: A Practical Approach	M. Mohitpour, K.K. Botros and T. Van Hardeveld (ASME 2008)
7	Natural Gas Compressors	PetroWiki (SPE International)

WEB REFERENCES

#	Organisation/Company	Website
1	NFPA	http://catalog.nfpa.org/NFPA-20-Standard-for-the-Installation-of-Stationary-Pumps-for-Fire-Protection-P1160.aspx
2	ABB	https://library.e.abb.com/public/7afdc73aa9256670c12578b5004a2ae5/COG%20brochure%20RevC_lowres.pdf
3	ATB Laurence Scott	http://www.laurence-scott.com/products/p-series/
4	Carmagen	http://www.carmagen.com/news/engineering_articles/news145.htm
5	GE Oil Nuovo Pignone	https://www.scribd.com/document/129868437/GE-Oil-Nuovo-Pignone
6	GE Induction Motors	http://www.gepowerconversion.com/product-solutions/induction-motors/mv-hv-cage-induction-motors
7	Jeumont Electric	http://www.jeumontelectric.com/produits/1055-2/?lang=en
8	Kawasaki Centrifugal Compressor	https://global.kawasaki.com/en/upload_pdf/pdf_machinery_A6_01.pdf
9	Siemens	http://www.industry.siemens.com/topics/global/en/pumps-fans-compressors/compressor-drives/Pages/Default.aspx

PART 2 CHAPTER 13

REFERENCES

#	Title	Author
1	RP 17A/ISO 13628-1:2005: Design and Operation of Subsea Production Systems—General Requirements and Recommendations	American Petroleum Institute (API)
2	ISO 13628-1:2005: Petroleum and natural gas industries – Design and operation of subsea production systems – Part 1: General requirements and recommendations	International Organization for Standardization (ISO)
3	ISO 13628-7:2005: Petroleum and natural gas industries – Design and operation of subsea production systems – Part 7: Completion/workover riser systems	
4	ISO 13628-6:2006: Petroleum and natural gas industries – Design and operation of subsea production systems – Part 6: Subsea production control systems	
5	ISO 13628-9:2000: Petroleum and natural gas industries – Design and operation of subsea production systems – Part 9: Remotely Operated Tool (ROT) intervention systems	
6	ISO 13628-15:2011: Petroleum and natural gas industries – Design and operation of subsea production systems – Part 15: Subsea structures and manifolds	

—Continued

#	Title	Author
7	U-001 Subsea production systems (Edition 4, October 2015)	NORSOK
8	Telemetry Systems Engineering (2002)	Frank Carden, Russ Jedlica (Artech House Communications Library)
9	Telemetry Systems Design (1995)	Frank Carden (Artech House Communications Library)
10	Cathodic Protection: Industrial Solutions for Protecting Against Corrosion	Volkan Cicek (Wiley 2013)
11	An Introduction to Impressed Current Cathodic Protection	J. Paul Guyer (Createspace 2014)

WEB REFERENCES

#	Organisation/Company	Website
1	ABB	http://new.abb.com/news/detail/1032/deep-water-dreams
2	Aker Solutions	http://akersolutions.com/what-we-do/products-and-services/subsea-trees/
3	Cameron Subsea	http://cameron.slb.com/products-and-services/subsea
4	Technip/FMC	http://www.fmctechnologies.com/en/SubseaSystems/Technologies/SubseaProductionSystems/SubseaTrees/EVDT.aspx
5	Emerson Subsea Measurement	http://www2.emersonprocess.com/en-US/brands/rosemount/subsea/Pages/index.aspx
6	GE Vetco Gray	https://www.geoilandgas.com/subsea-offshore/subsea-wellheads-specialty-connectors-pipe/subsea-wellheads
7	HMC Subsea	http://subsea-controls.com/
8	Siemens Subsea	http://w3.siemens.com/markets/global/en/oil-gas/Pages/subsea.aspx

PART 2 CHAPTER 14

REFERENCES

#	Title	Author
1	CAP 437: 8th Edition 9th December, 2016	UK Civil Aviation Authority
2	FAA Advisory Circular—Heliport Design: 150/5930	US FAA
3	SLL Lighting Handbook	CIBSE
4	Application Guide on Lighting in Hostile and Hazardous Environments (Currently withdrawn)	CIBSE
5	The offshore installations (prevention of fire and explosion, emergency response) regulations 1995 (PFEER)	UK Health & Safety Executive
5	Recommendation on The Marking of Man-Made Offshore Structures: O-139	IALA
6	DNV Flameproof Luminaire Standard MR-E010	DNV-GL
7	IEC60079-1:2007: "Explosive atmospheres – Part 1: Equipment protection by flameproof enclosures 'd'"	International Electrotechnical Commission
8	IEC 60598-1: Luminaires	

WEB REFERENCES

#	Organisation	Website
1	UK Civil Aviation Authority	http://publicapps.caa.co.uk/modalapplication.aspx?appid=11&mode=detail&id=523
2	US Federal Aviation Authority	https://www.faa.gov/airports/engineering/design_standards/
3	CIBSE	http://www.cibse.org/knowledge/cibse-publications
4	UK Health & Safety Executive	http://www.hse.gov.uk/offshore/index.htm
5	IALA-AISM	http://www.iala-aism.org/about-iala/basic documents/
6	Glamox	http://glamox.com/gmo/products/explosion-proof-lights-ul
7	Eaton CEAG/Crouse-Hinds	https://www.crouse-hinds.de/en/
8	Hubbell Victor Lighting	http://victorlightings.com/
9	Hubbell Chalmit Lighting	http://www.chalmit.com/Hazardous_Guide.html

PART 2 CHAPTER 15
REFERENCES

#	Title	Website
1	SI 2015 398: Offshore Installations (Offshore Safety Directive) (Safety Case etc.) Regulations 2015	http://www.legislation.gov.uk/uksi/2015/398/pdfs/uksi_20150398_en.pdf
2	Directive 2013/30/EU of the European Parliament and of The Council of 12 June 2013 on safety of offshore oil and gas operations and amending Directive 2004/35/EC	http://eur-lex.europa.eu/legal-content/en/ALL/?uri=CELEX%3A32013L0030
3	The Offshore Installations (Offshore Safety Directive) (Safety Case etc.) Regulations 2015. Guidance on Regulations	http://www.hse.gov.uk/pubns/books/l154.htm
4	Offshore Installations (Prevention of Fire and Explosion, and Emergency Response) Regulations 1995. Approved Code of Practice and guidance	http://www.hse.gov.uk/pubns/books/l65.htm
5	IEC 61508-1:2010: Functional safety of electrical/electronic/programmable electronic safety-related systems – Part 1: General requirements	https://webstore.iec.ch/publication/5515&preview=1
6	IEC 61511-1:2016: Functional safety – Safety instrumented systems for the process industry sector – Part 1: Framework, definitions, system, hardware and application programming requirements	https://webstore.iec.ch/publication/24241
7	ISA-TR84.00.02-2002: Part 1 Safety Instrumented Functions (SIF) – Safety Integrity Level (SIL) Evaluation Techniques Part 1: Introduction	https://www.isa.org/pdfs/microsites195/tr-8402p1/
8	HSE Guidance: Control Systems	http://www.hse.gov.uk/comah/sragtech/techmeascontsyst.htm
9	IEC Guidance: Functional safety and IEC 61508 A basic guide	http://www.ida.liu.se/~simna73/teaching/SCRTS/IEC61508_Guide.pdf

WEB REFERENCES

#	Author	Title	Website
1	B. Allen	Piper Alpha: condolences are not enough	https://www.healthandsafetyatwork.com/piper-alpha-lessons
2	F. Redmill	An Introduction to the Safety Standard IEC 61508	http://homepages.cs.ncl.ac.uk/felix.redmill/publications/4B.IEC%2061508%20Intro.pdf
3	G. Wild	API RP 14C has been updated	https://www.linkedin.com/pulse/api-rp-14c-has-been-updated-george-wild
4	Oil & Gas UK	Piper Alpha: Lessons Learnt, 2008	http://oilandgasuk.co.uk/wp-content/uploads/2015/05/HS048.pdf
5	ABB	Advanced Process Control, Optimization & Simulation	http://www.abb.com/industries/db0003db004061/4df43ae2d423f4c3c12573a7004d45d3.aspx
6	Emerson Process Management	DeltaV Digital Automation System System Overview	http://www2.emersonprocess.com/siteadmincenter/PM%20DeltaV%20Documents/Brochures/DeltaV-System-Overview-v11-Brochure.pdf
7	Honeywell Process Solutions	Offshore Oil and Gas Solutions Reduce Risk. Optimize ROI.	https://www.honeywellprocess.com/library/marketing/brochures/HPS_Discover_Oil_and_Gas_Offshore_Solutions.pdf

PART 2 CHAPTER 16

REFERENCES

#	Title	Author
1	BS EN 50588: Medium power transformers 50 Hz, with highest voltage for equipment not exceeding 36 kV	BSI
2	BS EN 60076-16:2011: Power transformers. Transformers for wind turbines applications	
3	DNVGL-ST-0076: Design of electrical installations for wind turbines	DNVGL
4	DNVGL-ST-0145: Offshore substations	
5	DNVGL-RP-0423: Manufacturing and commissioning of offshore substations	
6	Notice No.1 Classification of Offshore Units	LR

—Continued

#	Title	Author
7	Recommended Practices for Design, Deployment, and Operation of Offshore Wind Turbines in the United States	AWEA
8	J. & P. Transformer Book, 13th Edition	Martin Heathcote (Elsevier)
9	Transformer Inspection and Testing for Process Plants (Process Plant Maintenance Book 1) Kindle Edition	Wayne Smith

WEB REFERENCES

#	Organisation/ Company	Website
1	BSI	http://shop.bsigroup.com/
2	DNVGL	https://www.dnvgl.com/energy/generation/index.html
3	AWEA	http://www.awea.org/
4	ABB Distribution Transformers	http://new.abb.com/products/transformers/distribution
5	ABB Wind Energy	http://new.abb.com/products/transformers/by-customer-segment/wind/turbine-generators
6	GE Transformers, Prolec	http://www.prolecge.com/index.php/en/products/overview
7	Brush Transformers	http://www.brush.eu/en/38/Offerings/Transformers
8	OTDS	http://www.otds.co.uk/products/1/transformers
9	R. Baker (Electrical) Ltd.	https://www.rbaker.co.uk/services/three-phase-transformers/
10	Siemens	http://www.energy.siemens.com/hq/en/power-transmission/transformers/distribution-transformers/distribution/
11	Schneider Electric	http://www.schneider-electric.co.uk/en/product-category/88215-medium-voltage-transformers/?filter=business-6-medium-voltage-distribution-and-grid-automation

PART 2 CHAPTER 17
REFERENCES

#	Title	Author
1	Prevention of fire and explosion, and emergency response on offshore installation (PFEER) – Approved Code of Practice and guidance	HSE UK (Third Edition 2016)
2	SOLAS Consolidated Edition (2014)	IMO
3	The MODU Code (2010)	
4	Safety Related Telecommunications Systems on Fixed Offshore Installations, Issue 1	Oil & Gas UK
5	BS EN 60079: Explosive atmospheres	BSI UK
6	PD CLC/TR 50427: Assessment of inadvertent ignition of flammable atmospheres by radio-frequency radiation. Guide	
7	Automation, safety and telecommunication systems	DNV-GL (January 2017)
9	Facilities on Offshore Installations (2014)	ABS (2014)
10	IEC 60079: Explosive atmospheres	IEC
11	API RP 505: Recommended Practice for Classification of Locations for Electrical Installation at Petroleum Facilities Classified as Class I, Zone 0, Zone 1 and Zone 2 (2002).	API
12	NFPA 497: Recommended Practice for the Classification of Flammable Liquids, Gases, or Vapors and of Hazardous (Classified) Locations for Electrical Installations in Chemical Process Areas	NFPA
13	NFPA 70: National Electrical Code	
14	Canadian Marine Safety Guidelines	Transport Canada

WEB REFERENCES

#	Organisation/ Company	Website
1	ABS	https://ww2.eagle.org/en/rules-and-resources/rules-and-guides/archives.html
2	DNV-GL	https://rules.dnvgl.com/ServiceDocuments/dnvgl/#!/home
3	UK HSE	http://www.hse.gov.uk/offshore/index.htm
4	UK Energy Institute	https://www.energyinst.org/technical/technical-publications
5	BSI UK	http://shop.bsigroup.com/SearchResults/?q=BS%20EN%2060079

—Continued

#	Organisation/ Company	Website
6	IEC	https://webstore.iec.ch/searchform&q=IEC%2060079
7	API	http://www.techstreet.com/api/standards/api-rp-14fz?product_id=1856869
8	Transport Canada	https://www.tc.gc.ca/eng/marinesafety/tp-menu-515.htm
9	NFPA	http://www.nfpa.org/codes-and-standards/all-codes-and-standards/list-of-codes-and-standards?mode=code&code=497
10	IMO (SOLAS Conventions)	http://www.imo.org/en/Publications/Pages/Home.aspx
11	Oil & Gas UK	http://oilandgasuk.co.uk/publicationssearch.cfm?page=result&keyword=Telecommunications

PART 3 CHAPTER 1
REFERENCES

#	Title	Author
1	Making Projects Critical (Management, Work and Organisations)	Damian Hodgson (Editor), Svetlana Cicmil (Palgrave Macmillan; 2006)
2	Simple Project Management for Businesses: How to design a simple, pragmatic project management framework for small and large businesses	Adrian Shaw (Book Publishing Academy 2015)
3	Effective Document and Data Management: Unlocking Corporate Content	Bob Wiggins (Routledge; 3rd edition, 2012)
4	Securing Intellectual Property: Protecting Trade Secrets and Other Information Assets (Information Security)	Butterworth-Heinemann (2008)
5	Engineering Documentation Control Handbook: Configuration Management and Product Lifecycle Management	Frank B. Watts (William Andrew; 4th edition 2012)
6	Project Planning Handbook	Paul Whatley (Matador Business, 2014)
7	Manual of Engineering Drawing: to British and International Standards	Colin H. Simmons and Dennis E. Maguire (Newnes; 2nd edition, 2003)

Continued

—Continued

#	Title	Author
8	Units & Symbols for Electrical & Electronic Engineers	IET
9	IEC 60617: Graphical Symbols for Diagrams	IEC
10	315-1975: IEEE Standard for Graphic Symbols for Electrical and Electronics Diagrams (Including Reference Designation Letters)	IEEE
11	BS EN ISO/IEC 27001:2017: Information technology. Security techniques. Information security management systems. Requirements	BSI
12	BS EN 61078:2016: Reliability block diagrams.	
13	BS ISO 1219-3:2016: Fluid power systems and components. Graphical symbols and circuit diagrams. Symbol modules and connected symbols in circuit diagrams	
14	BS ISO 15519-2:2015: Specifications for diagrams for process industry. Measurement and control	
15	BS ISO 14084-2:2015: Process diagrams for power plants. Graphical symbols	
16	PD ISO/TR 16310:2014: Symbol libraries for construction and facilities management.	
17	BS ISO 128-24:2014: Technical drawings. General principles of presentation. Lines on mechanical engineering drawings	
18	BS EN ISO 10628-2:2012: Diagrams for the chemical and petrochemical industry. Graphical symbols	
19	PD IEC/TR 61352:2006: Mnemonics and symbols for integrated circuits.	
20	BS ISO 19018:2004: Ships and marine technology. Terms, abbreviations, graphical symbols and concepts on navigation.	
21	BS ISO 14617-1:2002: Graphical symbols for diagrams. General information and indexes	
22	BS ISO 14617-2:2002: Graphical symbols for diagrams. Symbols having general application	
23	BS ISO 14617-3:2002: Graphical symbols for diagrams. Connections and related devices	
24	BS ISO 14617-4:2002: Graphical symbols for diagrams. Actuators and related devices	
25	BS ISO 14617-5:2002: Graphical symbols for diagrams. Measurement and control devices	
26	BS ISO 14617-6:2002: Graphical symbols for diagrams. Measurement and control functions	
27	BS ISO 14617-7:2002: Graphical symbols for diagrams. Basic mechanical components	
28	BS ISO 14617-8:2002: Graphical symbols for diagrams. Valves and dampers	
29	BS ISO 14617-9:2002: Graphical symbols for diagrams. Pumps, compressors and fans	
30	BS ISO 14617-10:2002: Graphical symbols for diagrams. Fluid power converters	

—*Continued*

#	Title	Author
31	BS ISO 14617-11:2002: Graphical symbols for diagrams. Devices for heat transfer and heat engines	
32	BS ISO 14617-12:2002: Graphical symbols for diagrams. Devices for separating, purification and mixing	
33	BS EN 1861:1998: Refrigerating systems and heat pumps. System flow diagrams and piping and instrument diagrams. Layout and symbols.	
34	BS 5070-4:1990: Engineering diagram drawing practice. Recommendations for logic diagrams	
35	BS 1646-3:1984: Symbolic representation for process measurement control functions and instrumentation. Specification for detailed symbols for instrument interconnection diagrams	
36	BS 1646-4:1984: Symbolic representation for process measurement control functions and instrumentation. Specification for basic symbols for process computer, interface and shared display/control functions.	
37	BS 1553-1:1977: Specification for graphical symbols for general engineering. Piping systems and plant	
38	BS 1553-2:1950: Specification for graphical symbols for general engineering. Graphical symbols for power generating plant	
39	ISO 14617-13:2004: Graphical symbols for diagrams – Part 13: Devices for material processing	ISO
40	ISO 14617-14:2004: Graphical symbols for diagrams – Part 14: Devices for transport and handling of material	
41	ISO 14617-15:2002: Graphical symbols for diagrams – Part 15. Installation diagrams and network maps	

WEB REFERENCES

#	Organisation/ Company	Webpage
1	BSI	http://shop.bsigroup.com/ProductDetail/?pid=000000000030347472
2	IEEE	http://standards.ieee.org/findstds/standard/315-1975.html
3	IET	http://www.theiet.org/resources/books/index.cfm
4	ISO	https://www.iso.org/search/x/query/14617
5	Microsoft	https://products.office.com/en-us/project/project-and-portfolio-management-software?tab=tabs-1
6	Easy Projects	https://www.easyprojects.net/?aff=rvws
7	Mindview Project	http://www.projectmanagesoft.com/software/mindview
8	Autodesk	http://www.autodesk.com/products/autocad/overview
9	Aveva	http://www.aveva.com/en/Your_Industry/Oil_and_Gas/

PART 4 CHAPTER 1
WEB REFERENCES

#	Protection Company/ Manufacturer	Website
1	ANSI device numbers	https://en.wikipedia.org/wiki/ANSI_device_numbers
2	ABB Numerical Relays	http://new.abb.com/medium-voltage/distribution-automation/numerical-relays
3	ABB Substation Automation	http://new.abb.com/substation-automation/products
4	Ashida Protection Relay	http://www.ashidaelectronics.com/index.php/protection-relays/
5	Automation & Protection GE Grid Solutions	http://www.gegridsolutions.com/automation_protection.htm
6	Basler Electric	https://www.basler.com/SiteMap/Products/Protective-Relay-Systems/
7	Bender	http://www.bender-us.com/products/voltage-relays.aspx
8	CEE Relays (ICE)	http://www.ceeitaliana.com/en/Products_All.html
9	Computers & Control	http://www.computers-and-control.com/products/protection-relays/
10	Deif Marine & Offshore	http://www.deif.com/products-documentation/products/marineoffshore
11	GE Multilin	https://www.gegridsolutions.com/multilin/catalog/protection_control.htm
12	GE – Instrument Transformer Basic Technical Information and Application	www.gegridsolutions.com/products/brochures/ITItechInfo.pdf
13	P&B Relays	http://www.pbsigroup.com/protection-relays/
14	Littelfuse	http://www.littelfuse.com/technical-resources/technical-centers/relays-and-controls-technical-center.aspx
15	Protective Relaying Information Site	http://powersystemprotectiverelaying.blogspot.co.uk/search/label/Protection
16	Selinc Relays	https://selinc.com/products/
17	Siemens Substation Automation	http://w3.siemens.com/smartgrid/global/en/products-systems-solutions/Protection/Pages/overview.aspx
18	Schneider Substation Automation	http://www.schneider-electric.com/en/product-category/4600-protection-relays-by-application/?filter=business-6-medium-voltage-distribution-and-grid-automation
19	Toshiba Protection & Control Systems	http://www.toshiba-tds.com/tandd/products/pcsystems/en/f_p_rly.htm

PART 4 CHAPTER 2
REFERENCES

#	Title	Author
1	Protection Of Distribution Transformers	Ashish Seth (Kindle Edition, 2017)
2	Fundamentals of Electrical Design – Transformers Practical Approach and Application: Transformers	David R. Carpenter PhD (Kindle Edition, 2016)
3	Transformer Protection Application Guide	Basler Electric
4	BS EN 60076-7: Oil-Filled Power transformers.	BSI UK [IEC equivalents IEC 60076 and IEC 50216]
5	BS EN 60076-11: Dry-Type [includes cast-resin] Power transformers.	
6	BS EN 50216-3:2002: Power transformer and reactor fittings. Protective relay for hermetically sealed liquid-immersed transformers and reactors without gaseous cushion	
7	BS EN 50216-2:2002: Power transformers and reactor fittings. Gas and oil actuated relay for liquid immersed transformers and reactors with conservator	
8	IEEE Standard C57.12.10-2010: Standard Requirements for Liquid-Immersed Power Transformers	IEEE
9	NEMA ST 20-2014: Dry Type Transformers for General Applications	NEMA

WEB REFERENCES

#	Organisation/Company	Website
1	BSI UK	http://shop.bsigroup.com/
2	IEEE	http://webstore.ansi.org/RecordDetail.aspx?sku=IEEE+Std+C57.12.10-2010
3	IEC	https://webstore.iec.ch/publication/591
4	NEMA	https://www.nema.org/Standards/Pages/Dry-Type-Transformers-for-General-Applications.aspx
5	Basler Electric	https://www.basler.com/Product/BE1-11t-Transformer-Protection-System
6	P & B Relays (Buchholz Relays)	http://www.pbsigroup.com/protection-relays/buchholz-relays/
7	Viat Instruments	http://viatin.com/pro.html
8	Schneider Electric	http://www.schneider-electric.co.uk/en/work/support/customer-care/contact-centre.jsp

PART 4 CHAPTER 3
REFERENCES

#	Title	Author
1	BS EN 60255-149:2013: Measuring relays and protection equipment. Functional requirements for thermal electrical relays	BSI UK
2	IEC 60255-149:2013: Measuring relays and protection equipment – Part 149: Functional requirements for thermal electrical relays	IEC
3	IEC 60050-448 Ed. 2.0 t:1995: International Electrotechnical Vocabulary – Chapter 448: Power system protection	
4	IEEE C37.96-2012: IEEE Guide for AC Motor Protection	IEEE
5	IEEE C37.235-2007: IEEE Guide for the Application of Rogowski Coils Used for Protective Relaying Purposes	
6	IEEE 3004.8-2016: IEEE Approved Draft Recommended Practice for Motor Protection in Industrial and Commercial Power Systems	
7	Motor Fault Detection Using a Rogowski Sensor Without an Integrator	Oscar Poncelas, Javier A. Rosero, Jordi Cusidó, Juan Antonio Ortega, and Luis Romeral, IEEE Members.
8	Practical Aspects of Rogowski Coil Applications to Relaying	IEEE Power Engineering Society:2010
9	Condition Monitoring of Rotating Electrical Machines, 2nd Edition	Peter Tavner, Li Ran, Jim Penman and Howard Sedding (IET: 2008)

WEB REFERENCES (FOR MOTOR PROTECTION RELAY MANUFACTURERS, SEE TABLE 4.3.1)

#	Organisation	Website
1	BSI UK	https://www.bsigroup.com/en-GB/standards/
2	IEEE	http://standards.ieee.org/findstds/index.html?utm_source=mm_link&utm_campaign=find&utm_medium=std&utm_term=find%20standards
3	IEC	https://webstore.iec.ch/?ref=menu
4	IET	http://www.theiet.org/resources/books/pow-en/condition-monitoring-2nd-edition.cfm

PART 4 CHAPTER 4

REFERENCES

#	Title	Author
1	Protection of Industrial Power Systems 2nd, Kindle Edition by T. Davies (Author)	2nd and Kindle Editions by T. Davies (Newnes/Butterworth-Heinemann 1996)
2	Resistance of Low Voltage Arcs:	Lawrence E. Fisher (IEEE Trans IGA-6 Nov. Dec. 1970).
3	Power System Protection	Edited by the Electricity Training Association (IET 1995)
4	Network Protection & Automation Guide (NPAG)	G.E. Alstom Grid Edition May 2011
5	BS IEC 60255-1:2010: Measuring relays and protection equipment. Common requirements	BSI UK
6	IEC 60255-1:2009: Measuring relays and protection equipment – Part 1: Common requirements	IEC
7	IEEE Standard C37.90-2005 (R2011): IEEE Standard for Relays and Relay Systems Associated with Electric Power Apparatus IEEE Standard C37.234-2009: IEEE Guide for Protective Relay Applications to Power System Buses	IEEE

WEB REFERENCES

#	Organisation/Company	Website
1	ABB Relays	http://new.abb.com/substation-automation/products
2	Basler Electric	https://www.basler.com/Product/BE1-11t-Transformer-Protection-System
3	Ashida Protection Relay	http://www.ashidaelectronics.com/index.php/protection-relays/
4	BSI UK	http://shop.bsigroup.com/ProductDetail/?pid=000000000030136364
5	CEE Relays	http://www.ceeitaliana.com/en/Products_All.html
6	Deif Marine & Offshore	http://www.deif.com/products-documentation/products/marineoffshore
7	GE Multilin	https://www.gegridsolutions.com/multilin/catalog/protection_control.htm

Continued

—Continued

#	Organisation/Company	Website
8	IEC	https://webstore.iec.ch/publication/1160
9	IEEE/ANSI	http://ieeexplore.ieee.org/document/827748/?reload=true
10	IET	http://www.theiet.org/resources/index.cfm
11	GE Alsthom Grid Solutions	http://www.alstom.com/grid
12	Littelfuse	http://www.littelfuse.com/technical-resources/technical-centers/relays-and-controls-technical-center.aspx
13	P&B Relays	http://www.pbsigroup.com/protection-relays/
14	Siemens Substation Automation	http://w3.siemens.com/smartgrid/global/en/products-systems-solutions/Protection/Pages/overview.aspx
15	Schneider Substation Automation	http://www.schneider-electric.com/en/product-category/4600-protection-relays-by-application/?filter=business-6-medium-voltage-distribution-and-grid-automation
16	Toshiba Protection & Control Systems	http://www.toshiba-tds.com/tandd/products/pcsystems/en/f_p_rly.htm

PART 4 CHAPTER 5
REFERENCES

#	Title	Author
1	Power System Analysis: Short-Circuit Load Flow and Harmonics, Second Edition	J.C. Das (CRC Press, 2016)
2	Understanding Symmetrical Components for Power System Modelling	J.C. Das (IEEE Press Series on Power Engineering, 2017)
3	Electrical Installation Calculations: Basic, 9th edition	C. Kitcher and A.J. Watkins (Routledge, 2013)
4	Higher Electrical Engineering	J. Shepherd (Longman, 1986)
5	BS IEC 61892: Mobile and fixed offshore units. Electrical installations.	BSI/IEC
6	BS IEC 60287: Electric cables. Calculation of the current rating.	
7	BS IEC 60092: Electrical installation in ships	

—Continued

#	Title	Author
8	E 848: IEEE Standard Procedure for the Determination of the Ampacity Derating Factor for Fire-Protected Cable Systems	IEEE
9	EEE 1580: IEEE Recommended Practice for Marine Cable for Use on Shipboard and Fixed or Floating Facilities	
10	EE 45.8: IEEE Recommended Practice for Electrical Installations on Shipboard--Cable Systems	

WEB REFERENCES

#	Company/Organisation	Website
1	BSI	http://shop.bsigroup.com/SearchResults/?q=61892
2	Electrical Engineering Portal	http://electrical-engineering-portal.com/download-center/electrical-software/cable-designing-program
3	IEC	https://webstore.iec.ch/searchform&q=Cables#
4	IEEE	http://www.techstreet.com/ieee/searches/16156246
4	IET Forum	http://www.theiet.org/forums/forum/messageview.cfm?catid=205&threadid=40074

PART 4 CHAPTER 7

REFERENCES

#	Title	Author
1	IEC 60909-0:2016: Short-circuit currents in three-phase a.c. systems – Part 0: Calculation of currents	IEC
2	IEC 61363-1:1998: Electrical installations of ships and mobile and fixed offshore units – Part 1: Procedures for calculating short-circuit currents in three-phase a.c.	
3	IEEE: 399-1997: IEEE Recommended Practice for Industrial and Commercial Power Systems Analysis (IEEE Brown Book)	ANSI/IEEE
4	IEEE: 242-2001: IEEE Recommended Practice for Protection and Coordination of Industrial and Commercial Power Systems (IEEE Buff Book)	
5	IEEE: C37.010-1999: IEEE Application Guide for AC High-Voltage Circuit Breakers Rated on a Symmetrical Current Basis	
6	IEEE: C37.010-2016: IEEE Approved Draft Application Guide for AC High-Voltage Circuit Breakers > 1000 Vac Rated on a Symmetrical Current Basis	

Continued

—Continued

#	Title	Author
7	Power System Control and Stability	P. Anderson/A. Fouad (IEEE Press 1993)
8	Power System Analysis	J. Grainger/W. Stevenson (McGraw-Hill 1994)
9	Advances in Power System Modelling, Control and Stability Analysis	Prof. Federico Milano (ed.) (IET 2016)

WEB REFERENCES

#	Organisation	Website
1	IEC	https://webstore.iec.ch/
2	IEEE	http://www.ieee.org/standards/index.html
3	ANSI	http://webstore.ansi.org/FindStandards.aspx?Action=displaydept&DeptID=135
4	IET	http://www.theiet.org/resources/books/index.cfm
5	SKM (PTW)	http://www.skm.com/Dapper.html
6	CEE Relays (PTW)	http://www.ceerelays.co.uk/ptwsoftware.htm [UK SKM Agent]
7	IPSA	http://www.ipsa-power.com/
8	ETAP	https://etap.com/sector/generation
9	ERACS	http://www.eracs.co.uk/index.php

PART 4 CHAPTER 8

REFERENCES

#	Title	Author(s)
1	Power management system in vessel. (conference paper)	Abdul Hafis A. Jaleel; Gloria Devassy; Melvin C. Vincent; Neethu Rose; Rinto Raphel; Sreechithra C.M. (IEEE, 2016)
2	Dynamic positioning power plant system reliability and design. (conference paper)	Kamal Garg; Lew Weingarth; Saurabh Shah (IEEE, 2011)
3	Today's Dynamic Positioning Systems	Nick Van Overdam and Andrew Phillips (Kongsberg, 2011)

—*Continued*

#	Title	Author(s)
4	BS IEC 60092-501:2013: Electrical installations in ships. Special features. Electric propulsion plant	BSI/IEC
5	BS IEC 60092-504:2016: Electrical installations in ships. Automation, control and instrumentation	
6	BS-IEC-61892-5: Mobile and fixed off-shore units. Electrical installations. Mobile units	BSI/IEC

WEB REFERENCES

#	Organisation/ Company	Website
1	ABB	http://new.abb.com/marine/systems-and-solutions/automation-and-marinesoftware/integrated-automation
2	BSI	http://shop.bsigroup.com/SearchResults/?q=BS%20IEC%2060092
3	IEC	https://webstore.iec.ch/searchform&q=IEC%2060092
4	IEEE	http://ieeexplore.ieee.org/document/5936973/
5	Kongsberg	https://www.km.kongsberg.com/ks/web/nokbg0240.nsf/AllWeb/E1C7040DC299F88CC12570A400338307?OpenDocument
6	Rolls Royce Marine	http://www.maritime-suppliers.com/supplier/products2.aspx?company=0632
7	Wärtsilä	http://www.wartsila.com/services/areas-of-expertise/services-catalogue/electrical-automation-services/integrated-automation-system-wias

PART 4 CHAPTER 9

REFERENCES

#	Title	Author
1	HARMONICS – Understanding the Facts	Richard P. Bingham
2	Power System Harmonics	Jos Arrillaga and Neville R. Watson (Wiley – Blackwell, 2003)
3	Power System Harmonics and Passive Filter Designs	J.C. Das (Wiley, 2015)

Continued

—Continued

#	Title	Author
4	Power Systems Harmonics: Computer Modelling and Analysis	Acha (Wiley, 2001)
5	Power System Harmonic Analysis Using ETAP	Mohammed Alsaaq (LAP LAMBERT Academic Publishing, 2014)
6	519-2014: IEEE Recommended Practice and Requirements for Harmonic Control in Electric Power Systems	IEEE
7	IEC 61000-2-12:2003 [BS EN 61000-2-12:2003]: Electromagnetic compatibility (EMC) – Part 2-12: Environment – Compatibility levels for low-frequency conducted disturbances and signaling in public medium-voltage power supply systems	BSI/IEC
8	IEC 61000-3-12:2011 [BS EN 61000-3-12:2011]: Electromagnetic compatibility (EMC). Limits. Limits for harmonic currents produced by equipment connected to public low-voltage systems with input current >16 A and ≤75 A per phase	BSI/IEC

WEB REFERENCES

#	Organisation	Website
1	BSI	http://shop.bsigroup.com/ProductDetail?pid=000000000019985310
2	IEC	https://webstore.iec.ch/searchform&q=IEC%2061000-3-12%3A2011
3	IEEE	https://standards.ieee.org/findstds/standard/519-2014.html
4	ABB	http://new.abb.com/uk/campaigns/hvac-drives/consultants/harmonics
5a	GE	http://www.gegridsolutions.com/HVMV_Equipment/catalog/gem-active.htm
5b	GE	https://www.gegridsolutions.com/HVMV_Equipment/metal_harmonic_filter.htm
6a	Schneider Electric – Electrical Installation Wiki	http://www.electrical-installation.org/enwiki/Total_harmonic_distortion_(THD)
6b	Schneider Electric – Oil & Gas solution: Offshore platform harmonics solution	http://www2.schneider-electric.com/sites/singapore/en/solutions/business_segments/oil-and-gas/oil-and-gas-upstream-extraction/offshore-platform-harmonics-solution.page

—Continued

#	Organisation	Website
7a	Siemens	http://w3.usa.siemens.com/building technologies/us/en/building-automation-and-energy-management/variable-frequency-drives/low-harmonics-technology/Pages/low-harmonics-technology.aspx
7b	Siemens	https://w3.siemens.com/powerdistribution/global/SiteCollectionDocuments/en/mv/switchgear/Whitepaper_Switching-of-filter-circuits-with-vacuum-switches.pdf

PART 5 CHAPTER 1
REFERENCES

#	Title	Author
1	API RP 14C: Analysis, Design, Installation, and Testing of Safety Systems for Offshore Production Facilities, Eighth Edition	American Petroleum Institute
2	A guide to the Offshore Installations (Safety Case) Regulations 2005	UK HSE (2005)
3	Guidance for the provision of accommodation on offshore installations	UK HSE (2015)
4	DNVGL-OS-D201: Edition January 2017 Electrical installations	DNV GL
5	DNVGL-OS-D202: Edition January 2017 Automation, safety and telecommunication systems	
6	DNVGL-ST-0076: Edition 2015 Design of electrical installations for wind turbines	
7	DNVGL-ST-0145: Edition 2016 Offshore substations	
8	Rules and Regulations for the Classification of Offshore Units July 2016	LR
9	Rules For Building And Classing Facilities On Offshore Installations	American Bureau of Shipping (ABS)
10	Ship-Shaped Offshore Installations: Design, Building, and Operation	Jeom Kee Paik (Cambridge University Press; Reissue edition, 2011)
11	Decommissioning Offshore Structures	D.G. Gorman & J. Neilson (Springer London, 1998)
12	The challenges of offshore power system construction-bringing power successfully to Troll A, one of the world's largest oil and gas platforms	L. Stendius; P. Jones (IEEE Conference Publication, 2006)

WEB REFERENCES

#	Organisation/Company	Website
1	ABS	http://ww2.eagle.org/en/rules-and-resources/rules-and-guides.html
2	API	http://www.api.org/products-and-services/standards/purchase
3	DNV GL	https://rules.dnvgl.com/ServiceDocuments/dnvgl/#!/industry/1/Maritime/9/DNV%20GL%20offshore%20standards%20(OS)
4	HSE	http://www.hse.gov.uk/offshore/index.htm
5	LR	http://www.lr.org/en/RulesandRegulations/

PART 5 CHAPTER 2

REFERENCES

#	Title	Author
1	Principals of Electrical Grounding	John C. Pfeiffer
2	Guidelines on Earthing/Grounding/Bonding in the Oil and Gas Industry	UK Energy Institute (2016)
3	IEEE 142 Recommended Practice for Grounding of Industrial and Commercial Power Systems	IEEE
4	Offshore earthing a different perspective	S.G. Lawton (IEEE)
5	BS EN/IEC 60092: Electrical installations in ships	BSI/IEC
6	BS EN/IEC 61892: Mobile and fixed offshore units – Electrical installations	
7	BS EN 62305-1: Protection against lightning. Part 1 General principles.	
8	BS EN 62305-2: Protection against lightning. Part 2. Risk management.	
9	NFPA 70: National Electrical Code	NFPA
10	NFPA 780: Standard for the Installation of Lightning Protection Systems	
11	UL 467: Grounding and Bonding Equipment	Underwriter's Laboratories
12	IEC 61400-24:2010: Wind turbines – Part 24: Lightning protection	IEC

WEB REFERENCES

1	BSI	http://shop.bsigroup.com/ProductDetail?pid=000000000030324821
2	IEC	https://webstore.iec.ch/publication/5437
3	IEEE	http://ieeexplore.ieee.org/search/searchresult.jsp?queryText=IEEE%20Std%20142-2007%20.LB.Revision%20Of%20IEEE%20Std%20142-1991.RB.%20-%20Redline&newsearch=true
4	UK Energy Institute	https://www.energyinst.org/home
5	NFPA	http://www.nfpa.org/codes-and-standards
6	Underwriters' Laboratories	https://standardscatalog.ul.com/

PART 5 CHAPTER 3

REFERENCES

#	Title	Author
1	Preservation of large motors and generators from weather on offshore platforms	V.S. Pai (IEEE Transactions on Industry Applications (Volume: 26, Issue: 5, Sep/Oct 1990)
2	Facilities on Offshore Installations (2017)	ABS
3	BS EN 60947-1:2007 + A2:2014: Low-voltage switchgear and controlgear. General rules	BSI UK [IEC 60947-1:2007]
4	BS EN 60034-5:2001, IEC 60034-5:2000: Rotating electrical machines. Degrees of protection provided by the integral design of rotating electrical machines (IP code). Classification	BSI/IEC
5	BS EN/IEC 60092: Electrical installations in ships	
6	BS EN/IEC 61892: Mobile and fixed offshore units – Electrical installations	
7	IEC 60092-201:1994: Electrical installations in ships – Part 201: System design – General	
8	841-2001: IEEE Standard for Petroleum and Chemical Industry – Severe Duty Totally Enclosed Fan-Cooled (TEFC) Squirrel Cage Induction Motors – Up to and Including 370 kW (500 hp)	IEEE
9	45-2002: IEEE Recommended Practice for Electric Installations on Shipboard	
10	P45.4: Recommended Practice for Electrical Installations on Shipboard – Marine Sectors and Mission Systems	

Continued

—Continued

#	Title	Author
11	DNVGL-OS-A201: Edition July 2015: Winterization for cold climate operations	DNV GL (Offshore Standards)
12	DNVGL-OS-D201: Edition January 2017: Electrical installations	
13	Rules and Regulations for the Classification of Offshore Units (July 2016)	LR

WEB REFERENCES

#	Organisation/Company	Website
1	ABS	http://ww2.eagle.org/en/rules-and-resources/rules-and-guides.html?q=Offshore
2	BSI	http://shop.bsigroup.com/ProductDetail/?pid=000000000030146272
3	DNV GL	https://rules.dnvgl.com/ServiceDocuments/dnvgl/#!/industry/1/Maritime/9/DNV%20GL%20offshore%20standards%20(OS)
4	IEEE	http://standards.ieee.org/findstds/index.html?utm_source=mm_link&utm_campaign=find&utm_medium=std&utm_term=find%20standards
5	IEC	https://webstore.iec.ch/publication/678
6	LR	http://www.lr.org/en/services/rules-for-offshore-units-in-oil-and-gas.aspx

PART 5 CHAPTER 4

REFERENCES

#	Title	Author
1	Dangerous substances and explosive atmospheres Dangerous Substances and Explosive Atmospheres Regulations 2002. Approved Code of Practice and guidance	UK HSE (2nd edition 2013)
2	The 'Blue Guide' on the implementation of EU product rules 2016	European Commission (2016)
3	EI Model code of safe practice Part 15: Area classification for installations handling flammable fluids	UK Energy Institute
4	Energy Institute Model Code Of Safe Practice, Part 21: Guidelines for the control of hazards arising from static electricity	

—Continued

#	Title	Author
5	BS EN 60079: Explosive atmospheres	BSI UK
6	PD CLC/TR 50427: Assessment of inadvertent ignition of flammable atmospheres by radio-frequency radiation. Guide	
7	BS EN ISO 80079-36:2016: Explosive atmospheres. Non-electrical equipment for explosive atmospheres. Basic method and requirements	
8	BS EN ISO 80079-37:2016: Explosive atmospheres. Non-electrical equipment for explosive atmospheres. Non-electrical type of protection constructional safety "c", control of ignition sources "b", liquid immersion "k"	
9	IEC 60079 Explosive atmospheres	IEC
10	API RP 14FZ: Design and Installation of Electrical Systems for Fixed and Floating Offshore Petroleum Facilities for Unclassified and Class I, Zone 0, Zone 1 and Zone 2 Locations, Second Edition	API
11	API RP 505: Recommended Practice for Classification of Locations for Electrical Installation at Petroleum Facilities Classified as Class I, Zone 0, Zone 1 and Zone 2 (2002).	
12	NFPA 497: Recommended Practice for the Classification of Flammable Liquids, Gases, or Vapors and of Hazardous (Classified) Locations for Electrical Installations in Chemical Process Areas	NFPA
13	NFPA 70: National Electrical Code	
14	Canadian Marine Safety Guidelines	Transport Canada

WEB REFERENCES

#	Organisation/Company	Website
1	UK HSE	http://www.hse.gov.uk/offshore/index.htm
2	European Commission	http://www.industry-finder.com/machinery-directive/new-blue-guide-dated-04-2014.html
3	UK Energy Institute	https://www.energyinst.org/technical/technical-publications
4	BSI UK	http://shop.bsigroup.com/SearchResults/?q=BS%20EN%2060079
5	IEC	https://webstore.iec.ch/searchform&q=IEC%2060079
6	API	http://www.techstreet.com/api/standards/api-rp-14fz?product_id=1856869
7	Transport Canada	https://www.tc.gc.ca/eng/marinesafety/tp-menu-515.htm
8	NFPA	http://www.nfpa.org/codes-and-standards/all-codes-and-standards/list-of-codes-and-standards?mode=code&code=497

Continued

—Continued

#	Organisation/Company	Website
9	R. Stahl	http://www.rstahl.co.uk/
10	MTL IS Barriers	https://www.mtl-inst.com/
11	Pepperl + Fuchs	https://www.pepperl-fuchs.us/usa/en/21.htm?PA=1
12	Weidmueller-Klippon	http://www.weidmueller.com/int/products/klippon--protect-enclosure-system

PART 6 CHAPTER 1

REFERENCES

#	Title	Author
1	Dynamic Positioning of Offshore Vessels	Max J. Morgan (Pennwell Books, 1978)
2	Requirements concerning Mobile Offshore Drilling Units	IACS (2012)
3	ABS Guide for Dynamic Positioning Systems	ABS (2012)
4	DNVGL-OS-C301: Stability and watertight integrity	DNV GL (2017)
5	DNVGL-RP-E306: Dynamic positioning vessel design philosophy guidelines	DNV GL (2015)
6	DNVGL-OS-D203: Integrated software dependent systems (ISDS)	DNV GL (2015)
7	Rules for Classification: Part 6 Additional class notations Chapter 3 Navigation, manoeuvring and position keeping	DNV GL (2017)
8	Rules and Regulations for the Classification of Ships, Part 7, Chapter 4: Dynamic Positioning Systems	LR (2016)

WEB REFERENCES

#	Organisation/Company	Website
1	ABS	http://ww2.eagle.org/en/rules-and-resources/rules-and-guides.html
2	DNV GL	https://rules.dnvgl.com/ServiceDocuments/dnvgl/#!/industry
3	LR	http://www.lr.org/en/rulesandregulations/
4	Kongsberg	https://www.km.kongsberg.com/ks/web/nokbg0240.nsf/AllWeb/F85938EB2AE9BD86C1256A48004161A8?OpenDocument
5	Marine Technologies	http://marine-technologies.com/dynamicpositioning/dp3.html
6	Praxis Automation	https://www.praxis-automation.nl/products/dynamic-positioning-system/dp-dynamic-positioning-system/#dp2-dp3

PART 6 CHAPTER 2

REFERENCES

#	Title	Author
1	Rules For Building And Classing Floating Production Installations	ABS (2017)
2	DNVGL-RU-OU-0102: Floating production, storage and loading units	DNV GL (2017)
3	Rules & Regulations for the Classification of Ships 2014 Part 4	LR

WEB REFERENCES

#	Organisation/Company	Website
1	ABS	http://ww2.eagle.org/en/rules-and-resources/rules-and-guides.html?q=Offshore#/content/dam/eagle/rules-and-guides/current/offshore/82_Floating_Production_Installations_2017
2	DNV GL	https://rules.dnvgl.com/ServiceDocuments/dnvgl/#!/industry/1/Maritime/8/DNV%20GL%20rules%20for%20classification:%20Offshore%20units%20(RU-OU)
3	LR	http://www.webstore.lr.org/products/744-rules-regulations-for-the-classification-of-ships-2014-part-4.aspx

PART 6 CHAPTER 3

REFERENCES

#	Title	Author
1	Rules For Building And Classing Mobile Offshore Drilling Units	ABS (2016)
2	DNVGL-RU-OU-0101: Offshore drilling and support units	DNV GL (2017)
3	DNVGL-RU-OU-0102: Floating production, storage and loading units	
4	IMO Code for the Construction and Equipment of Mobile Offshore Drilling Units, 2009	IMO (2009)
5	Rules & Regulations for the Classification of Ships 2014 Part 4	LR

WEB REFERENCES

#	Organisation/Company	Website
1	ABS	https://ww2.eagle.org/en/rules-and-resources/rules-and-guides/archives.html
2	DNV GL	https://rules.dnvgl.com/ServiceDocuments/dnvgl/#!/industry/1/Maritime/8/DNV%20GL%20rules%20for%20classification:%20Offshore%20units%20(RU-OU)
3	IMO	http://www.safety4sea.com/imo-2009-modu-code/
4	LR	http://www.webstore.lr.org/default.aspx

PART 6 CHAPTER 4
REFERENCES

#	Title	Author
1	Guide For Building And Classing Mobile Offshore Units	ABS (2017)
2	DNVGL-RU-OU-0104: Self-elevating units, including wind turbine installation units and liftboats	DNV GL (2017)
3	IEC 61892:2015 Mobile and fixed offshore units – Electrical installations	IEC (2015)
4	Requirements concerning Mobile Offshore Drilling Units	IACS (2012)
5	Rules and Regulations for the Classification of Offshore Units July 2016	LR (2016)

WEB REFERENCES

#	Organisation/Company	Website
1	ABS	http://ww2.eagle.org/en/rules-and-resources/rules-and-guides.html?q=Offshore#/content/dam/eagle/rules-and-guides/current/offshore/6_rules_building_classing_mobile_offshore_drilling_units
2	DNV GL	https://rules.dnvgl.com/ServiceDocuments/dnvgl/#!/industry/1/Maritime/8/DNV%20GL%20rules%20for%20classification:%20Offshore%20units%20(RU-OU)
3	IACS	https://www.iacs.org.uk/document/public/Publications/Resolution_changes/PDF/UR_D4_Rev3_pdf1773.pdf
4	IEC	https://webstore.iec.ch/publication/6083
5	LR	http://www.lr.org/en/services/rules-for-offshore-units-in-oil-and-gas.aspx

PART 6 CHAPTER 5
REFERENCES

#	Title	Author
1	Guide For Building And Classing Mobile Offshore Units	ABS (2017)
2	DNV-OS-J101: Design of Offshore Wind Turbine Structures	DNV GL (2013)
3	DNVGL-OS-D201: Electrical installations	DNV GL (2017)
4	DNVGL-ST-0076: Design of electrical installations for wind turbines	DNV GL (2015)
5	Service Specification DNVGL-SE-0077: Certification of fire protection systems for wind turbines	DNV GL (2015)
6	IEC 61892:2015: Mobile and fixed offshore units – Electrical installations	IEC (2015)
7	IEC 61400-3:2009: Wind turbines – Part 3: Design requirements for offshore wind turbines	IEC (2009)
8	Requirements concerning Mobile Offshore Drilling Units	IACS (2012)
9	Rules and Regulations for the Classification of Offshore Units July 2016	LR (2016)

WEB REFERENCES

#	Organisation/Company	Website
1	ABS	http://ww2.eagle.org/en/rules-and-resources/rules-and-guides.html?q=Offshore#/content/dam/eagle/rules-and-guides/current/offshore/6_rules_building_classing_mobile_offshore_drilling_units
2	DNV GL	https://rules.dnvgl.com/ServiceDocuments/dnvgl/#!/industry/1/Maritime/8/DNV%20GL%20rules%20for%20classification:%20Offshore%20units%20(RU-OU)
3	IACS	https://www.iacs.org.uk/document/public/Publications/Resolution_changes/PDF/UR_D4_Rev3_pdf1773.pdf
4	IEC	https://webstore.iec.ch/publication/6083
5	LR	http://www.lr.org/en/services/rules-for-offshore-units-in-oil-and-gas.aspx

PART 6 CHAPTER 6

REFERENCES

#	Title	Author
1	Offshore Wind: A Comprehensive Guide to Successful Offshore Wind Farm Installation	Kurt Thomsen (Elsevier 2011)
2	Offshore Wind Farms: Technologies, Design and Operation	Chong Ng and Li Ran (Woodhead Publishing 2016)
3	MGN 543: (M+F) Safety of Navigation: Offshore Renewable Energy Installations (OREIs) – UK Navigational Practice, Safety and Emergency Response.	UK Maritime and Coastguard Agency
4	MGN 543: Check list for developers	
5	Offshore Renewable Energy Installations: Requirements, Guidance and Operational Considerations for Search and Rescue and Emergency Response	
6	HSE L144: Managing health and safety in construction – Construction (Design and Management) Regulations 2007 Approved Code of Practice.	HSE 2007
7	Offshore Wind and Marine Energy Health and Safety Guidelines (Issue 2 2014)	RenewableUK
8	IEC/BS EN 61400-3:2009: Wind turbines – Part 3: Design requirements for offshore wind turbines	IEC/BSI (2009)
9	DNVGL-ST-0076: Design of electrical installations for wind turbines	DNV GL (2015)
10	Service Specification DNVGL-SE-0077: Certification of fire protection systems for wind turbines	DNV GL (2015)
11	DNVGL-ST-0145: Offshore substations	DNV GL (2016)
12	EU Council Directive 92/57/EEC– Temporary and Mobile Construction Sites.	EU Council

WEB REFERENCES

#	Organisation/Company	Website
1	DNV GL	https://rules.dnvgl.com/ServiceDocuments/dnvgl/#!/industry/1/Maritime/8/DNV%20GL%20rules%20for%20classification:%20Offshore%20units%20(RU-OU)
2	IEC	https://webstore.iec.ch/publication/6083
3	RenewableUK	http://www.renewableuk.com/page/HealthSafety
4	UK Maritime and Coastguard Agency	https://www.gov.uk/government/publications/mgn-543-mf-safety-of-navigation-offshore-renewable-energy-installations-oreis-uk-navigational-practice-safety-and-emergency-response

PART 7 CHAPTER 1
REFERENCES

#	Title	Author
1	Power System Commissioning and Maintenance Practice (Energy Engineering)	Keith Harker (Editor) (IET 1997)
2a	BS EN 60255: Measuring relays and protection equipment.	BSI (UK)
2b	IEC 60255: Measuring relays and protection equipment.	IEC
3a	IEEE C37.102: IEEE Guide for AC Generator Protection	IEEE
3b	IEEE C37.103: IEEE Guide for Differential and Polarizing Relay Circuit Testing	
3c	IEEE C37.106: IEEE Guide for Abnormal Frequency Protection for Power Generating Plants	
3d	IEEE C37.112: IEEE Standard Inverse-Time Characteristic Equations for Overcurrent Relays	
3e	IEEE C37.117: IEEE Guide for the Application of Protective Relays	
4	EIS: Inspection Testing and Commissioning	Malcom Doughton and John Hooper (Cengage Learning 2012)
5	Chemical and Process Plant Commissioning Handbook: A Practical Guide to Plant System and Equipment Installation and Commissioning	Martin Killcross (Elsevier – Butterworth-Heinemann 2011)
6	Managing Industrial Controls	N.E. Battikha (2014 International Society of Automation)
7	Commissioning of a Type Approved PLC – ISA TECH 1997	Ekkehard Pofahl
8	Loop Checking: A Technician's Guide	H.M Jeffery (International Society of Automation)
9	Start-up: A Technician's Guide, Second Edition	Diane R. Barkin (International Society of Automation)
10	Wiring Diagram Book	Groupe Schneider, Square D Book (File 0140 – pdf format)

WEB REFERENCES

#	Organisation/Company	Website
1	ABB	http://new.abb.com/control-systems/industry-specific-solutions/oil-gas-and-petrochemicals
2	ACS	http://www.acsprocess.com/
3	BSI	http://shop.bsigroup.com/
4	EEP: Industrial automation – engineering pocket guide	http://electrical-engineering-portal.com/download-center/books-and-guides/automation-control/industrial-automation-pocket-guide
5	Intech	http://www.intechww.com/oil-and-gas-applications/offshore-production/
6	Honeywell	https://www.honeywellprocess.com/en-US/innovations/Pages/latest-experion-pks.aspx
7	Rockwell Silvertech	http://www.silvertech-me.com/what-we-do/process-control-systems
8	Kongsberg	https://www.km.kongsberg.com/ks/web/nokbg0240.nsf/AllWeb/E1C7040DC299F88CC12570A400338307?OpenDocument
9	Siemens	http://w3.siemens.com/mcms/automation/en/process-control-system/Pages/Default.aspx
10	Yokogawa	https://www.yokogawa.com/solutions/solutions/safe-operations/advanced-process-control/

PART 7 CHAPTER 5

REFERENCES

Text references – See PART 7 Chapter 1

Control and Protection Circuits – see PART 7 Chapter 4

WEB REFERENCES

#	Organisation	Website
1	BS EN 60034: Rotating Electrical Machines	http://shop.bsigroup.com/ProductDetail/?pid=000000000030184921
2	IEC 60034: Rotating Electrical Machines	https://webstore.iec.ch/?ref=menu
3	43-2000: IEEE Recommended Practice for Testing Insulation Resistance of Rotating Machinery	http://standards.ieee.org/findstds/standard/43-2000.html
4a	A guide to Diagnostic Insulation Testing Above 1 kV	https://uk.megger.com/

—*Continued*

#	Organisation	Website
4b	The Complete Guide to Electrical Insulation Testing	https://www.instrumart.com/assets/Megger-insulationtester.pdf
5	Monition	http://www.monition.com/electric-motor-winding-testing.html
6	ABB Technical Guide No. 102: Effects of AC Drives on Motor Insulation	https://duckduckgo.com/?q=Motor+Insulation+Testing&t=hb&atb=v55-4_a&ia=web
7	Powerpoint Engineering	http://powerpoint-engineering.com/technical-support/technical-guides/insulation-testing-technical-guides/
8	IEEE Power Engineering Society: IEEE Standard Test Procedure for Polyphase Induction Motors and Generators	https://engineering.purdue.edu/~dionysis/EE452/Lab12/IEEEstd_112.pdf

PART 7 CHAPTER 6
REFERENCES

#	Title	Author
1	The safe isolation of plant and equipment	UK Health and Safety Executive
2	Start-up: A Technician's Guide, Second Edition	Diane R. Barkin (ISA)
3	Process Plant Commissioning	D.M.C. Horsley (IChemE, 1998)
4	Chemical and Process Plant Commissioning Handbook 1st Edition	Martin Killcross (Elsevier 2011)
5	IEC 62337:2012 BS EN 62337:2012 Commissioning of electrical, instrumentation and control systems in the process industry. Specific phases and milestones	IEC BSI UK
6	ANSI/NETA ECS-2015: Standard for Electrical Commissioning of Electrical Power Equipment and Systems	ANSI/NETA
7	NFPA 3-2012: Recommended Practice on Commissioning and Integrated Testing of Fire Protection and Life Safety Systems, 2012 Edition	ANSI/NFPA
8	Installation, Commissioning, Validation etc…	Safeprod

WEB REFERENCES

#	Title/Organisation	Website
1	Pocket Guide on Industrial Automation (S Medida, IDC Technologies, 2008)	http://www.pacontrol.com/download/Industrial-Automation-Pocket-Guide.pdf
2	BSI UK	http://shop.bsigroup.com/ProductDetail/?pid=000000000030277236
3	IEC	https://webstore.iec.ch/publication/6871
4	Safeprod	http://www.sp.se/sv/index/services/functionalsafety/Documents/Safetylife%20cycles%20after%20and%20including%20installation%20guideline%20process.pdf
5	ANSI/NETA	http://webstore.ansi.org/RecordDetail.aspx?sku=ANSI%2FNETA+ECS-2015
6	ANSI/NFPA	http://webstore.ansi.org/RecordDetail.aspx?sku=NFPA+3-2012

PART 8 CHAPTER 1

REFERENCES

#	Title	Author(s)
1	Reliability Technology	A.E. Green and A.J. Bourne
2	Statistical Tables (sixth edition) 1980	R.A. Fisher and F. Yates
3	Offshore Wind Turbines: Reliability, Availability and Maintenance	IET http://www.theiet.org/resources/books/
4	Power System Reliability Evaluation	Roy Billinton (Routledge: January 1, 1970)
5	Looking Forward – Reliability of Safety Critical Control Systems on Offshore Drilling Vessels	Jon Espen Skogdalen Øyvind Smogeliii Deepwater Horizon Study Group Working Paper – January 2011
6	Reliability of Power Electronic Converter Systems (Energy Engineering)	Henry Shu-Hung Chung (Editor), Huaiqing Wang (Editor), Frede Blaabjerg (Editor): IET 2015
7	Power Distribution System Reliability: Practical Methods and Applications	Ali Chowdhury, Don Koval (Wiley Blackwell:2009)
8	Reliability Block Diagram Modeling – Comparisons of Three Software Packages	IEEE http://ieeexplore.ieee.org/Xplore/home.jsp
9	IEC 60812 Procedure for failure mode and effects analysis (FMEA)	IEC https://webstore.iec.ch/?ref=menu
10	IEC 61025: Fault tree analysis (FTA)	
11	IEC 61078: Analysis techniques for dependability – Reliability block diagram method	

WEB REFERENCES

#	Organisation/Company	Website
1	Arms Reliability	https://www.armsreliability.com/our-services/
2	Item Software	http://www.itemsoft.com/products.html
3	ALD	http://aldservice.com/Download/download-reliability-and-safety-software.html
4	Isograph	https://www.isograph.com/software/reliability-workbench/fault-tree-analysis/
5	Barringer and Associates, Inc.	http://www.barringer1.com/raptor.htm

PART 8 CHAPTER 2

REFERENCES

#	Title	Author
1	Preventive Maintenance, 3rd Edition	J.D. Patton, Jr. (The International Society of Automation, 2005)
2	IEC/BS EN 13460:2009: Documentation for maintenance	IEC/BSI
3	IEC/BS EN 60079-17:2013: Explosive atmospheres – Part 17: Electrical Installations inspection and maintenance	
4	NFPA 70B: Recommended Practice for Electrical Equipment Maintenance, 2016 edition	NFPA (2016)
5	DNVGL-CG-0058: Maintenance of safety equipment	DNV GL (2016)

WEB REFERENCES

#	Organisation/Company	Website
1	BSI	http://shop.bsigroup.com/ProductDetail/?pid=000000000030160768
2	DNV GL	https://www.dnvgl.com/rules-standards/
3	IEC	https://webstore.iec.ch/searchform&q=IEC%2013460%3A2009
4	The International Society of Automation	https://www.isa.org/standards-publications/isa-publications/isa-books/
5	NFPA	http://catalog.nfpa.org/NFPA-70B-Recommended-Practice-for-Electrical-Equipment-Maintenance-P1196.aspx?icid=B484

PART 9 CHAPTER 1
REFERENCES

#	Title	Website
1	Directive 2013/30/EU of the European Parliament And Of The Council of 12 June 2013, on safety of offshore oil and gas operations and amending Directive 2004/35/EC	https://www.ecolex.org/details/legislation/directive-201330eu-of-the-european-parliament-and-of-the-council-on-safety-of-offshore-oil-and-gas-operations-and-amending-directive-200435ec-lex-faoc125458/
2	Offshore Installations and Wells (design and construction, etc) Regulations 1996 [DCR]	http://www.hse.gov.uk/offshore/publications.htm
3	The Offshore Installations (Offshore Safety Directive) (Safety Case etc.) Regulations 2015	http://www.legislation.gov.uk/uksi/2015/398/contents/made
4	The offshore installations (prevention of fire and explosion, emergency response) regulations 1995 [PFEER]	http://www.hse.gov.uk/foi/internalops/hid_circs/enforcement/spcenf155.htm
5	The offshore installations and pipeline works (management and administration) regulations 1995 [MAR]	http://www.hse.gov.uk/foi/internalops/hid_circs/enforcement/spcenf153.htm
6	The pipelines safety regulations 1996 [PSR]	http://www.hse.gov.uk/foi/internalops/hid_circs/enforcement/spcenf156.htm
7	Electrical standards and approved codes of practice Please note: • Electrical standards relating to offshore topics such as ATEX and DSEAR are covered in the appropriate sections of this book. • The effect of BREXIT on electrical standards is presently beyond the scope of this book **DISCLAIMER** This Chapter is an introductory guide, and although it has been extensively reviewed, only *current* statutory and HSE documents must be used on live projects.	http://www.hse.gov.uk/electricity/standards.htm Contains a useful but not exhaustive list of commonly used electrical standards and approved codes of practice. The specifier must select and apply the appropriate standards whether listed here or not. You should also ensure that the standards you are using are current. The standards are organised into a number of topic areas and are ordered with the lowest number at the top of each table: Electrical and Power Electrical Appliances Electromagnetic Compatibility Flammable Atmospheres Machinery

Index

Note: 'Page numbers followed by "f" indicate figures, "t" indicate tables'.

E

I

J

K

L

M

P

Q

R

S

T

U

V

Printed in the United States
By Bookmasters